Measure Theory and Probability

Second Edition

FEW EXAMPLES PER CHAPTER
[NO SOLUTIONS TO EXERCISES (PROBLEMS AT END OF CHAPT.)
[NO ANSWERS

A.K. BASU
Formerly Faculty, Department of Statistics
University College of Science
University of Calcutta
Kolkata

Reviewed and Revised by
Dr. Antar Bandyopadhyay
Indian Statistical Institute
New Delhi

PHI Learning Private Limited
New Delhi-110001
2012

₹ 225.00

MEASURE THEORY AND PROBABILITY, Second Edition
A.K. Basu

ISBN-978-81-203-4385-6

The export rights of this book are vested solely with the publisher.

Fifth Printing (Second Edition) ... ... **April, 2012**

Published by Asoke K. Ghosh, PHI Learning Private Limited, M-97, Connaught Circus, New Delhi-110001 and Printed by Rajkamal Electric Press, Plot No. 2, Phase IV, HSIDC, Kundli-131028, Sonepat, Haryana.

To
My Parents

Contents

3. Distribution Functions   38–49

4. Measurable Functions   50–59

5. Integration Theory and Expectation   60–74

6. Types of Convergence and Limit Theorems   75–83

7. Independence   84–94

8. Law of Large Numbers and Associated Limit Theorems   95–111

Editor's Preface

Due to the sudden demise of the author, Dr. A.K. Basu, the publisher, PHI Learning, had to take the assistance of a senior faculty, Dr. Antar Bandyopadhyay, from Indian Statistical Institute (ISI), New Delhi to review the book.

The publisher is ever grateful to Dr. Bandyopadhyay for the efforts he has put in for reviewing the book. As per his suggestions, some long chapters have been split into smaller ones for easy understanding and presentation. For example, old Chapter 3 has been split into Chapters 3 and 9, and old Chapter 11 has been split into Chapters 11, 12 and 13.

The publisher hopes that with these changes, the book will become more student friendly and accessible.

Preface to the First Edition

This text is the outgrowth of the several lectures I have delivered for advanced undergraduate course (honours) at Laurentian University, Canada, a graduate course at the University of Brasilia, and a graduate course at Calcutta University.

It develops the Lebesgue theory of measure and integration, using probability theory as the motivating force. This is in contrast to conventional approaches in which probability theory is usually developed after thorough exposure to the theorems and techniques of measure and integration. Due to simultaneous development of measure and probability theory, many overlaps are avoided and due to their interaction, both the fields become well illustrated and better understood. The book provides a rigorous treatment of mathematical theory and finely blends these with their applications.

Almost all theorems are illustrated by examples and applications, making them better understood and often making the reader feel that it is worth reading the long and sometimes tedious proofs of many theorems. A large number of worked-out problems ranging from easy to difficult, make the book ideal for self-study. For the benefit of the students, all the exercises are solved, and this appears as a separate section to help the not so bright students. The exercises are not only an aid for the students to understand the subject better but they complement the text. They contain a good number of examples and counter examples, illuminating the subtleties and finer points of the general theory.

A novel feature of this book is a section on large sample theory of statistics (including order in probability) just after Chapter 10 on CLT and a section on large deviation theory in the Appendix. A section on Stieltjes integration before the chapter on characteristic function and distribution function makes further reading easier. Occasionally, a different proof is given instead of a classical one (e.g. Inversion theorem for characteristic function and Bochner's theorem).

The only prerequisites for understanding the text are a sound knowledge of advanced calculus and an elementary knowledge of probability theory.

This book is suitable for one-year (or two-semester) courses offered to postgraduate students in Statistics and Mathematics (as well as for courses in operation research) of Indian universities and advanced undergraduate and first year graduate students of US universitites in Statistics, Mathematics, Operations Research and Engineering.

In a work of this kind, one has to depend on many sources. In this context, the books by Y.S. Chow and H. Tiecher, H. Tucker, K.L. Chung and M. Loéve have been of great help to me. My warm thanks and appreciation go to Mr. Lahiri and Ms. Dynacmp for their expert typing of the manuscript. I am indebted to my students and friends for the enthusiasm and encouragement they have shown while I was writing this book. Finally, I thank my ex-students, and my colleagues Dr. S. Sen Roy and Dr. D. Bhattacharyya for their assistance in checking the manuscript.

Any constructive comments for improving the contents will be warmly appreciated.

A.K. BASU

List of Symbols and Abbreviations

$\in$	element of
$\notin$	not an element of
$\Rightarrow$	logical implication
$\Leftrightarrow$	logical equivalence
iff	if and only if
$\ni$	such that
$\exists$	there exists
$\forall$	for all
$\because$	since
$\sigma(\mathcal{G})$	σ-field generated by the class $\mathcal{G}$
$\sigma(\times)$	σ-field generated by the r.v. $\times$
r.v./r.vs.	random variable(s)
$\Vert \times \Vert_p$	p-norm of $\times$, i.e. $(E \mid \times \mid^P)^{1/P}$
$c(F)$	continuity set of the function F
$\xrightarrow{\ P \text{ or } \mu\ }$	convergence in probability or μ-measure
$\xrightarrow{\ d\ }$	convergence in distribution
$\xrightarrow{\ \ \ } \text{a.s. or a.e.}$	convergence almost surely or almost everywhere
$\xrightarrow{\ L_p\ }$	convergence in pth mean
$\mathcal{B}^{\mathrm{r}}$	class of r-dimensional Borel sets or field
ch.f.	characteristic function
d.f. (c.d.f)	distribution function (cumulative distribution function)
i.i.d.	independently identically distributed
M.C.	Markov Chain
u.i.	uniformly integrable
i.o.	infinitely often
C.L.T.	Central Limit Theorem
WLLN	Weak Law of Large Numbers

SLLN	Strong Law of Large Numbers
m.g.f.	moment generating function
G.F.	Generating function
P.G.F.	Probability generating function
$\wedge$	minimum of
$\vee$	maximum of
$m(x)$	median of x
$\varlimsup\limits_{n\to\infty}$ or $\lim$	limit superior
$\varliminf\limits_{n\to\infty}$ or $\underline{\lim}$	limit inferior

$A \triangle B$ — SYMMETRIC DIFFERENCE BETWEEN SETS A AND B. P. 2

$$A \triangle B = (A - B) \cup (B - A)$$

$A \backslash B = A - B$ — DIFFERENCE OF SETS A AND B. P. 2

$f/A : A \to N$ — f RESTRICTED TO THE SET A INTO THE NATURAL NUMBERS. P. 2

1

Introduction

Sets, Indicator Functions, and Classes of Sets

1.1 SETS AND SEQUENCES OF SETS

Definition 1.1 A set is a collection of arbitrary elements. An empty set is a set of no elements and is denoted by ϕ. A set of sets is called a *class,* e.g.

1. $\{1, 2, \ldots\}$ = the set of natural numbers is a set.
2. $\{\omega : \omega = n, n$ is a positive integer$\}$ is a set.
3. $\{\omega : \omega = m/n, n = \pm 1, \pm 2, \pm 3, \ldots, m = 0, 1, 2, \ldots\}$ the set of all rational number is a set.

An important class of sets is the class of *intervals.* There are three types of intervals:

$[a, b] = \{\omega : a \leq \omega \leq b\}$ and $(a, b) = \{\omega : a < \omega < b\}$ are respectively *closed* and *open* intervals, while $[a, b) = \{\omega : a \leq \omega < b\}$ or $(a, b] = \{\omega : a < \omega \leq b\}$ are *semi-closed* intervals.

Let Ω be a non-empty set called a *space* or in statistics a *sample space.* An element of Ω is called a *point* and is denoted by ω. Let A and B be two sets of points. A is said to be a sub-set of B or included in B if all points of A are points of B and we write $A \subset B$ or $B \supset A$. The class of all subsets of Ω is called the power set of Ω and is denoted by $S(\Omega)$. ϕ is a *null* set, i.e. a set with no element. $\omega \in A$ ($\omega \notin A$) means that ω is (is not) a point of A. Obviously, $\phi \subset \Omega$. $A \subset B$, $B \subset C \Rightarrow A \subset C$.

Definition 1.2 If $A \subset B$ and $B \subset A$, then we say A and B are equal and write $A = B$.

Definition 1.3 A set A is said to be *countable,* if there exists a 1–1 (one-to-one) correspondence between A and a subset of the set of all positive integers. If A is not countable, then A is said to be *non-countable* or *non-denumerable* or simply *uncountable.*

1

EXAMPLE 1.1   Prove that if A is non-countable and $B \supset A$, then B is also non-countable.

Proof. $A \subset B$ and A is non-countable.

Suppose if possible B is countable. Then $\exists f : B \to \mathbb{N}$ which is one-to-one. But then $f/A : A \to \mathbb{N}$ which is one-to-one. This contradicts to the assumption that A is non-countable. Thus, B is non-countable.

1.1.1 Set Operations

Definition 1.4   Let A and B be two sets of points. The *difference* $A\backslash B$ is the set of all points of A which do not belong to B. $\Omega\backslash A$ is called the *complement* of A and is denoted by A^C (or A').

The *intersection* $A \cap B$ or simply AB (or $A \cdot B$) is the set of all points common to A and B. The *union* $A \cup B$ is the set of all points which belong to at least one of the sets A or B.

If $AB = \phi$, then A and B are said to be *disjoint*. $A \Delta B = (A - B) \cup (B - A)$ is called *symmetric difference* between A and B.

Simple consequences 1.

$(A \cup B) \cup C = A \cup (B \cup C)$, $(AB)C = A(BC)$

$A \cup B = B \cup A$, $AB = BA$ (commutative)

$A \cup A = A$, $A \cap A = A$ (reflexive)

$(A \cup B)C = AC \cup BC$ (distributive)

and $A \cup B = A$ iff $B \subset A$,

$A \subset B \Leftrightarrow A^C \supset B^C$, $AA^C = \phi$, $A \cup A^C = \Omega$, $\Omega^C = \phi$ and $\phi^C = \Omega$, $(A^C)^C = A$, $A - B = AB^C = A - AB$, $(A \cup B)^C = A^C B^C$, $(AB)^C = A^C \cup B^C$ (these are known as De-Morgan Laws)

$A \cup (B \cap C) = (A \cup B) \cap (A \cup C)$ (distributive).

EXAMPLE 1.2   Prove that $(A \cup B)(C \cup D) = (AC) \cup (AD) \cup (BC) \cup (BD)$ and $(A \cup B)(A \cup C) = A \cup (BC)$.

Definition 1.5   Let $\{A_t, t \in T\}$ be a class of sets (of points), where T is an arbitrary (index) set. The intersection (and similarly union) of all sets $\{A_t, t \in T\}$ is defined to be the set of all those points which belong to every (at least one) A_t and is denoted by $\underset{t \in T}{\cap} A_t \,(\underset{t \in T}{\cup} A_t)$.

If all sets of a class $\{A_t, t \in T\}$ are pairwise disjoint, then $\{A_t, t \in T\}$ is said to be disjoint or a disjoint class.

1.1.2 Limits of Sets

Definition 1.6 $\underset{k \geq n}{\inf} A_k = \overset{\infty}{\underset{k=n}{\cap}} A_k$

$$\underset{k \geq n}{\sup} A_k = \overset{\infty}{\underset{k=n}{\cup}} A_k$$

$$\liminf_{n\to\infty} A_n = \liminf_{n\to\infty}_{k\ge n} A_k = \bigcup_{n=1}^{\infty} \bigcap_{k=n}^{\infty} A_k$$

$$\limsup_{n\to\infty} A_n = \limsup_{n\to\infty}_{k\ge n} A_k = \bigcap_{n=1}^{\infty} \bigcup_{k=n}^{\infty} A_k$$

If $\liminf_{n\to\infty} A_n = \limsup_{n\to\infty} A_n = A$, say, then A is said to be the limit of A_n as $n \to \infty$ and is denoted by $\lim_{n\to\infty} A_n$ and the sequence of sets $\{A_n\}$ is said to be convergent to A. Hence, we write $A_n \to A$ as $n \to \infty$ or $\lim_{n\to\infty} A_n = A$. The following results are always true:

$$\liminf_{n\to\infty} a_n \le \limsup_{n\to\infty} a_n \text{ for numerical sequence } \{a_n\} \text{ and } \liminf_{n\to\infty} A_n \subset$$
$\limsup_{n\to\infty} A_n$ for arbitrary sequence of sets.

Proof. $A_n \supset \bigcap_{k=n}^{\infty} A_k$ for each $n \ge 1$, $\bigcup_{k=n}^{\infty} A_k \supset \bigcap_{k=m}^{\infty} A_k$ for $n \ge 1$ and $m \ge 1$, because for $i \ge \max(m, n)$ $\bigcap_{k=m}^{\infty} A_k \subset A_i \subset \bigcup_{k=n}^{\infty} A_k$. Taking union over m and intersection over n we get $\bigcap_{n=1}^{\infty} \bigcup_{k=n}^{\infty} A_k \supset \bigcup_{m=1}^{\infty} \bigcap_{k=m}^{\infty} A_k$.

Alternate proof. Let $A_* = \liminf_{n\to\infty} A_n = \{\omega : \omega \in A_n \text{ for all but a finite number of indices } n\}$ and $A^* = \limsup_{n=\infty} A_n = \{\omega : \omega \in A_n \text{ for infinitely many } n\}$.

Employing the abbreviation i.o. to designate "infinitely often" we can write $A^* = \{\omega : \omega \in A_n \text{ i.o.}\}$.

By De-Morgan's rule $A^{*C} = (\liminf A_n^C) = \{\omega : \omega \text{ belonging to only finite number of } A_n\text{'s}\}$. So A^* is the set of all points which belong to infinitely many A_n's.

Then $\omega \in A^*$ iff for every positive integer n, $\omega \in \bigcup_{k=n}^{\infty} A_n$, i.e. iff $\omega \in \bigcap_{n=1}^{\infty} \bigcup_{k=n}^{\infty} A_k$.

Therefore, $A_* \subset A^*$, since if ω belongs to all but a finite number of A_n's, it belongs to infinitely many of them.

We may use notations $\underline{\lim} A_n = \liminf_{n\to\infty} A_n$ and $\overline{\lim} A_n = \limsup_{n\to\infty} A_n$ in future.

Definition 1.7 $\{A_n\}$ is said to be monotone if it is either non-decreasing $(A_n\uparrow)$ i.e. $A_1 \subset A_2 \subset \ldots \subset A_k$ or non-increasing $(A_n\downarrow)$, i.e. $A_1 \supset A_2 \supset \ldots \supset A_k$.

Simple consequences 2.

$$A_n \uparrow \Rightarrow \lim_{n\to\infty} A_n = \bigcup_{n=1}^{\infty} A_n; \quad A_n \downarrow \Rightarrow \lim_{n\to\infty} A_n = \bigcap_{n=1}^{\infty} A_n.$$

Therefore every monotone sequence of sets is convergent.

Proof.   Since

$$A_n \uparrow \Rightarrow \bigcap_{k=n}^{\infty} A_k = A_n \Rightarrow \lim_{n \to \infty} \inf A_n = \bigcup_{n=1}^{\infty} A_n$$

By definition,

$$\lim_{n \to \infty} \sup A_n = \bigcap_{n=1}^{\infty} \bigcup_{k=n}^{\infty} A_k \subset \bigcup_{k=1}^{\infty} A_k = \lim_{n \to \infty} \inf A_n. \tag{1}$$

However,

$$\lim_{n \to \infty} \inf A_n \subset \lim_{n \to \infty} \sup A_n \tag{2}$$

From (1) and (2), $\underline{\lim} \, A_n = \overline{\lim} \, A_n$. Hence $\lim A_n$ exists and is equal to $\bigcup_{n=1}^{\infty} A_n$. Proof is similar for decreasing sequence $\{A_n\}$.

1.2   SEQUENCE OF INDICATOR FUNCTIONS

Definition 1.8   The indicator function of a set A is a function of ω defined by

$$I_A(\omega) = \begin{cases} 1 & \text{if } \omega \in A \\ 0 & \text{if } \omega \in A^C \end{cases}$$

1.2.1   Some Simple Properties of Indicator Function

1. $A \subset B \Rightarrow I_A \leq I_B$

2. $I_{A^C} = 1 - I_A$, $I_{A \cup B} = \max\{I_A, I_B\}$
        $\leq I_A + I_B$ (and equality iff $AB = \phi$)
     $I_{AB} = I_A I_B = \min(I_A, I_B)$.

3. $I_{B \backslash A}(\omega) = I_B(\omega) - I_A(\omega) I_B(\omega)$,
     $I_{A \cup B} = I_A + I_B - I_{AB}$ ($\omega \in (B \backslash A) \Rightarrow \omega \in B$ but not in A)

4. $I_{\inf_{n \geq k} A_n}(\cdot) = \inf_{n \geq k} I_{A_n}(\cdot)$, $I_{\sup_{n \geq k} A_n}(\cdot) = \sup_{n \geq k} I_{A_n}(\cdot)$ $I_{\underset{1}{\cup A_n}} \leq \sum_{n=1}^{\infty} I_{A_n}$ and   equality iff

$A_n A_m = f$ for $n \, {}^1 \, m$ (pairwise disjoint).

$I_{\lim \inf}(\cdot) = \lim_{n \to \infty} \inf I_{A_n}(\cdot)$ and $I_{\lim \sup A_n}(\cdot) = \lim_{n \to \infty} \sup I_{A_n}$.

Proof. $I_{\inf_{n \geq k} A_n} = \inf_{n \geq k} I_{A_n}$.

Now

$$I_{\inf A_n}(\omega) = 1 \Leftrightarrow \omega \in \inf_{n \geq k} A_n = \bigcap_{n=k}^{\infty} A_n$$

$$\Leftrightarrow I_{A_n}(\omega) = 1 \text{ for } n \geq K$$

$$\Leftrightarrow \inf_{n \geq k} I_{A_n}(\omega) = 1.$$

EXAMPLE 1.3   Prove that

If
$$A_n \subset \Omega, n \geq 1, I_{\underset{1}{\overset{\infty}{\cup}} A_n}(\cdot) = \sup_{n \geq 1} I_{A_n}(\cdot)$$

and
$$I_{\underset{1}{\overset{\infty}{\cap}} A_n}(\cdot) = \inf_{n \geq 1} I_{A_n}(\cdot)$$

Define
$$S_1 = \sum_{j=1}^{n} I_{A_j}, S_2 = \sum_{1 < j_1 < j_2 \leq n} I_{A_{j_1} A_{j_2}}, ..., S_n =$$
$$\sum_{1 < j_1 < .. < j_n \leq n} I_{A_{j_1} ... A_{j_n}} = I_{A_1 A_2 ... A_n}$$

Then
$$I_{\underset{1}{\overset{n}{\cup}} A_j} = S_1 - S_2 + S_3 + ... + (-1)^{n-1} S_n. \tag{1}$$

Proof of (1).   If for some ω, $I_{\underset{1}{\overset{n}{\cup}} A_j}(\omega) = 0$, then $S_k(\omega) = 0$ for $1 \leq k \leq n$. Hence (1) holds.

On the other hand if
$$I_{\underset{1}{\overset{n}{\cup}} A_n}(\omega) = 1,$$

then $\omega \in A_j$ for at least one j, $1 \leq j \leq n$. Suppose that ω belongs to exactly m of the sets, $A_j, ..., A_n$. Then
$$S_1(\omega) = m, S_2 = \binom{m}{2}, ..., S_m = \binom{m}{m}, S_{m+1} + 0, ..., S_n(\omega) = 0.$$

Hence
$$S_1 - S_2 + S_3 + ... + (-1)^{n-1} S_n = \binom{m}{1} - \binom{m}{2} + ... + (-1)^{m-1} \binom{m}{m}$$
$$= 1 - \sum_{k=0}^{m} (-1)^k \binom{m}{k}$$
$$= 1 - (1 + (-1))^m = 1 = I_{\underset{1}{\overset{n}{\cup}} A_j}$$

EXAMPLE 1.4   Prove that if B_m is the set of points belonging to exactly $m(1 \leq m \leq n)$ or $A_1, ..., A_n$, then $I_{Bm} = S_m - \binom{m+1}{m} S_{m+1} + \binom{m+2}{m} S_{2+m} + ... +$
$$(-1)^{n-m} \binom{n}{m}.$$

De-Morgan rule.   For an arbitrary class of sets $\{A_t, t \in T\}$

    (a) $(A \cup B)^C = A^C \cap B^C$

         $(A \cap B)^C = A^C \cup B^C$

    (b) $(\underset{t \in T}{\cup} A_t)^C = \underset{t \in T}{\cap} A_t^C, (\underset{t \in T}{\cap} A_t)^C = \underset{t \in T}{\cup} A_t^C.$

Duality rule.   Every valid relation between sets, obtained by taking complements, unions and intersections is transformed into a valid relation if "=" and "c" remaining unchanged $\cap$, $\subset$ and ϕ interchanged with $\cup$, $\supset$ and Ω and conversely, e.g.

$$A(B \cup C) = (AB) \cup (AC), \; A \cup (BC) = (A \cup B)(A \cup C)$$

$$\Omega^C = \phi. \text{ and conversely } \phi^C = \Omega.$$

Definition 1.9   A class of sets α is said to be closed under a set operation (e.g. arbitrary union, countable or finite) union, intersection and complement if the sets obtained by performing these operations on sets of α are in α, e.g. α is closed under union means $A, B \in \alpha \Rightarrow A \cup B \in \alpha$.

Thus $S(\Omega)$, the class of all subsets of Ω and (ϕ, Ω) are closed under every set operations.

1.3   FIELDS AND σ-FIELDS

Definition 1.10   A *field (algebra)* is a non-empty class of sets (in Ω) closed under all finite set operations, i.e. finite union, finite intersection and complementation.

A *σ-field* (or *σ-algebra* or *Borel-field*) is a non-empty class of sets (in Ω) closed under all countable set operations.

EXAMPLE 1.5 $S(\Omega)$ and (ϕ, Ω) are examples of both fields and σ-fields.

(a) α is a field $\Rightarrow \phi \in \alpha$, $\Omega \in \alpha$,

For $B \in \alpha$, $B \backslash B = \phi \in \alpha$, therefore $\phi^C = \Omega \in \alpha$.

(b)   A field α is a σ-field if Ω is finite, e.g. $\Omega = \{1, 2, 3, 4, 5, 6\}$

(c)   If α is closed under complementation and finite (countable) unions $\Rightarrow \alpha$ is a field (σ-field).

Proof. $A_i \in \alpha$ is closed under complementation and countable unions, then

$$\bigcup_{i=1}^{\infty} A_i^C \in \alpha . \text{ By De-Morgan rule, } \bigcup_{i=1}^{\infty} A_i = (\bigcup_{i=1}^{\infty} A_i^C)^C \in \alpha \Rightarrow \bigcup_{i=1}^{\infty} A_i \in \alpha.$$

Definition 1.11   *The σ-field generated by a given class of sets α.*

Let "0" be a set operation and each α_t, $t \in T$ be closed under "0". Then $\bigcap\limits_{t \in T} \alpha_t = \alpha$ (say), is closed under "0".

Let "0" be countable union, i.e. $0(A_1, A_2, \ldots) = \bigcup_1^{\infty} A_i$ and $A_i \in \alpha_t$, $i = 1, 2, 3, \ldots$ and each $t \in T$. Let $B_i \in \alpha_t$ for $i = 1, 2, 3, \ldots$ then each $B_i \in \alpha$, $t \in T$.

Then $\bigcup_1^{\infty} B_i \in \alpha_t$ for each $t \in T$, therefore $\bigcup_1^{\infty} B_i \in \bigcap_1^{\infty} \alpha_t = \alpha$. Therefore, α is closed under countable union.

We can conclude that arbitrary intersection of fields (or σ-fields) is again a field (or σ-field).

Definition 1.12   Let α be a class of sets in Ω.

The *minimal field* (or σ-field) α' over α (i.e. α' containing α) is a field (or σ-field), $\alpha' \supset \alpha$ and if $\alpha'' \supset \alpha$ is another field (or σ-field), then $\alpha'' \supset \alpha'$.

The minimal field (or σ-field) generated by a given class α is called the *field (or σ-field) generated by α.*

The *σ-field generated by α is denoted by $\mathscr{B}(\alpha)$.*

Theorem 1.1   There is a unique minimal field (σ-field) over a class α.

Proof.   Put $\alpha = \{\mathscr{B} \mid \mathscr{B} \supset \alpha, \mathscr{B}$ is a field$\}$. Then α is a non-empty class of sets, since $S(\Omega)$ is such a class. Let $\alpha' = \underset{\mathscr{B}\in\alpha}{\cap} \mathscr{B}$.

Then α' is a field and it contains α. Suppose that if α'' is a field and it contains α.

Then $\alpha' \subset \alpha''$ and α' is the minimal one.

EXAMPLE 1.6 Ω is any infinite set.

$\alpha = \{$All finite subsets of Ω and those subsets of Ω whose complements are finite$\}$ is a field.

For example, take $\Omega = \{1, 2, 3, 4, 5, \ldots\}$. The α is an algebra but not a σ-algebra, because if we teke $A_i = \{2i\}$, then

$$\overset{\infty}{\underset{i=1}{\cup}} A_i = \{2, 4, 6, \ldots\} \notin \alpha.$$

EXAMPLE 1.7 Ω = Real line, $\alpha = \{$Finite disjoint unions of a semiclosed intervals of the type $(a, b]\}$. (a, ∞) is by conventions right semiclosed for $-\infty \le a < b < \infty$, since $(-\infty, a] \in \alpha \Rightarrow (a, \infty) \in \alpha$.
Hence that is a field.

EXAMPLE 1.8 $R^2 : \alpha = \{$Finite disjoint union of semiclosed rectangles, $(a, b] \times (c, d]\}$. In this case also similar conventions are followed. This is a field.
NOTE.   In Example 1.6 all finite subsets of Ω are not an algebra (unless Ω is finite).

EXAMPLE 1.9 Ω = An uncountable set.

$\mathscr{B} = \{$All countable subsets of Ω and those subsets of Ω whose complements are countable$\}$ is a σ-field.

1.4   MONOTONE CLASS

1.4.1   Monotone Class or Monotone System of Sets

Definition 1.13   A class $\mathscr{M}$ of subsets of Ω is said to be a monotone class if (1) $\mathscr{M}$ is non-empty (2) $A_n \in \mathscr{M}$ for $n \ge 1$ and $A_n \uparrow$ or $\downarrow$ then $A = \lim A_n$ (exists) and $A \in \mathscr{M}$.

Proposition 1.1   Given a class of sets α, there is the smallest monotone class $\mathscr{M}(\alpha)$ which contains α (is called the monotone class generated by α).

Proof.   First we prove, if $\mathscr{M}_\alpha$ is any collection of monotone classes, then, $\mathscr{M} = \underset{\alpha}{\cap} \mathscr{M}_\alpha$ is a monotone class provided $\mathscr{M}$ is non-empty. Suppose $\mathscr{M}$ is non-empty, let $A_n \in \mathcal{M}$, and $A_n \uparrow$ or $\downarrow$. Then $A_n \in \mathcal{M}_\alpha$ for each $\alpha \Rightarrow \lim A_n$ exists and $\lim A_n \in \mathcal{M}_\alpha$ for each $\alpha \Rightarrow \lim A_n \in \underset{\alpha}{\cap} \mathcal{M}_\alpha = \mathcal{M}$.

Define $\mathcal{M}(\alpha) = \underset{\alpha}{\cap} [\mathcal{M}_\alpha \mid \mathcal{M}_\alpha \supset \alpha$ and $\mathcal{M}_\alpha$ is a monotone class] is a non-empty class of sets, since $S(\Omega) \supset \alpha$ and is a monotone class. Obviously $\mathcal{M}(\alpha)$ is the required monotone class.

Theorem 1.2   If α is an algebra, then $\mathcal{M}(\alpha) = \mathcal{B}(\alpha)$, where $\mathcal{M}(\alpha)$ is the monotone class, generated by α and $\mathcal{B}(\alpha)$ is the σ-field generated by α.

EXAMPLE 1.10 Ω = any uncountable set.

$\mathcal{M}$ = class of all countable subsets of Ω is a monotone class but not a σ-field. Take $A_n \in \mathcal{M}$ and $A_n \uparrow$. Then their union is countable and hence is in $\mathcal{M}$ and $\mathcal{M}$ is closed under monotone limit. So $\mathcal{M}$ is a monotone class.

Let A be a countable subset of Ω then A^C is uncountable. Therefore $A^C \notin \mathcal{M}$. Hence $\mathcal{M}$ is not a σ-field (since $\mathcal{M}$ is not closed under complementation). Note that a σ-field is a monotone class. For if $\mathcal{F}$ be a σ-field and taking any $A_n \uparrow$ sets from $\mathcal{F}$ ($A_n \in \mathcal{F}$, $n \geq 1$),

$$\lim A_n = \overset{\infty}{\underset{1}{\cup}} A_n \in \mathcal{F}$$

Hence $\mathcal{F}$ is a monotone class.

EXAMPLE 1.11   Prove that a σ-field is a monotone field (monotone class and also a field) and conversely.

Proof (of theorem 1.2).

    (a) $\mathcal{M}(\alpha)$ is a monotone class and

    (b) $\mathcal{M}(\alpha) \supset \alpha$.

We have proved that if $\mathcal{G} \supset \alpha$ and $\mathcal{G}$ is a monotone class, the $\mathcal{G} \supset \mathcal{M}(\alpha)$.

We also have seen that $\mathcal{B}(\alpha)$ is a monotone class and $\mathcal{B}(\alpha) \supset \alpha$ and hence $\mathcal{B}(\alpha) \supset \mathcal{M}(\alpha)$. So it is enough to prove $\mathcal{B}(\alpha) \subset \mathcal{M}(\alpha)$, i.e. it is enough to prove $\mathcal{M}(\alpha)$ is a σ-field.

Now to prove $\mathcal{M}(\alpha)$ is closed under countable union, we first assume that $\mathcal{M}(\alpha)$ is closed under finite union. Let

$$B_n = \overset{n}{\underset{1}{\cup}} A_i, \, A_i \in \mathcal{M}(\alpha), i \geq 1.$$

Then

$$B_n \uparrow \overset{\infty}{\underset{1}{\cup}} A_i.$$

Now $\mathcal{M}(\alpha)$ is a monotone class and hence

$$\underset{n \to \infty}{\lim} B_n \in \mathcal{M}(\alpha)$$

Therefore,

$$\overset{\infty}{\underset{1}{\cup}} A_i \in \mathcal{M}(\alpha).$$

Now we need to prove that if $A \in \mathcal{M}(\alpha)$, then $A^C \in \mathcal{M}(\alpha)$.

Let $\mathcal{K} = \{A \mid A \in \mathcal{M}(\alpha), A^C \in \mathcal{M}(\alpha)\}$. Then it is non-empty, since ϕ, Ω $\in \mathcal{M}(\alpha)$. Now if $A \in \alpha$, then $A^C \in \alpha$ and therefore $A \in \mathcal{K}$. Also $\mathcal{K} \supset \alpha$. If $A_n \downarrow A$ and each $A_n \in \mathcal{K}$, then $A_n \in \mathcal{M}(\alpha)$ and $A_n^C \in \mathcal{M}(\alpha)$ and therefore $A \in \mathcal{M}(\alpha)$ and $A^C \in \mathcal{M}(\alpha)$ and that implies $A \in \mathcal{K}$. Similarly, taking an increasing sequence of sets from $\mathcal{K}$, the limit set must be in $\mathcal{K}$, i.e. $\mathcal{K}$ is a monotone class. Since $\mathcal{K} \supset \alpha$, $\mathcal{K} \supset \mathcal{M}(\alpha)$. Therefore $A \in \mathcal{M}(\alpha) \Rightarrow A \in \mathcal{K}$ and $A^C \in \mathcal{M}(\alpha)$. Hence $\mathcal{M}(\alpha)$ is closed under complement. To prove $\mathcal{M}(\alpha)$ is closed under finite union we define

$\mathcal{D}_B = \{A \mid A \cup B \in \mathcal{M}(\alpha) \text{ for all } B \in \alpha\}$. If $A_n \in \mathcal{D}_B$ and $A_n \downarrow A$, then $A = \overset{\infty}{\underset{n=1}{\cap}} A_n$ and $A \cup B \in \mathcal{M}(\alpha)$, for $(A_n \cup B) \in \mathcal{M}(\alpha)$ and $\{A_n \cup B\} \downarrow$ implies $A \cup B = \overset{\infty}{\underset{n=1}{\cap}} (A_n \cup B) \in \mathcal{M}(\alpha)$ and hence $\lim A_n \in \mathcal{D}_B$. Similarly for $A_n \uparrow$ in $\mathcal{D}_B$, $\lim A_n \in \mathcal{D}_B$. So $\mathcal{D}_B$ is a monotone class and $\mathcal{D}_B \supset \alpha$. Therefore by minimality of $\mathcal{M}(\alpha)$, $\mathcal{M}(\alpha) \subset \mathcal{D}_B$. Define $\mathcal{D} = \{B \mid A \cup B \in \mathcal{M}(\alpha) \text{ for all } A \in \mathcal{M}(\alpha)\}$. Then $\mathcal{D}$ is a monotone class, since if $B_n \uparrow B$ and $B_n \in \mathcal{D}$, $(A \cup \lim B_n) = \lim (A \cup B_n) \in \mathcal{M}(\alpha)$. Now from $\mathcal{M}(\alpha) \subset \mathcal{D}_B$ and $B \in \alpha \Rightarrow A \cup B \in \mathcal{M}(\alpha)$ and in other words $B \in \alpha \Rightarrow B \in \mathcal{D}$. Thus $\alpha \subset \mathcal{D}$. By minimality $\mathcal{M}(\alpha) \subset \mathcal{D}$ and hence $A, B \in \mathcal{M}(\alpha) \Rightarrow A \cup B \in \mathcal{M}(\alpha)$.

1.5 BOREL SETS IN A REAL LINE

The set $R = (-\infty, \infty)$ of all finite numbers is called a real line. The basic class of sets in R are the four types of intervals

$$(a, b), (a, b], [a, b) \text{ and } [a, b].$$

The Borel-field on R is the σ-field (minimal) generated by the finite intervals. The members of the Borel-field are called the Borel sets of the real line. Note that any type of interval can be written as countable operations of any other type. For example

$$[a, b] = \overset{\infty}{\underset{n=1}{\cap}} \left(a - \frac{1}{n}, b + \frac{1}{n} \right), \{a\} = \overset{\infty}{\underset{n=1}{\cap}} \left(a - \frac{1}{n}, a + \frac{1}{n} \right)$$

$$(a, b] = \overset{\infty}{\underset{1}{\cap}} \left(a, b + \frac{1}{n} \right)$$

$$[a, b) = \overset{\infty}{\underset{1}{\cap}} \left(a - \frac{1}{n}, b \right).$$

Also note that every open set on the real line can be written as countable union of open intervals. The Borel field in R is also the minimal σ-field over the class of infinite intervals of the form $(-\infty, x)$, $-\infty < x < \infty$, since any finite interval can be written as $[a, b) = (-\infty, b) - (-\infty, a)$.

PROBLEMS AND COMPLEMENTS

1.1   Let $x \in \mathbb{R}^2$ and A_n = interior of a circle with centre $\left(\dfrac{(-1)^n}{n}, 0 \right)$ and radius unity. Find $\limsup A_n$, $\liminf A_n$ and $\lim A_n$ if it exists.

1.2   Find $\limsup A_n$ and $\liminf A_n$ of the following sequence of sets:

(a) $A_n = \begin{cases} [x:0 \le x < 1] & \text{if } n \text{ is odd} \\ [x:1 \le x \le 2] & \text{if } n \text{ is even} \end{cases}$

(b) $E_n = \begin{cases} [x:-n \le x \le 0] & \text{if } n \text{ is odd} \\ [x:1/n \le x \le n] & \text{if } n \text{ is even} \end{cases}$

1.3   Give an example of a class of sets closed under finite unions and finite intersections but not under complementation.

1.4   Show that the following set relations hold:

(a) $E \triangle F = F \triangle E$

(b) $(E \triangle F) \triangle G = E \triangle (F \triangle G)$

(c) $(E \triangle F) \triangle (G \triangle H) = (E \triangle G) \triangle (F \triangle H)$

(d) $E \triangle F = \phi$ iff $E = F$

(e) $(E \triangle F)^C \cup (E^C \triangle F^C) = \Omega$

(f) $(E \triangle F) \subset (E \triangle G) \cup (G \triangle F)$

(g) $E \triangle F = E^C \triangle F^C$, $G = E \triangle F$ iff $E = F \triangle G$

(h) $(\bigcup_{i=1}^{\infty} E_i) \triangle (\bigcup_{i=1}^{\infty} F_i) = \bigcup_{i=1}^{\infty}(E_i \triangle F_i)$

(i) $(\bigcup_{i=1}^{\infty} E_i) \triangle (\bigcup_{i=1}^{\infty} F_i) \subset \bigcup_{i=1}^{\infty}(E_i \triangle F_i)$

(j) $(\bigcap_{i=1}^{\infty} E_i) \triangle (\bigcap_{i=1}^{\infty} F_i) \subset \bigcap_{i=1}^{\infty}(E_i \triangle F_i)$

1.5   Let $\Omega = [0, \infty)$ and $\mathcal{F}_1$ the class of all intervals of the type $[a, b)$ or $[a, \infty)$ where $0 \le a < b < \infty$. Let $\mathcal{F}_2$ be the class of all finite disjoint unions of intervals of $\mathcal{F}_1$. Show that $\mathcal{F}_1$ is not a field and $\mathcal{F}_2$ is a field but not a σ-field.

1.6   Show that

(a) if $\mathcal{F}$ is a field such that for every disjoint sequence $\{F_n, n \ge 1\}$ of sets in $\mathcal{F}$, $\bigcup_{n=1}^{\infty} F_n \in \mathcal{F}$, then $\mathcal{F}$ is a σ-field.

(b) if $\mathcal{F}_n$ is an increasing sequence of σ-fields, then $\mathcal{F} = \bigcup_{n=1}^{\infty} \mathcal{F}_n$ is a field.

Give a counter-example to show that $\mathcal{F}$ may not be a σ-field.

(c) the σ-field generated by a countable class of disjoint, non-empty sets whose union is equal to the whole set Ω is the class of all unions of these sets.

1.7 Let $\{A_i, i = 1, \ldots, n\}$ be a partition of Ω. Find the minimal field $\alpha(\{A_i\})$ generated by $\{A_i\}$. What is the cardinality of this field?

1.8 Let

$$A_n = \{(x, y) : 0 \le x \le n, 0 \le y < 1/n\}, (x, y) \in \mathbb{R}^2.$$

Show that A_n is not monotone but $\lim_{n \to \infty} A_n$ exists. Also find $\lim_{n \to \infty} A_n$.

1.9 (a) Show that any sequence of disjoint sets converges to the null set.

(b) Show that $\{E_n\}$ is a convergent sequence iff there is no point x of X such that each $x \in E_n$, $x \in X - E_n$ hold for infinitely many n.

(c) Let $E_n = \begin{cases} \left[x : 0 < x < 1 - \dfrac{1}{n} \right] & \text{if } n \text{ is odd} \\[2ex] \left[x : \dfrac{1}{n} \le x < 1 \right] & \text{if } n \text{ is even} \end{cases}$

Prove that $\{E_n\}$ converges but is not monotone.

2

Measure Space and Probability Space

2.1 DEFINITION AND FUNDAMENTAL PROPERTIES OF MEASURE

Definition 2.1

1. By a *set function* μ we mean a function which assigns an extended real number to certain class of sets.

2. By a *measurable space* we mean $(\Omega, \mathcal{B})$ consisting of a set Ω and a σ-algebra $\mathcal{B}$ of subsets of Ω. A subset A of Ω is called measurable (or $\mathcal{B}$-measurable if $A \in \mathcal{B}$.

3. By a *measure* μ on a measurable space $(\Omega, \mathcal{B})$ we mean a non-negative set function defined for all sets $\mathcal{B}$ and satisfying $\mu(\phi) = 0$,

$$\mu(\bigcup_1^\infty E_i) = \sum_1^\infty \mu(E_i)$$

   for any sequence $\{E_i\}$ of disjoint measurable sets, i.e.

$$E_i \cap E_j = \phi, \ E_i \in \mathcal{B}. \tag{1}$$

4. $(\Omega, \mathcal{B}, \mu)$ is called a *measure space.*

NOTE. Property 1 is called countable additivity (σ-additivity).

If
$$\mu(\bigcup_1^n E_i) = \sum_1^n \mu(E_i) \text{ for disjoint } E_i \in \mathcal{B} \tag{2}$$

then the set function μ is called finitely additive. Countably additive $\Rightarrow$ finitely additive but the converse is not true.

2.2 DIFFERENT EXAMPLES OF MEASURES

EXAMPLE 2.1 *Lebesgue measure.* Lebesgue measure space $(R, \mathcal{B}, m)$, where $R = (-\infty, \infty)$ is the real line, $\mathcal{B}$ = Borel sets of the real line, and m = Lebesgue measure. Lebesgue measure is *translation invariant* i.e. $m(A + x) = m(A)$ where

x is a point of the real line and $m(I) = l(I)$, where I is an interval of the real line and $l(I)$ is the length of the interval I.

EXAMPLE 2.2 *Counting measure.* $(R, S(R), r)$, where $S(R)$ is the class of all subsets of real line R. r = counting measure defined by

$$r(E) = \begin{cases} \infty \text{ if } E \text{ is an infinite set} \\ \text{number of elements of } E \text{ if } E \text{ is a finite set} \end{cases}$$

EXAMPLE 2.3 Probability measure. Let $(\Omega, \mathcal{F}, P)$ be a probability space,

Ω = any space, $\mathcal{F}$ = a σ-algebra of the subsets of Ω.
P = probability measure, where P is a set function defined on Ω and satisfies

$$0 \le P(E) \le 1, \text{ for } E \in \mathcal{F}, \text{ and } P(\Omega) = 1.$$

P is a measure $\Rightarrow P$ is non-negative and $P(\phi) = 0$.
We know that $P(A \cup B) = P(A) + P(B)$ if $A \cap B = \phi$.

The following analogies exist between different terminologies in probability theory and measure theory.

$$\text{Events} \equiv \text{Measurable sets}, E_i \cap E_j \ne \phi, i \ne j, P(\overset{r}{\underset{1}{\cup}} E_i) = P(\Omega) = 1$$

means $E_1, \ldots, E_r$ are mutually exclusive and exhaustive.

Mutually exclusive events $\equiv$ Disjoint sets and E is an impossible event means $P(E) = 0$.

EXAMPLE 2.4 Ω = any uncountable set
$\mathcal{B}$ = the family of those subsets which are countable or complement of a countable set

$$\mu(E) = \begin{cases} 0 & \text{if } E \text{ is countable} \\ 1 & \text{if } E^C \text{ is countable} \end{cases}$$

Note that all of the above measures satisfy the fundamental properties of measure listed in Section 2.1.

Further Properties of Measure

Proposition 2.1 If $A \in \mathcal{B}, B \in \mathcal{B}$ and $A \subset B$ then $\mu(A) \le \mu(B)$ (monotone) and

$\mu(\overset{\infty}{\underset{1}{\cup}} E_n) \le \overset{\infty}{\underset{1}{\Sigma}} \mu(E_n), E_i \in \mathcal{B}$ (countably subadditive). In particular,

$$\mu(\overset{k}{\underset{1}{\cup}} E_n) \le \overset{k}{\underset{1}{\Sigma}} \mu(E_n), \ E_i \in \mathcal{B} \text{ (finitely subadditive)}$$

Proof. Since $A \subset B, B = A \cup (B \cap A^C)$ (disjoint union), then $\mu(B) = \mu(A) + \mu(BA^C) \ge \mu(A)$.

$\overset{\infty}{\underset{n=1}{\cup}} E_n = \overset{\infty}{\underset{1}{\cup}} B_n,$ where $B_n = E_n \cap E_1^C \cap E_2^C \cap \ldots \cap E_{n-1}^C$ and B_n's are disjoint. So

$$\mu(\overset{\infty}{\underset{1}{\cup}} E_n) = \mu(\overset{\infty}{\underset{1}{\cup}} B_n) = \overset{\infty}{\underset{1}{\Sigma}} \mu(B_n) = \overset{\infty}{\underset{1}{\Sigma}} \mu(E_n \cap E_1^C \cap \ldots \cap E_{n-1}^C) \le \overset{\infty}{\underset{1}{\Sigma}} \mu(E_n).$$

Proposition 2.2   Let $(\Omega, \mathcal{B}, \mu)$ be a measure space. If $E_i \in \mathcal{B}$, $\mu(E_1) < \infty$ and $E_{i+1} \subset E_i$ (i.e. decreasing sequence of sets) then $\mu(\overset{\infty}{\underset{1}{\cap}} E_i) = \lim_{n \to \infty} \mu(E_n)$. This is known as *continuity theorem of measure*.

Proof.   Set $E = \overset{\infty}{\underset{i=1}{\cap}} E_i$, $E_1 = E \cup \overset{\infty}{\underset{1}{\cup}} (E_i - E_{i+1})$ (disjoint union) and $(E_i - E_{i+1}) \in \mathcal{B}$. Hence

$$\mu(E_1) = \mu(E) + \overset{\infty}{\underset{i=1}{\Sigma}} \mu(E_i - E_{i+1}) \tag{1}$$

Note the $E_i = (E_{i+1}) \cup (E_i - E_{i+1})$ is a disjoint union. Therefore

$$\mu(E_i) = \mu(E_{i+1}) + \mu(E_i - E_{i+1}) \text{ (since } \mu(E_{i+1}) \le \mu(E_i) < \infty)$$
$$\Rightarrow -\mu(E_{i+1}) + \mu(E_i) = \mu(E_i - E_{i+1}) \tag{2}$$

From (1) and (2),

$$\mu(E_1) = \mu(E) + \overset{\infty}{\underset{i+1}{\Sigma}} [\mu(E_i) - \mu(E_{i+1})]$$

(note that $\mu(E_i) - \mu(E_{i+1}) \ge 0$)
or

$$\mu(E_1) = \mu(E) + \lim_{n \to \infty} \overset{n-1}{\underset{i=1}{\Sigma}} [\mu(E_i) - \mu(E_{i+1})]$$
$$= \mu(E) + \mu(E_1) - \lim_{n \to \infty} \mu(E_n)$$
$$\Rightarrow \mu(\overset{\infty}{\underset{1}{\cap}} E_i) = \lim_{n \to \infty} \mu(E_n)$$
$$\Rightarrow \mu(E) = \lim_{n \to \infty} \mu(E_n) \text{ (since } \mu(E_1) < \infty)$$

(Note that the proposition is true if μ is replaced by P without any condition of finiteness.)

2.3   RANDOM VARIABLE AND MEASURABLE TRANSFORMATION

Definition 2.2

1. A measure μ is called *finite* if $\mu(\Omega) < \infty$.

2. A measure μ is called *σ-finite* if there is a sequence $\{A_n\}$ of set in $\mathcal{B}$ such that $\Omega = \underset{n}{\cup} A_n$ and $\mu(A_n) < \infty$. Lebesgue measure on $R = (-\infty, \infty)$ is a fine example of a σ-finite measure. Any probability measure is a finite measure. The counting measure on an uncountable set is a measure which is not σ-finite.

3. A set E is said to be of *finite measure* if $E \in \mathcal{B}$ and $\mu(E) < \infty$. A set E is said to be of *σ-finite measure* if E is the union of a countable collection of measurable sets (i.e. $E_i \in \mathcal{B}$) and each E_i is of finite measure.

NOTE: If μ is σ-finite, then every measurable set is of σ-finite measure (by monotonic property of measure).

4.   A measure space $(\Omega, \mathcal{B}, \mu)$ is said to be complete if $\mathcal{B}$ contains all subsets of sets of measure zero, i.e. if $B \in \mathcal{B}$,

$$\mu(B) = 0 \text{ and } A \subset B \Rightarrow A \in \mathcal{B}.$$

Definition 2.3   Let X be an extended real valued function on Ω. If $\{\omega \mid X(\omega) < a\} = [X < a] \in \mathcal{B}$, i.e. $X^{-1}[-\infty, a) \in \mathcal{B}$ for every $-\infty < a < \infty$, then X is said to be a measurabole function or $\mathcal{B}$-measurable function. Note that $X(\omega)$ is a mapping from Ω to $\bar{R} = [-\infty, \infty]$. In a probability space, finite measurable functions are random variables (r.vs.), i.e. $P[\mid X \mid = \infty] = 0$ and X is measurable.

2.3.1   Measurable Transformation

Let Ω_1 and Ω_2 be two reference sets and X is a function on Ω_1 with values in Ω_2 i.e. $X: \Omega_1 \to \Omega_2$.

For every subset A of Ω_2 and class $\mathcal{G}$ of subsets of Ω_2, define, $X^{-1}(A) = \{\omega : \omega \in \Omega_1, X(\omega) \in A\}$ and $X^{-1}(\mathcal{G}) = \{X^{-1}(A) : A \in \mathcal{G}\}$.

The set $X^{-1}(A)$ and $X^{-1}(\mathcal{G})$ are called respectively the inverse images of the sets A and of the class $\mathcal{G}$. Note that

$$X[\cup_{\alpha} A_{\alpha}] = \cup_{\alpha}[X(A_{\alpha})], \text{ but } X[\cap_{\alpha} A_{\alpha}] \subset \cap_{\alpha} X(A_{\alpha}).$$

Note also that inverse image is closed under all set operations, viz.

$$X^{-1}[\cup_{\alpha} A_{\alpha}] = \cup_{\alpha}[X^{-1}(A_{\alpha}), X^{-1}[\cap_{\alpha} A_{\alpha}] = \cap_{\alpha} X^{-1}(A_{\alpha}),$$

$$X^{-1}(A^C) = [X^{-1}(A)]^C$$

and that

$$X[X^{-1}(A)] \subset A, X^{-1}(X(A)) \supset A.$$

If follows that $X^{-1}(\mathcal{G})$ of σ-algebra $\mathcal{G}$ on Ω_2 is a σ-algebra on Ω_1 and the class $\{B : B \subset \Omega_2, X^{-1}(B) \in \mathcal{K}\}$ is a σ-algebra on Ω_2 if $\mathcal{K}$ is a σ-algebra on Ω_1. If $(\Omega_1, \mathcal{G}_1)$ and $(\Omega_2, \mathcal{G}_2)$ are measurable space and $X : \Omega_1 \to \Omega_2$ then X is said to be a measurable transformation from $(\Omega_1, \mathcal{G}_1)$ to $(\Omega_2, \mathcal{G}_2)$ provided $X^{-1}(\mathcal{G}_2) \subset \mathcal{G}_1$.

2.4   DISCRETE SAMPLE SPACE

A finite family of events $(A_n)_{n=1}^{N}$ is said to be independent if $P(A_{i_1} \ldots A_{i_k}) = P(A_{i_1})P(A_{i_2}) \ldots P(A_{i_1})$, $1 \leq i_1, \ldots, i_k \leq N, k \leq N$. A finite family of r.vs. $(X_n)_{n=1}^{N}$ is said to be independent if

$$P[X_{i_1} \leq a_{i_1}, \ldots, x_{i_k} \leq a_{i_k}] = P[X_{i_1} \leq a_{i_1}] \ldots P[x_{i_k} \leq a_{i_k}]$$

for all $a_1, \ldots, a_N$ of real numbers and $i \leq i_1, \ldots, i_k \leq N$. A family of events $(A_t)_{t \in T}$ (in case of r.vs. $(X_t)_{t \in T}$) is said to be independent if for every finite subset $T^* \subset T$ (or $(X_t)_{t \in T}^*$) is independent, e.g. $T^* = \{t_1, \ldots, t_n\} \subset T = [0, 1]$.

Definition 2.4   A r.v. X is said to be *discrete* if there exists a countable set $\{r_n\}_{n=1}^{\infty}$ of real numbers such that $P[\omega : X(\omega) \notin \{r_n\}_{n=1}^{\infty}] = 0$ and in the case r_n's are said to be the possible values of the r.v. X. If r is a possible value of X then $P[X = r] > 0$.

Definition 2.5   Two r.vs. X and Y are said to be identically distributed if $P[X \leq a] = P[Y \leq a]$ for all real numbers a. Two countable families of r.vs. $(X_n)_1^{\infty}$ and $(Y_n)_1^{\infty}$ are said to have the same joint distribution if for every finite positive integer $N = 1, 2, 3, \ldots$

$$[X_1 \leq a_1, \ldots, X_N \leq a_N] = P[Y_1 \leq a_1, \ldots, Y_N \leq a_N]$$

for all $a_1, \ldots, a_N$ of real numbers. Let Ω be a countable set. More specifically $\Omega = \{\omega_j, j \in J\}$, where J is a countable index set and let $\mathscr{F} = S(\Omega)$. Choose any sequence of real numbers $\{p_j, j \in J\}$ satisfying for all

$$j \in J, \, p_j \geq 0, \, \sum_{j \in J} p_j = 1 \tag{1}$$

and define a set function P on $\mathscr{F}$ as follows. For all $E \in \mathscr{F}$. $P(E) = \sum_{\omega_j \in E} p_j$. In words we assign p_j the value of the "probability" of the singleton $\{\omega_j\}$ and for an arbitrary set of ω_j's we assign as its probability the sum of all probabilities assigned to its elements. Clearly axioms (i) for all $E \in \mathscr{F}$, $P(E) \geq 0$, (ii) if $\{E_j\}$ is a countable collection of (pairwise) disjoint events in $\mathscr{F}$, then $P[\cup_j E_j] = \sum_j P(E_j)$, (iii) $p(\Omega) = 1$ are satisfied. Hence $P(.)$ so defined is a probability measure. Conversely, let any such P as given on $\mathscr{F}$. Since $\{\omega_j\} \in \mathscr{F}$ for every j, $P(\{\omega_j\})$ is defined, let its value be $p_j \geq 0$. Then (i) is satisfied. We have thus exhibited all the probability measures on Ω or rather the pair $(\Omega, \mathscr{F})$. This is the alternative way (the constructive way) to define a discrete sample space.

2.4.1   A Mathematical Model for Tossing a Coin N Times

In order to construct a mathematical model for the above experiment, let $\Omega = \{\omega = (\omega_1, \ldots, \omega_n) \mid \omega_j = 0, 1 \text{ for } j = 1, \ldots N\}$ and $\mathscr{F} = S(\Omega)$. Then Ω has 2^N points and $\mathscr{F}$ has 2^{2^N} sets. For a typical $\omega = (\omega_1, \ldots, \omega_n)$ put

$$p_\omega = p^{\omega_1 + \omega_2 + \ldots + \omega_n} q^{N - \sum_1^N \omega_i} \quad (1), \quad P(A) = \sum_{\omega \in A} P_\omega, A \in \mathscr{F} = S(\Omega) \tag{2}$$

Now $p_\omega = p^{\sum \omega_i} q^{N - \sum_1^N \omega_i} \geq 0.$ Obviously $P(\Omega) = 1.$

$$P(A \cup B) = \sum_{\omega \in A \cup B} p_\omega = \sum_{\omega \in A} p_\omega + \sum_{\omega \in B} p_\omega \text{ if } A \text{ and } B \text{ are disjoint} = P(A) + P(B)$$

Obviously, P is additive and since $\mathscr{F}$ is finite, P is σ-additive also.

$$P(\Omega) = \sum_{\omega \in \Omega} p_\omega = \Sigma_{\omega_1} \Sigma_{\omega_2} \ldots \Sigma_{\omega_n} p^{\sum_1^N \omega_i} q^{N - \sum_1^N \omega_i}, \quad p + q = 1$$

$$= \Sigma_{\omega_1} \Sigma_{\omega_2} \ldots \sum_{\omega_{N-1}} p^{\sum_1^{N-1} \omega_i} q^{N-1-\sum_1^N \omega_i} \sum_{\omega_N = 0.1} p^{\omega_n} q^{1-\omega_n}$$

$$= \Sigma_{\omega_1} \ldots \sum_{\omega_{N-1}} p^{\sum_1^{N-1} \omega_i} q^{N-1-\sum_1^{N-1} \omega_i} \ldots = \sum_{\omega_1} p^{\omega_1} q^{1-\omega_1} = (p+q) = 1$$

Define $X_j(\omega) = \omega_j$, $1 \le j \le N$ for $\omega \in \Omega$. X_j's are called coordinate r.vs. in this model.

2.5 COMBINATORIAL ASPECTS OF SET FUNCTIONS

2.5.1 Combinatorial Formulae for Additive Set Function

Theorem 2.1(a) Let α be any class of sets and ϕ be a σ-additive set function on α. If $\{A_n\}_1^\infty \subset \alpha$ is disjoint and

$$\bigcup_1^\infty A_n \in \alpha \quad \text{and} \quad \left| \phi\left(\bigcup_1^\infty A_n\right) \right| < \infty \tag{1}$$

then

$$\sum_1^\infty |\phi(A_n)| < \infty.$$

Note that for a real sequence $|\Sigma a_n| < \infty \not\Rightarrow \Sigma |a_n| < \infty$.

Proof. By σ-additivity of ϕ, $\phi\left(\bigcup_1^\infty A_n\right) = \sum_1^\infty \phi(A_n) < \infty$ by (1). For any permutation $\{n_k, k \ge 1\}$ of $\{n, n \ge 1\}$, $\bigcup_{k=1}^\infty A_{n_k} = \bigcup_1^\infty A_n$ (because for union of sets the choice of its order is immaterial). Hence

$$\sum_{k=1}^\infty \phi(A_{n_k}) = \phi\left(\bigcup_{k=1}^\infty A_{n_k}\right) = \phi\left(\bigcup_1^\infty A_n\right) = \sum_1^\infty \phi(A_n) < \infty.$$

(By Weierstrass theorem of rearrangements of terms Sank $\Sigma a_{n_k} < \infty$ for all rearrangements of the terms a_n implies $\Sigma |a_n| < \infty$)

Theorem 2.1(b) Let ϕ be additive on a field α and $A_i \in \alpha$ for $i \ge 1$.

  (i) If $A_1 \subset A_2$ and $\phi(A_2) < \infty$, then $\phi(A_1) < \infty$.

  (ii) If $\phi \ge 0$, then ϕ is non-decreasing, i.e. $\phi(A_2) \le \phi(A_1)$ if $A_1 \subset A_2$.

  (iii) If $\phi \ge 0$ and ϕ is σ-additive, then $\phi\left(\bigcup_1^\infty A_n\right) \le \sum_1^\infty \phi(A_n)$ if $\bigcup_1^\infty A_n \in \alpha$
    (σ-subadditive).

Proof. The proof follows similarly as for measures and hence is omitted.

Theorem 2.1(c) Let ϕ be additive on a field α and $A_k \in \alpha$ for $k = 1, \ldots, n$. If moreover $| \phi(\overset{n}{\underset{1}{\cup}} A_n) | < \infty$, then

$$\phi(\overset{n}{\underset{1}{\cup}} A_k) = \overset{\infty}{\underset{1}{\Sigma}} \phi(A_k) - \underset{1 \le k < k' \le n}{\Sigma} \phi(A_k A_{k'}) + \ldots + (-1)^{n-1} \phi(A_1 A_2 \ldots A_n).$$

Proof. Put $\phi(A_k) = \phi_k$, $\phi(A_{k_1} A_{k_2}) = \phi_{k_1 k_2}, \ldots, \phi(A_{k_1} \ldots A_{k_r}) = \phi_{k1 \ldots k_r}$, $r \ge 2$. Obviously, (c) holds for $n = 1$.

Assume (c) holds for $n = m$. Then

$$\phi(\overset{m+1}{\underset{1}{\cup}} A_k) = \phi(\overset{m}{\underset{1}{\cup}} A_k) + \phi(A_{m+1} - \overset{m}{\underset{1}{\cup}} A_k) \text{ (disjoint unions)}$$

and $\phi(A_{m+1}) = \phi(A_{m+1} - \overset{m}{\underset{1}{\cup}} A_k) + \phi(A_{m+1} \cap (\overset{m}{\underset{1}{\cup}} A_k))$ $[A = (A - B) \cup (AB)$

disjoint union]

$$\Rightarrow \phi(A_{m+1} - \overset{m}{\underset{1}{\cup}} A_k) = \phi(A_{m+1}) - \phi(\overset{m}{\underset{k=1}{\cup}} A_{m+1} A_k)$$

Therefore,

$$\phi(\overset{m+1}{\underset{1}{\cup}} A_k) = \overset{m}{\underset{k=1}{\Sigma}} \phi_k - \underset{1 \le k_1 < k_2 \le m}{\Sigma} \phi_{k_1 k_2} + \ldots + (-1)^{m-1} \phi_{1 2 \ldots m}$$

$$+ \phi_{m+1} - [\underset{1 \le k \le m}{\Sigma} \phi_{k,m+1} - \underset{1 \le k_1 < k_2 \le m}{\Sigma} \phi_{k_1 k_2 m+1} + \ldots + (-1)^{m-1} \phi_{1 2 \ldots m \, m+1}]$$

$$\Rightarrow \phi(\overset{m+1}{\underset{1}{\cup}} A_k) = \overset{m+1}{\underset{1}{\Sigma}} \phi_k - \underset{1 \le k_1 < k_2 \le m+1}{\Sigma} \phi_{k_1 k_2} + \ldots + (-1)^{(m+1)-1} \phi_{1 2 \ldots m \, m+1}.$$

Definition 2.6 Events $A_1, A_2, \ldots, A_n$ of a probability space $(\Omega, \mathscr{F}, P)$ are called interchangeable (exchangeable) if for all choices of

$$1 \le i_1 < i_2 < \ldots < i_j \le n \text{ and all } 1 \le j \le n, P(A_{i_1} A_{i_2} \ldots A_{i_j}) = p_j \ge 0 \quad (2)$$

This is due to Haag (1924–28) and De Finetti (1937).

Corollary 1 of 2.1(c) If $A_1, \ldots, A_n$ are exchangeable events of a probability space $(\Omega, \mathscr{F}, P)$ with p_j defined by (2) then

$$P(\overset{n}{\underset{1}{\cup}} A_j) = np_1 - \binom{n}{2} p_2 + \binom{n}{3} p_3 + \ldots + (-1)^{n-1} p_n.$$

Corollary 2 of 2.1(c) (*Poincaré formula*). If $A_1, \ldots, A_n$ are events in a probability space $(\Omega, \mathscr{F}, P)$ and $T_k = \underset{1 \le j_1 < j_2 \ldots < j_k \le n}{\Sigma} P(A_{j_1} \ldots A_{j_k})$ then

$$P(\bigcup_{j=1}^{n} A_j) = \sum_{k=1}^{n} (-1)^{k-1} T_k.$$

Theorem 2.1(d)   Let ϕ be an additive set function on a field α and $A_k \in \alpha$, $k = 1, \ldots, n$.

If $|\phi(\bigcup_1^n A_k)| < \infty$ and $B_r = \{\omega \mid \omega \in A_k$ for exactly r of the events $A_k \in \alpha$, $1 \le k \le n.\}$, $0 \le r \le n$, then $\phi(AB_0) = \phi(A) - \sum_{1 \le k \le n} \phi(A_k A) + \sum_{1 \le k_1 < k_2 \le n} \phi(A_{k_1} A_{k_2} A)$
$- \ldots + (-1)^n \phi(A_{k_1} A_{k_2} \ldots A_{k_n} A)$ for all $A \in \alpha$.

NOTE.   If $\phi = P$, $A = \Omega$, $B_0^n = \{$None of the events $A_1, \ldots, A_n$ occurs$\}$. Then $P(B_0^n)$ = Probability of occurrence of none of the events = 1 − Probability of occurrence of at least one event.

$$\phi(B_r) = \sum_{1 \le k_1 < \ldots < k_r \le n} \phi(A_{k_1} \ldots A_{k_r}) - \binom{r+1}{r} \sum_{1 \le k_1 < \ldots < k_{r+1} \le n}$$

$$\phi(A_{k_1} \ldots A_{k_{r+1}}) + \binom{r+2}{r} \sum_{1 \le k_1 < \ldots < k_{r+2} \le m} \phi(A_{k_1} \ldots A_{k_{r+2}}) - \ldots$$

$$+ (-1)^{n-r} \binom{n}{r} \phi(A_1 \ldots A_n)$$

$$= \sum_{i=r}^{n} \binom{i}{r} (-1)^{i-r} \sum_{1 \le k_1 < \ldots < k_1 \le n} \phi(A_{k_1} \ldots A_{k_i}).$$

Proof. $A = (A \cap B_0) \cup (A \cap \bigcup_1^n A_k)$ (disjoint union), $B_0^C = \bigcup_1^n A_k$,

$$B_0 = (\bigcup_1^n A_k)^C = A_1^C \ldots A_n^C$$

$$\phi(A) = \phi(A \cap B_0) + \phi(A \cap \bigcup_1^n A_k).$$

Hence

$$\phi(AB_0) = \phi(A) - \phi(\bigcup_1^n AA_k)$$

$$= \phi(A) - \sum_{1 \le k \le n} \phi(A_k A) + \sum_{1 \le k_1 < k_2 \le n} \phi(A_{k_1} A_{k_2} A) - \ldots$$

$$+ (-1)^n \phi(A_1 \ldots A_n A) \quad \text{(from (c))}$$

If $A = \Omega$, then we get $\phi(B_0)$. In general, for $1 \le r \le n$, $B_r = \bigcup_{1 \le k_1 < \ldots < k_r \le n}$

$(A_{k_1} A_{k_2} \ldots A_{k_r} A_{j_1}^C \ldots A_{j_{n-r}}^C)$ is a disjoint union, where $1 \le j_1 < j_2 < \ldots < j_{n-r} \le n$ are the remaining integers in $\{1, 2, \ldots, n\}$. Since ϕ is additive,

$$\phi(B_r) = \sum_{1 \le k_1 < \ldots < k_r \le n} \phi(A_{k_1} \ldots A_{k_r} A_{j_1}^C \ldots A_{j_{n-r}}^C)$$

Let

$$B_0^n = \{\text{none of } A_1 \ldots A_n \text{ occurs}\}.$$

Put

$$A = A_{k_1}, \ldots, A_{k_r}, \; B_0^{n-r} = A_{j_1}^C \ldots A_{j_{n-r}}^C = \left(\bigcup_{i=1}^{n-r} A_{j_i} \right)^C$$

Now

$$\phi(B_r) = \sum_{1 \le k_1 < \ldots < k_r \le n} \phi(AB_0^{n-r}) = \sum_{1 \le k_1 < k_2 < \ldots < k_r \le n} \phi(AA_{j_1}^C \ldots A_{j_{n-r}}^C)$$

$$\Rightarrow \phi(B_r) = \sum_{1 \le k_1 < k_2 < \ldots < k_r \le n} \phi(A) - \sum_{1 \le k \le n-r} \phi(AA_{j_k}) + \ldots + (-1)^{n-r}$$

$$\phi(A_{j_1} \ldots A_{j_{n-r}} A)$$

$$\left(\text{since} \binom{n-r}{1}\binom{n}{r} = \binom{n}{r+1}\binom{r+1}{r}, \binom{n-r}{2}\binom{n}{r} = \binom{n}{r+2}\binom{r+2}{r}, \ldots, \right.$$

$$\left. \binom{n-r}{j}\binom{n}{r} = \binom{n}{r+j}\binom{r+j}{r} \right)$$

$$= \sum_{1 \le k_1 < \ldots < k_r \le n} \phi(A_{k_1} \ldots A_{k_r}) - \binom{r+1}{r} \sum_{1 \le k_1 < k_2 < \ldots < k_{r+1} \le n} f(A_{k_1} \ldots A_{k_r}) + \ldots$$

$$+ (-1)^{n-r} \binom{n}{r} \phi(A_1 \ldots A_n).$$

EXAMPLE 2.5   Let a deck of N cards, marked from 1 to *N*, be put into a random order. If a card occupies the place as its number, we say a match will occur. Let $P_{[m]}^N$ be the probability that exactly *m* matches occur $(0 \le m \le N)$. Prove that

(a) $P_{[0]}^N = 1 - 1 + \dfrac{1}{2!} - \dfrac{1}{3!} + \ldots + (-1)^N \, 1/N! = 1 - P_1^N$

(b) $|P_{[0]}^N - e^{-1}| \le \dfrac{1}{(N+1)!}$

(c) $P_{[m]}^N = \dfrac{1}{m!}\left[1 - 1 + \dfrac{1}{2!} - \dfrac{1}{3!} + \ldots + (-1)^{N-m} \dfrac{1}{(N-m)!} \right] \to \dfrac{1}{m!}e^{-1}$

(d) Define $P_m^N = P_{[m]}^N + P_{[m+1]}^N + \ldots + P_{[N]}^N, \; 0 \le m \le N$

$\qquad\qquad = $ probability of at least *m* matches.

Prove that $P_2^N \to e^{-1}\left(\dfrac{1}{2!} - \dfrac{1}{3!} + \ldots \right)$

(e) $|P_{[m]}^N - e^{-1}/m!| < \dfrac{1}{m!(N-m+1)!}$

Theorem 2.1(e)   If $T_k = \sum\limits_{1 \le j_1 < \ldots < j_k \le n} P(A_{j_1} \ldots A_{j_k})$, then $P(B_r) = P[\text{exactly } r \text{ of}$ $A_1, \ldots, A_n \text{ occurs}]$

$$= T_r - \binom{r+1}{1} R_{r+1} + \ldots + (-1)^{n-r} \binom{n}{r} T_n \tag{3}$$

In case of interchangeable events (3) simplifies considerably and may be expressed in terms of the classical difference operator defined by

$$\Delta p_n = p_{n+1} - p_n, \Delta^k p_n = \Delta(\Delta^{k-1} p_n), k > 1.$$
$$= \sum_{j=0}^{k} (-1)^j \binom{k}{j} p_{n+j}$$

Corollary of 2.1(e)   If $A_1, \ldots, A_n$ are interchangeable events of the probability space $(\Omega, \mathscr{F}, P)$ with $p_j = P(A_1 A_2 \ldots A_j)$, $1 \le j \le r$, then setting $p_0 = 1$, for any integer r in $[0, n]$.

$$q_r = p \,[\text{exactly } r \text{ of } A_1, \ldots, A_n \text{ occur}] = \sum_{i=r}^{n} (-1)^{i-r} \binom{n}{i}\binom{i}{r} p_i$$

$$= \binom{n}{r} \sum_{i=0}^{n-r} (-1)^i \binom{n-r}{i} p_{r+i}$$

$$= \binom{n}{r} (-1)^{n-r} \Delta^{n-r} p_r \tag{4}$$

NOTE. $\Omega = \bigcup\limits_{r=0}^{n} B_r : B_r$ are $n+1$ mutually exclusive and exhaustive events. In case of interchangeable events, $p_j \ge 0$, $\sum\limits_{j} p_j = 1$. So, $\{p_j\}$ are not any arbitrary non-negative numbers but they constitute a probability distribution.

EXAMPLE 2.6   Take

$$P_j = \frac{\binom{N-j}{m-j}}{\binom{N}{m}}, j = 0, 1, \ldots, m, m \le N.$$

Then (4) becomes,

$$q_r = \binom{n}{r}\binom{N-n}{m-r} \Big/ \binom{N}{m} \ge 0, 0 \le r \le n, \sum_{r} q_r = 1 = p_0$$

are the hypergeometric probabilities. Let $S_n = \sum_{j=1}^{n} I_{A_j}$ for each $n \geq 1$ and $A_1, \ldots, A_n$ are interchangeable events, then S_n is a r.v.

Theorem 2.2 $P[S_n = m] = \sum_{j=0}^{n-m} (-1)^j \binom{n}{m+j} \binom{m+j}{m} p_{m+j}^{(n)} + P(B_m)$, say

$$= \sum_{j=0}^{n-m} (-1)^j \binom{n}{m} \binom{n-m}{j} p_{m+j}, \; 0 \leq m \leq n,$$

where

$$P_{m+j}^{(n)} = p_{m+j} = P(A_{i_1} \ldots A_{i_{m+j}}).$$

Assume for some $0 \leq \lambda_0 < \infty$ and $0 \leq \lambda_n < \infty$, $0 < n(n-1) \ldots (n-k+1)$ $p_k^{(n)} \leq \lambda_0^k$ for $1 \leq k \leq n$ and $n^k p_k^{(n)} = \lambda_n^k + o(1)$, $k \geq 1$, as $n \to \infty$. Then

$$\left(P(S_n = m) - \frac{\lambda_n^m e^{-\lambda_n}}{m!} \right) \to 0, \text{ as } n \to \infty \text{ for each } m \geq 0.$$

The proof is given later.

2.5.2   Occupancy Problem: *r* Balls in *n* Cells

Put balls randomly into n cells with each arrangement having probability n^{-r}. A mathematical model or probability space describing such an experiment is: $\Omega = \{\omega \mid \omega = (\omega_1, \ldots, \omega_r), \omega_j = 0, 1, \ldots, n \text{ for } j = 1, \ldots, r\}$. $\mathscr{F} = S(\Omega)$. For each typical point $\omega_0 \in \Omega$, put $p_{\omega_0} = P[\omega = \omega_0] = n^{-r}$. For $A \in \mathscr{F}$, define $P(A) = \sum_{\omega \in A} p_\omega = $ (number of elements in A). (n^{-r}).

Obviously, $(\Omega, \mathscr{F}, P)$ is a probability space. Let $X_j(\omega) = \omega_j$ for, $j = 1, \ldots, r$. Then X_j is the number of the cell containing the jth ball. These r.vs. X_j's are called coordinate r.vs. (projection mapping). Then $X_1, \ldots, X_n$ are i.i.d. r.vs., since

$$P[(X_1, \ldots, X_r) = \omega^0] = P[x_1 = \omega_1^0, \ldots, x_r = \omega_r^0] = n^{-r}, \text{ where } \omega^0 = (\omega_1^0, \ldots,$$
$\omega_r^0)$. $P[X_j = \omega_j^0] = P[\omega \mid \omega_j = \omega_j^0] = $ [Number of all points such that $\omega_j = \omega_j^0$]/

$$n^r = \frac{n^{r-1}}{n^r} = \frac{1}{n}.$$

Therefore,

$$P[(X_1, \ldots, X_r) = \omega^0] = P[X_1 = \omega_1^0] \ldots P[X_r = \omega_r^0].$$

We have n-cells and we give them numbers $1, \ldots, n$, and r balls are also numbered $l, \ldots, r$.

Put $A_k = (X_1 \neq k, \ldots, X_r \neq k)$ for $k = 1, \ldots, n$.

P[no balls in the kth cell, i.e. kth cell is empty] $= P(A_k) = P(X_1 \neq k) \ldots$

$$P(X_r \neq k) = \left(1 - \frac{1}{n} \right)^r, \; k = 1, \ldots, n.$$

Now $P(A_{k_1} A_{k_2}) = p(k_1\text{th cell and } k_2\text{th cell both are empty}) = \left(1 - \dfrac{2}{n}\right)^r$. In general,

$$P(A_{k_1} \ldots A_{k_j}) = \left(1 - \dfrac{j}{n}\right)^r, 1 \le k_1 < \ldots < k_j \le n \text{ and } j = 1, \ldots, n.$$

NOTE. $A_1, \ldots, A_n$ are interchangeable events. Put $P_m(n, r) = P$ (exactly m empty cells), $m = 0, \ldots, r$. So by previous formulae,

$$P_0(r, n) = \sum_{i=0}^{n} (-1)^i \binom{n}{i} \left(1 - \dfrac{1}{n}\right)^r \text{ and for } m \ge 1, m \le n$$

$p_m(r, n) = $ Probability that exactly m cells are empty.

$$= \sum_{i=m}^{n} (-1)^{i-m} \binom{n}{i}\binom{i}{m}\left(1 - \dfrac{i}{n}\right)^r = \binom{n}{m}\sum_{j=0}^{n-m} (-1)^j \binom{n-m}{j}\left(1 - \dfrac{j+m}{n}\right)^r$$

Theorem 2.3 (*von Mises*, 1939).   Let $r(n) = r_n$ balls be randomly distributed into n-cells. Each arrangement has probability n^{-r}. Let $p_m(r, n)$ be the probability of the event that there are exactly m-empty cells and if $r_n \to \infty$ as $n \to \infty$ such that

$$ne^{-r_n/n} = \lambda_n, \lambda_n \le \lambda_0 < \infty \text{ for all } n \ge 1.$$

Then

$$\left(p_m(r,n) - \dfrac{e^{-\lambda_n}\lambda_n^m}{m!}\right) \to 0 \text{ as } n \to \infty, \text{ for each fixed } m.$$

In particular, if $\lambda_n = \lambda(1 + o(1))$, then $\{p_m(r, n), m \ge 0\}$ tends, as $n \to \infty$, to the Poisson distribution with parameter λ.

Proof of Theorem 2.2   The first part is obvious and follows from Eq. (4) in corollary of 2.1(e). We have

$$\lambda_n^m e^{-\lambda_n}/m! = \sum_{j=0}^{\infty} (-1)^j \dfrac{\lambda_n^{m+j}}{m!\,j!}$$

and given $\qquad\qquad 0 \le n(n-1)\ldots(n-k+1)p_k^{(n)} \le \lambda_0^k \qquad\qquad (5)$

and for every fixed k,

$$n^k p_k^{(n)} = \lambda_n^k + o(1) \text{ as } n \to 0. \qquad\qquad (6)$$

We have

$$P(S_n = m) = \sum_{j=0}^{n-m} (-1)^j \binom{n}{m+j}\binom{m+j}{m} p_{m+j}^{(n)}, 0 \le m \le n$$

$$= \sum_{j=0}^{n-m} (-1)^j \dfrac{1}{m!\,j!} \dfrac{n!}{(n-m-j)!} p_{m+j}^{(n)}$$

Hence for $m \geq 0$, fixed n_0 and for all large $n > n_0$.

$$\left(P(S_n = m) - \frac{\lambda_n^m e^{-\lambda_n}}{m!} \right) = \sum_{j=0}^{n_0} \frac{(-1)^j}{m!\, j!} \left[\frac{n!}{(n-m-j)!}\, p_{m+j}^{(n)} - \lambda_n^{m+j} \right]$$

$$+ \sum_{j=n_0+1}^{n-m} (-1)^j \frac{n!}{m!\, j!(n-m-j)!}\, p_{m+j}^{(n)} - \sum_{j=n_0+1}^{n-m} (-1)^j \frac{n^{m+j}}{m!\, j!}$$

$$= I_1 + I_2 + I_3 \quad \text{(say)}$$

Since $\lambda_n + o(1) = n\, p_1^{(n)} \leq \lambda_0$, there exists $\lambda > \lambda_0$ such that $\lambda_n < \lambda$ for all n. For any $\varepsilon > 0$ and fixed m, choose n_0 to satisfy

$$\sum_{j=n_0+1}^{\infty} \frac{\lambda^{m+j}}{m!\, j!} < \varepsilon \Rightarrow |I_3| < \varepsilon$$

$$|I_2| \leq \sum_{j=n_0+1}^{n-m} \frac{n(n-1)\ldots(n-m-j+1)}{m!\, j!}\, p_{m+j}^{(n)} \leq \sum_{j=n_0+1}^{n-m} \lambda_0^{m+j} /(m!\, j!)$$

(From (5) taking $k = m + j$)

$$\leq \sum_{j=n_0+1}^{\infty} \frac{\lambda_0^{m+j}}{m!\, j!} < \varepsilon \text{ (choosing } n_0 \text{ properly)}$$

Now for $1 \leq m + j \leq m + n_0$.

$$\frac{n!}{(n-m-j)!}\, p_{m+j}^{(n)} - \lambda_n^{m+j} \leq n^{m+j}\, p_{m+j}^{(n)} - \lambda_n^{m+j}$$

$$= \lambda_n^{m+j} + o(1) - \lambda_n^{m+j} + o(1) \quad \text{as } n \to \infty \text{ (from (6))}$$

Also,

$$\frac{n!}{(n-m-j)!}\, p_{m+j}^{(n)} - \lambda_n^{m+j} \geq (n-m-j)^{m+j}\, p_{m+j}^{(n)} - \lambda_n^{m+j} = \left[n\left(1 - \frac{m+j}{n}\right) \right]^{m+j}$$

$$p_{m+j}^{(n)} - \lambda_n^{m+j} = \left(1 - \frac{m+j}{n}\right)^{m+j} (\lambda_n^{m+j} + o(1)) - \lambda_n^{m+j} = o(1) \text{ as } n \to \infty.$$

Hence

$$|I_1| = o(1) \text{ as } n \to \infty.$$

Proof of von Mises theorem: r balls in n cells. Now write $r = r_n = r(n)$. Let

$$p_j^{(n)} = (1 - j/n)^{r(n)} \text{ and } p_m(r_n, n) = \sum_{j=0}^{n-m} (-1)^j \binom{n}{m+j}\binom{m+j}{m} p_{m+j}^{(n)}.$$

We have to check that conditions (5) and (6) of the previous theorem are satisfied in this special case. Since $1 - x \leq e^{-x}$ for $0 < x < 1$, $n(n-1) \ldots (n-k+1)\, p_k^{(n)} \leq n^k e^{-r(n)k/n}$ (from conditions of theorem) $= \lambda_n^k \leq \lambda_0^k$. So (5) is satisfied.

For fixed $k \geq 1$, if $r(n) \leq n^{3/2}$, then

$$n^k p_k^{(n)} = n^k \left(1 - \frac{k}{n}\right)^{r(n)} = n^k [e^{-k/n} + 0(n^{-2})]^{r(n)}$$

$$= n^k e^{-kr(n)/n} [1 + 0(n^{-2}) e^{k/n}]^{r(n)}$$

$$= \lambda_n^k [1 + o(1)] = \lambda_n^k + o(1)$$

If $r(n) > n^{3/2}$, then

$$n^k p_k^{(n)} = n^k \left(1 - \frac{k}{n}\right)^{r(n)} \le n^k \left[\left(1 - \frac{k}{n}\right)^n\right]^{\sqrt{n}} \le n^k e^{-k\sqrt{(n)}} = o(1)$$

and $\qquad \lambda_n^k = n^k e^{-kr(n)/n} \le n^k e^{k\sqrt{(n)}} = o(1) \Rightarrow n^k p_k^{(n)} = \lambda_n^k + o(1)$ as $n \to \infty$.

So condition (6) is also satisfied.

2.6 INDUCED MEASURE AND DISTRIBUTION FUNCTION

Consider a measurable transformation X from a measure space $(\Omega, \mathcal{F}, \mu)$ to a measurable space $(\tilde{\Omega}, \tilde{\mathcal{F}}, \mu)$ i.e. $X: \Omega \to \tilde{\Omega}$ and $X^{-1}(\tilde{\mathcal{F}}) \subset \mathcal{F}$. The measure on $\mathcal{F}$ induces in a very natural fashion a measure $\tilde{\mu}$ on $\tilde{\mathcal{F}}$, known as the *induced measure,* denoted by μ_x and defined for all $\tilde{F} \in \tilde{\mathcal{F}}$ by

$$\mu_X(\tilde{F}) = \tilde{\mu}(\tilde{F}) = \mu[X^{-1}(\tilde{F})]$$

Obviously μ_x is a measure on $\tilde{F}$ and consequently that $(\tilde{\Omega}, \tilde{\mathcal{F}}, \mu_x)$ is a measure space. In particular, a r.v. X on a probability space $(\Omega, \mathcal{F}, P)$ induces a new probability space $(R, \mathcal{B}, P_X)$, where $\mathcal{B}$ is the class of all Borel sets of R. Since X is our primary interest we restrict the former to $(\Omega, \mathcal{B}(X), P)$, where $\mathcal{B}(X)$ is σ-field generated by X (See Section 7.1). Associated with any r.v. X on a probability space $(\Omega, \mathcal{F}, P)$ is a real function F_X called the distribution function of the r.v. X and defined by

$$F_X(x) = P(\omega : X(\omega) \le x), x \in R$$

(For details see Section 3.1.)

2.7 PROBABILITY GENERATING FUNCTION AND DISCRETE CONVOLUTION

Definition 2.7 Let $a_0, a_1, \ldots$ be a sequence of real numbers. If $A(s) = a_0 + a_1 s + a_2 s^2 + \ldots$ converges in some interval $-s_0 < s < s_0$, then $A(s)$ is called the *generating function* of the sequence $\{a_n\}$.

EXAMPLE 2.7 If X is the number scored in a throw of a perfect die, the probability distribution of X has the generating function $(s + s^2 + \ldots + s^6)/6$ (here $a_0 = 0$, $a_i = 1/6$, $i = 1, \ldots, 6$).

Let
$$p_j = P(X = j), \; q_j = P(X > j) = p_{j+1} + p_{j+2} + \cdots$$

$$P(s) = p_0 + p_1 s + p_2 s^2 + \cdots \text{ and } Q(s) = q_0 + q_1 s + q_2 s^2 + \cdots$$

As $P(1) = 1$, the series $P(s)$ converges absolutely at least for $-1 \le s \le 1$, $Q(s)$ also converges for $-1 < s < 1$ by comparison with G.P. series.

Theorem 2.4   For $-1 < s < 1$, $Q(s) = \dfrac{1 - P(s)}{1 - s}$.

Proof.   The coefficient of s^n in $(1 - s)Q(s) = q_n - q_{n-1} = -p_n$ for $n \ge 1$, $q_0 = p_1 + p_2 + \cdots + \cdots = 1 - p_0$

$$(1 - s)\, Q(s) = q_0 + (q_1 - q_0)s + (q_2 - q_1)s^2 + \cdots + (q_n - q_{n-1})s^n + \cdots$$

$$= q_0 - p_1 s - p_2 s^2 - \cdots - p_n s^n - \cdots$$

$$= 1 - p_0 - p_1 s - P_2 s^2 - \cdots = 1 - P(s) \Rightarrow Q(s) = \dfrac{1 - P(s)}{1 - s}.$$

Consider the Generating Function

$$P(s) = \sum_{k=0}^{\infty} p_k s^k$$

of a probability distribution $\{p_k\}_0^{\infty}$. This is a power series in s and can be differentiated term by term. $P'(s) = \sum_{k=1}^{\infty} k p_k s^{k-1}$ converges for $-1 \le s \le 1$ if $E(X) < \infty$. This follows from

$$|P'(s)| \le \sum_{k=1}^{\infty} k p_k \text{ if } |s| \le 1.$$

Theorem 2.5   If $E(X)$ exists (may be ∞), then $E(X) = \sum_{j=1}^{\infty} j p_j$

$$= \sum_{k=0}^{\infty} q_k = P'(1) = Q(1),$$

where
$$q_i = P(X > 1) \text{ and } Q(s) \text{ is the G.F. of } \{q_n\}.$$

Proof.   If $E(X) < \infty$, then $\sum_{k=1}^{\infty} k p_k$ converges. If $\sum_{k=1}^{\infty} k p_k$ diverges, then $P'(s) \to \infty$ as $s \to 1$ and $P'(1) = E(X) = \infty$. Now $P'(s)$ is a continuous function in the closed interval $-1 \le s \le 1$ if $E(X) < \infty$. Therefore, by mean value theorem.

$$P(s) = P(1)(1 - s)P'(\theta), \text{ where } s < \theta < 1.$$

$$= 1 - (1 - s)p'(\theta).$$

Hence

$$P'(\theta) = \dfrac{1 - P(s)}{1 - s} = Q(s) \text{ (by Theorem 2.1).}$$

Since $P'(s)$ and $Q(s)$ are monotonically increasing functions of s and have the same limit finite or infinite. Therefore

$$\lim_{s \to 1} P'(s) = \lim_{s \to 1} Q(s), \text{ i.e. } P'(1) = Q(1).$$

Theorem 2.6   var $(X) = P''(1) + P'(1) - [P'(1)]^2 = 2Q'(1) + Q(1) - Q^2(1)$. If var $(X) = \infty$, then $p''(s) \to \infty$ as $s \to 1$.

Proof.   If var $(X) < \infty$, then $P'(s) = \sum_{k=1}^{\infty} k p_k s^{k-1}$ and

$$P''(1) = \sum_{k=1}^{\infty} k(k-1) p_k = E[X(X-1)]$$

var $(X) = E(X^2) - E^2(X) = E(X(X-1)) + E(X) - E^2(X) = P''(1) + P'(1) - [P'(1)]^2$

By Theorem 2.4 $(1-s) Q(s) = 1 - P(s)$ and differentiating both sides with respect to s,

$$-Q(s) + (1-s) Q'(s) = -P'(s) \text{ or } P'(s) = Q(s) - (1-s)Q'(s).$$

Again differentiating, we get

$$P''(s) = Q'(s) + Q'(s) - (1-s) Q''(s)$$

$$= 2Q'(s) - (1-s) Q''(s)$$

Taking limit $s \to 1$ on both sides, $P''(1) = 2Q'(1)$.

EXAMPLE 2.8   Let X be a r.v. with G.F. $P(s)$. Find the G.F. of $X + 1$ and $2X$.

Answer. $p_j = P[X = j]$, $P[X + 1 = j] = P[X = j - 1] = p_{j-1}$, the coefficient of s^j.

$$P(s) = \sum_{k=0}^{\infty} p_k s^k \Rightarrow sP(s) = sp_0 + s^2 p_1 + s^3 p_2 + \ldots + s^j p_{j-1} + \ldots$$

So the G.F. of $Y = x + 1$ is $sP(s)$. If

$$Z = 2X, P(Z = j) = P\left[x = \frac{j}{2}\right] = p_{j/2}, \text{ the coefficient of } s^j.$$

So the G.F. of Z is $p_0 + p_1 s^2 + p_2 s^4 + \ldots = P(s^2)$.

EXAMPLE 2.9   Find the G.F. of the sequence $\{a_n\} = P(X \le n)$ in terms of $P(s)$, the G.F. of $\{p_j\}$.

Answer.   Let $a_n = P[X \le n] = p_0 + p_1 + \ldots + p_n$

Then the G.F. of $\{a_n\}$ is

$$p_0 + (p_0 + p_1)s + (p_0 + p_1 + p_2) s^2 + \ldots + (p_0 + p_1 + \ldots + p_n)s^n$$

$$= P(s)(1 + s + s^2 + \ldots + s^n + \ldots) = \frac{P(s)}{1-s}, \ |s| < 1$$

EXAMPLE 2.10   Find the G.F. of the following sequences:

(a) $b_n = P[X < n]$, *Answer.* $\dfrac{s}{1-s} P(s)$

(b) $c_n = P[X \geq n]$, *Answer.* $\dfrac{1-sP(s)}{1-s}$

(c) $d_n = P[X > n + 1]$, *Answer.* $-\dfrac{p_0}{s} + \dfrac{1 - P(s)/s}{1-s}$

EXAMPLE 2.11   Let u_n be the probability that the combination SF occurs for the 1st time at trials number, $n - 1$ and n with probability $P(S) = p$, $P(F) = q$, p, $q > 0$, $p + q = 1$. Find the G.F. of $\{u_n\}$ and also its mean and variance.

Answer. $pqs^2(1 - ps)^{-1} (1 - qs)^{-1}$, $1/pq$, $\dfrac{1 - 3pq}{p^2 q^2}$

2.7.1   Discrete Convolution

If a r.v. X takes only non-negative integral values, then s^X is a well-defined r.v.

Probability G.F. of X is $E(s^X) = \sum\limits_{k=1}^{\infty} p_k s^k$. If X and Y are two independent discrete r.vs. then, $E(s^{X+Y}) = E(s^X)(s^Y)$. If $P[X = j] = a_j$, $P[Y = j] = b_j$ are the probability distributions of X and Y respectively, then $P[X = j, Y = j] = a_j b_j$.

Now $S = (X + Y)$ is a new r.v. and the event

$$[S = r] = [X = 0, Y = r] \cup [X = 1, \ \ Y = r - 1] \ldots \cup [X = r, Y = 0]$$

is the union of mutually exclusive events. Therefore,

$$c_r = P[S = r] = a_0 b_r + a_1 b_{r-1} + \ldots + a_r b_0 \tag{1}$$

Definition 2.8   Let $\{a_k\}$ and $\{b_k\}$ be any two numerical sequences (not necessarily probability distribution). The sequence $\{c_r\}$ defined by (1) is called the convolution of $\{a_k\}$ and $\{b_k\}$ and will be denoted by $\{c_k\} = \{a_k\} * \{b_k\}$.

EXAMPLE 2.12 $a_k = k$, $b_k = 1$, then $c_k = 1 + 2 + \ldots + k = \dfrac{k(k + 1)}{2}$

Theorem 2.7   If $\{a_k\}$ and $\{b_k\}$ are numerical sequences with G.F. $A(s)$ and $B(s)$ respectively and $\{c_k\}$ is their convolution, then the G.F. $C(s) = \sum\limits_{k} c_k s^k$ is the product $A(s)\, B(s)$.

Proof.   The sequence $\{a_k\}$ and $\{b_k\}$ have G.F.

$$A(s) = \sum_{0}^{\infty} a_k s^k \quad \text{and} \quad B(s) = \sum_{0}^{\infty} b_k s^k.$$

The product $A(s)B(s)$ is obtained by term-wise multiplication of power series $A(s)$ and $B(s)$ (i.e. by the Cauchy product rule), i.e.

$$(\sum_j a_j s^j)(\sum_k b_k s^k) = \sum_{k=0}^{\infty} (\sum_{j=0}^{\infty} a_j b_{k-j})s^k = \sum_{k=0}^{\infty} c_k s^k.$$

NOTE.   Convolution operation has both associative and commutative properties because, $\{a_k\} * \{b_k\} = \{b_k\} * \{a_k\}$ and $\{a_k\} * \{b_k\} * \{c_k\} * \{d_k\}$ is the sequence of numbers whose G.F. is $A(s)B(s)C(s)D(s)$.

Definition 2.9   If $\{a_j\}$ is the common probability distribution of a sequence of i.i.d. r.vs. $\{X_n\}$ then the probability distribution of $S_n = X_1 + \cdots + X_n$ will be given by $\{a_j\}^{n*}$. Thus $\{a_j\}^{2*} = \{a_j\} * \{a_j\}$, $\{a_j\}^{n*}$ is the sequence of numbers whose G.F. is $[A(s)]^n$ where $A(s)$ is the G.F. of $\{X_n\}$.

EXAMPLE 2.13   **(a) Binomial distribution.**   G.F. of $b(k; n, p) = \binom{n}{p} p^k(1 - p)^{n-k}$

is $(E(s)^X) = \sum_{k=0}^{n} \binom{n}{k} (ps)^k q^{n-k} = (q + ps)^n$. It is the nth power of $(q + ps)$ so that $b(k; n, p)$ is the distribution of the sum, $S_n = X_1 + \cdots + X_n$ of n i.i.d. r.vs. with common G.F. $(q + ps)$ where each X_j assumes value 1 with probability p and 0 with probability $q = (1 - p)$. Therefore $\{b(k; n, p)\} * \{b(k; m, p)\} = \{b(k; m + n, p)\}$. By Theorem 2.5,

$$E(S_n) = P'(1) = np(q + ps)^{n-1} \big|_{s=1} = np$$

$$V(S_n) = P''(1) + P'(1) - [P'(1)]^2 = n(n - 1)p^2 + np - (np)^2 = npq.$$

(b) Poisson distribution.   The G.F. of $p(k; \lambda) = e^{-\lambda} \dfrac{\lambda^k}{k!}$ is

$$E(s^X) = \sum_{k=0}^{\infty} e^{-\lambda} \frac{(\lambda s)^k}{k!} = e^{-\lambda + \lambda s} = e^{\lambda(s-1)}.$$

Also $\{p(k; \lambda_1) * \{p(k; \lambda_2)\} = \{p(k; \lambda_1 + \lambda_2)\}$ proves the sum of two independent Poisson r.vs. is a Poisson r.v. with parameter as the sum of the parameters.

(c) Geometric distribution. $P[X = k] = q^k p$, $k = 0, 1, \ldots, p + q = 1, p$, $q \geq 0$. G.F. is

$$\sum_{k=0}^{\infty} s^k q^k p = p(1 - qs)^{-1}, |qs| < 1. \; E(X) = P'(1) = \frac{q}{p}, V(X) = q/p^2.$$

(d) Negative binomial distribution.   Let X be the waiting time for the 1st success in a sequence of independent Bernoulli trials. In general, suppose X_k be the number of failures following the $(k - 1)$th and preceding the kth success, then $S_r = X_1 + \cdots + X_r$ is the total number of trials upto and including the rth success.

If X_k's are independent with same distribution then S_r has G.F. $[p/(1 - qs)]r$, for coefficient of s^k is

$$p^r \binom{-r}{k}(-q)^k = \binom{r+k-1}{k} p^r q^k = f(k; r, p), k = 0, 1, 2, \ldots$$

$$\left(\text{for } r > 0, \binom{-r}{k} = (-1)^k \binom{r+k-1}{k}\right).$$

Therefore,

$$P(S_r = k) = f(k;\, r,\, p)$$

and this is the r-fold convolution of the geometric distribution with itself. Symbolically, $f(k;\, r,\, p) = \{q^k p\}^{r*}$.

Definition 2.10 For arbitrary fixed $r > 0$ (not necessarily positive integer) and $0 < p < 1$, the sequence of probability distributions $f(k;\, r,\, p)$ is called *negative binomial distribution* (since $\displaystyle\sum_{k=0}^{\infty} \binom{-r}{k} (-q)^k = (1 - q)^{-r} = p^{-r}$). By Theorems 2.2 and 2.3 and differentiating the G.F., we get $E(S_r) = rq/p$, var $(S_r) = rq/p^2$ and negative binomial distribution has multiplicative property.

$$f(k;\, r_1,\, p) * f(k;\, r_2,\, p) = f(k;\, r_1 + r_2,\, p).$$

If r is a positive integer, $f(k;\, r,\, p)$ is also called Pascal distribution. Note that probability distribution like binomial, Poisson, negative binomial with multiplicative property is called *factor closed distribution.*

2.7.2 Bivariate Generating Function

Let $P(X = j,\, Y = k) = p_{jk}$, The G.F. of two variables is

$$P(s_1, s_2) = \sum_{j,k} p_{jk}\, s_1^j s_2^k$$

Theorem 2.8 (a) The G.F. of the marginal distribution $P[X = j]$ and $P[Y = k]$ are $A(s) = P(s,\, 1)$ and $B(s) = P(1,\, s)$ respectively.

 (a) The G.F. of $X + Y$ is $P(s,\, s)$.

 (b) X and Y are independent iff $P(s_1,\, s_2) = A(s_1)B(s_2),\ |\, s_1\, | < 1,\ |\, s_2\, | < 1$.

Proof. $P(s,\, 1) = \displaystyle\sum_j \sum_k p_{jk} s^j = \sum_j p_{j0} s^j = A(s)$ and $B(s) = \displaystyle\sum_j p_{0k} s^k$

$$P(s,\, s) = \sum_j \sum_k p_{jk} s^{j+k} = E(s^{X+Y}) \text{ is the G.F. of } X + Y.$$

EXAMPLE 2.14 Bivariate Poisson distribution has G.F.

$$P(s,\, s) = \exp\,(-\lambda_1 - \lambda_2 - \mu + \lambda_1 s_1 + \lambda_2 s_2 + \mu s_1 s_2)$$

$$= \exp\,[-\lambda_1(1 - s_1) - \lambda_2(1 - s_2) - (1 - s_1 s_2)\mu]$$

with probability mass function.

$$p(k,\, j;\, \lambda_1,\, \lambda_2,\, \mu) = \exp\,(-\lambda_1 - \lambda_2 - \mu) \sum_{i=0}^{\min(j,k)} \mu^i/i!\ \lambda_1^{j-i}/(j-i)!\,\lambda_2^{k-i}/(k-i)!$$

$$A(s) = P(s, 1) = e^{(\lambda_1 + \mu)(1-s)}, \quad B(s) = P(1, s) = e^{(\lambda_2 + \mu)(1-s)},$$

$(X + Y)$ has G.F. $P(s, s) = \exp\left[-(\lambda_1 + \lambda_2)(1 - s) - \mu(1 - s^2)\right]$

which is a compound Poisson distribution.

(e) Multinomial distribution. Consider a sequence of n independent trials, each of which results in E_0, E_1 and E_2 with probabilities p_0, p_1, $p_2 \geq 0$, and $p_0 + p_1 + p_2 = 1$. Then we have trinomial distribution whose G.F. is given by $(p_0 + s_1 p_1 + s_2 p_2)^n$.

(f) Compound distribution. Sums of a random number of r.vs. Let X_k be a sequence of i.i.d r.vs. with common distribution $P(X_k = j) = p_j$ and has G.F. $P(s) = \sum_i p_i s^i$. We are interested in the sums $S_N = X_1 + X_2 + \ldots + X_N$, where the number of terms is a r.v. independent of X_j's and with probability distribution $P(N = n) = g_n$, $n = 0, 1, 2, \ldots$ Let $G(s) = \sum_{n=0}^{\infty} g_n s^n$ be the G.F. of $\{g_n\}$.

Theorem 2.9   The G.F. of $\{S_N\}$ is the compound function $G(P(s))$.

Proof.   Let the distribution of S_N be denoted by

$$h_j = P(S_N = j) = \sum_{n=0}^{\infty} P(N = n)P(S_n = j), \text{ since}$$

$$h_j = \sum_{n=0}^{\infty} P(S_n = j) \mid (N = n)P(N = n) = \sum_{n=0}^{\infty} P(N = n)P(S_n = j)$$

G.F. of $\{h_j\}$ is

$$\sum_{j=0}^{\infty} h_j s^j = \sum_j s^j \left(\sum_n g_n \{p_j\}^{n*}\right), \qquad p_i = P[X_j = i].$$

Note that the distribution of S_n is the n-fold convolution of $\{p_j\}$ denoted by $\{p_j\}^{n*}$. Interchanging the order of summations, we get

$$\sum_{j=0}^{\infty} h_j s^j = \sum_{n=0}^{\infty} s_n \sum_{n=0}^{\infty} s^j \{p_j\}^{n*} = \sum_{n=0}^{\infty} g_n [P(s)]^n = G(P(s)),$$

since the G.F. of $\{p_j\}^{n*}$ is $[P(s)]^n$.

EXAMPLE 2.15   Take $P(N = n) = e^{-\lambda} \dfrac{\lambda^n}{n!} = g_n$, $\lambda > 0$. Suppose that X_j has common distribution with G.F. $P(s)$. Then $G(s) = e^{-\lambda + \lambda P(s)}$, which is a compound Poisson distribution.

EXAMPLE 2.16   (a) Prove that var $(X) = E(\text{var }(X/Y) + \text{var }(E(X/Y))$, where var (X/Y) is the conditional variance given by $E(X^2/Y) - E^2(X/Y)$.

(b) Let N be an integer valued r.v. independent of $\{X_1\}$, where X_i are i.i.d. discrete r.vs. Then show that

$$\text{var} \left(\sum_{1}^{N} X_i \right) = E(N) \, \text{var} \, (X) + E^2(X) \, \text{var} \, (N).$$

(Do it either by conditioning or by G.F.)

(c) $E(\sum_{1}^{N} X_i) = E(N)E(X)$. This is known as Wald's identity.

2.8   CONTINUITY THEOREM FOR ADDITIVE SET FUNCTIONS AND APPLICATION

Theorem 2.10   Let Φ be an additive set function on a field α. Then

(a) Φ is σ-additive $\Rightarrow$ Φ is continuous
(b) Φ is continuous from below $\Rightarrow$ Φ is σ-additive.
(c) Φ is finite and continuous at ϕ (null set) $\Rightarrow$ Φ is σ-additive.

Proof.   (a) Let $A \in \alpha$, $A_n \in \alpha$, for all $n = 1, 2, 3, \ldots$ and $A_n \uparrow A$, $A = \bigcup_{n=1}^{\infty} A_n$. Now $A = A_1 \cup (A_2 \, A_1^C) \cup (A_3 \, A_2^C \, A_1^C) \cup \ldots$ is a disjoint union. Therefore

$$\Phi(A) = \Phi(A_1) + \Phi(A_2 \, A_1^C) + \Phi(A_3 A_2^C \, A_1^C) + \ldots \quad (\Phi \text{ being } \sigma\text{-additive})$$

$$= \lim_{n \to \infty} [\Phi(A_1) + \Phi(A_2 A_1^C) + \ldots + \Phi(A_n A_1^C \ldots A_{n-1}^C)]$$

$$= \lim_{n \to \infty} \Phi\left(\bigcup_{n=1}^{n} A_k \right) \left(\text{since } \bigcup_{k=1}^{n} A_k = A_1 \cup (A_2 A_1^C) \cup \ldots \cup (A_n A_1^C \ldots A_{n-1}^C) \right)$$

$$= \lim_{n \to \infty} \Phi(A_n), \text{ since } A_n \uparrow A.$$

Hence Φ is continuous from below. Now assume that $| \Phi(A_n) | < \infty$ for some n. Let $A \in \alpha$, $A_n \in \alpha$ for all $n \geq 1$. Suppose $A_n \downarrow A$. To prove that $\lim_{n \to \infty} \Phi(A_n) = \Phi(A)$, define $B_n = (A_1 - A_n) \in \alpha$ (since α is a field), $B_n \uparrow (A_1 - A) \in \alpha$ and

$A_1 - A = \bigcup_{n=1}^{\infty} B_n$. Therefore

$$\Phi(A_1 - A) = \lim_{n \to \infty} \Phi(B_n) \text{ (using last result).}$$

Hence $\Phi(A_n) < \infty$ for some $n \Rightarrow \Phi(A_1) - \Phi(A) = \lim_{n \to \infty} (\Phi(A_1) - \Phi(A_n)) \Rightarrow \Phi(A)$

$$= \lim_{n \to \infty} \Phi(A_n).$$

(b) Let $A_n \in \alpha$, take $A = \bigcup_{n=1}^{\infty} A_n \in \alpha$ and A_n's are disjoint.

Put $B_m = \bigcup_{n=1}^{\infty} A_n$ (since α is a field) and $B_m \uparrow A$ as $m \to \infty$. Then

$$\Phi(A) = \lim_{m \to \infty} \Phi(B_m) \text{ (since } \Phi \text{ is continuous from below)}$$

$$= \lim_{m \to \infty} [\Phi(A_1) + \ldots + \Phi(A_m)] \text{ (}\Phi \text{ is additive)}$$

$$= \sum_{i=1}^{\infty} \Phi(A_i), \text{ i.e. } \Phi \text{ is } \sigma\text{-additive.}$$

(c) Let $A_n \in \alpha$, A_n's are disjoint and $\bigcup_n A_n = A \in \alpha$

Put $B_m = (A - \bigcup_{n=1}^{m} A_n) \downarrow \phi$ (null set) as $m \to \infty$.

Since Φ is finite

$$\Phi(\phi) = 0 = \lim_{m \to \infty} \Phi(B_m) = \Phi(A) - \sum_{i=1}^{\infty} \Phi(A_i).$$

Therefore, Φ is σ-additive. Note that $\Phi(\phi) = \Phi(\phi \cup \phi) = \Phi(\phi) + \Phi(\phi)$, since Φ is additive. Hence $\Phi(\phi) = 0$. Recall that a measure μ on a σ-field $\mathcal{B}$ of subsets of Ω is a non-negative, σ-additive, set function defined on $\mathcal{B}$.

Definition 2.11 If $A \in \mathcal{B}$ and $\mu(A) = 0$, then A is called a *μ-null set*.

Corollary. Let $(\Omega, \mathcal{B}, \mu)$ be a measure space.

(a) If $\mu(\Omega) < \infty$, then μ is bounded.

(b) $A_n \uparrow A$, $A_n \in \mathcal{B} \Rightarrow \mu(A_n) \uparrow \mu(A)$.

(c) If $\mu(\Omega) < \infty$, $A_n \downarrow A$, $A_n \in \mathcal{B} \Rightarrow \mu(A_n) \downarrow \mu(A)$.

(d) (Fatou's Lemma) $A_n \in \mathcal{B} \Rightarrow \lim_{n \to \infty} \inf \mu(A_n) \geq \mu(\lim_{n \to \infty} \inf (A_n))$ and A_n

$\in \mathcal{B}$, μ is finite $\Rightarrow \lim_{n \to \infty} \sup \mu(A_n) \leq \mu(\lim_{n \to \infty} \sup A_n)$. Hence $A_n \in \mathcal{B}$,

μ is finite and $A_n \to A \Rightarrow \mu(A_n) \to \mu(A)$.

Proof of (a), (b) and (c) have already been given in the previous sections.

Proof of Corollary (d).

$$\mu (\lim \inf A_n) = \mu(\bigcup_{m=1}^{\infty} \bigcap_{n=m}^{\infty} A_n) = \lim_{n \to \infty} \mu(\bigcap_{n=m}^{\infty} A_n)$$

$$(\text{since } B_m = \bigcap_{n=m}^{\infty} A_n \uparrow) = \lim_{n \to \infty} \inf \mu(\bigcap_{n=m}^{\infty} A_n) \leq \lim_{m \to \infty} \inf \mu(A_m)$$

$$(\text{since } \bigcap_{n=m}^{\infty} A_m \subset A_m \text{ and } \mu \text{ is non-decreasing}).$$

If $A_n \in \mathcal{B}$, μ is finite, them $\mu(\lim_{n \to \infty} \sup A_n) \geq \lim \sup \mu(A_n)$ (this follows by taking complements). Now $\lim A_n \to A \Rightarrow \lim \inf A_n = \lim \sup A_n$. Combining them,

$$\liminf_{n \to \infty} \mu(A_m) \geq \mu(\limsup_{n \to \infty} A_n)$$

However,

$$\liminf_{n \to \infty} \mu(A_n) \leq \limsup_{n \to \infty} \mu(A_n) \text{ is always true.}$$

Therefore,

$$\liminf_{n \to \infty} \mu(A_n) = \limsup_{n \to \infty} \mu(A_n)$$

and hence $\lim \mu(A_n)$ exists and by previous arguments,

$$\lim \mu(A_n) = \mu(A).$$

EXAMPLE 2.17   Prove that a countable union of μ-null sets is a μ-null set.

Definition 2.12   Let $\{X_n\}$ be a sequence of r.vs. on a probability space $(\Omega, \mathscr{B}, P)$.

If $P[\lim_{n \to \infty} X_n \text{ exists}] = 1$, we say $\lim_{n \to \infty} X_n$ exists almost surely (a.s.).

If $P[\lim_{n \to \infty} X_n = X] = 1$ for a measurable function X on Ω, then we say $\lim_{n \to \infty}$ $X_n = X$ a.s. Recall that if X is a r.v., then the following properties hold.

(1) X is a measurable function

(2) It is finite almost surely, i.e. $[P \mid X \mid < \infty] = 1$ or $P[|X| = \infty] = 0$ are satisfied.

Theorem 2.11 (*First Borel-Cantelli lemma*).   Let $(\Omega, \mathscr{B}, P)$ be a probability space and $A_n \in \mathscr{B}$ for each $n = 1, 2, \ldots$

If $\sum_{n=1}^{\infty} P(A_n) < \infty$, then $P[A_n \text{ infinitely often}] = P[\overline{\lim} \, A_n] = 0$.

(Recall the definition

$$\limsup_{n \to \infty} A_n = [A_n \text{ i.o.}] = [A_n \text{ occur infinitely often}] = \bigcap_{m=1}^{\infty} \bigcup_{n=m}^{\infty} A_n)$$

Proof. $P[A_n \text{ i.o.}] = P[\bigcap_{m=1}^{\infty} \bigcup_{n=m}^{\infty} A_k] = \lim_{n \to \infty} P[\bigcup_{k=n}^{\infty} A_k].$

By continuity theorem of probability measure and noting

$$B_n = \bigcup_{k=n}^{\infty} A_k \downarrow, \, P[A_n \text{ i.o.}] \leq \lim_{n \to \infty} \sum_{k=n}^{\infty} P(A_k).$$

Now,

$$\sum_{1}^{\infty} P(A_n) < \infty \Rightarrow \sum_{k=n}^{\infty} P(A_k) < \varepsilon, \text{ if } n \geq n_0(\varepsilon).$$

Hence

$$P[A_n \text{ i.o.}] \leq \varepsilon \text{ for every } \varepsilon > 0.$$

PROBLEMS AND COMPLEMENTS

2.1 If S denotes the set of all irrational numbers in a finite interval $[a, b]$, prove that S is Lebesgue measurable with $\mu(S) = b - a$.

2.2 Let $(\Omega, \mathcal{F})$ be a measurable space and $B \in \mathcal{F}$, $B \neq \phi$. Define $B \cap \mathcal{F} = \{B \cap \mathcal{F}, A \in \mathcal{F}\}$. Show that $B \cap \mathcal{F}$ is a σ-field of subsets of B.

2.3 Let $(\Omega, \mathcal{F}, \mu)$ be a measure space. Show that the class of sets of the forms $E \cup N$ and $E \Delta N$ with $E \in \mathcal{F}$ and $N \subset A \in \mathcal{F}$ with $\mu(A) = 0$ form a σ-field.

2.4 (a) If E_1 and E_2 are subsets of X with a complete measure space $(X, \mathcal{F}, \mu)$, $\mu(E_1 \Delta E_2) = 0$ and E_1 is measurable, then prove that E_2 is also measurable and $\mu(E_2) = \mu(E_1)$.

(b) If A_1 and A_2 are measurable subsets of $[0, 1]$ and m is the Lebesgue measure on $[0, 1]$ with $m(A_1) = 1$, then show that $m(A_1 \cup A_2) = m(A_2)$.

***2.5** Let Q be the set of all rational, $\mathcal{F}_0$ the field of finite disjoint union of right semi-closed intervals of the form $\{x \in Q: a < x \leq b\}$, a, b rationals (consider $(0, \infty)$ and Q itself as right semi-closed intervals). Let $\mathcal{B} = \mathcal{B}\,(\mathcal{F}_0)$ be the σ-field generated by $\mathcal{F}_0$. Then prove that

(a) $\mathcal{B}$ consists of all subsets of Q.

(b) If $\mu(A)$ is the number of points in A (i.e. μ is counting measure), then μ is σ-finite on $\mathcal{B}$ but not on $\mathcal{F}_0$.

(c) There are sets $A \in B$ of finite measure that cannot be approximated by set in $\mathcal{F}_0$, i.e. there is no sequence $A_n \in \mathcal{F}_0$ with $\mu\,(A \Delta A_n) \to 0$.

******(d) If $\lambda = 2\mu$, then $\lambda = \mu$ on $\mathcal{F}_0$ but not on $\mathcal{B}$ (this shows that the Carathéodory extension and approximation fail if the measure μ on the field is not σ-finite).

2.6 Let $(\Omega, \mathcal{F}, \mu)$ be a measure space. If the measure μ is totally finite, what would be the probability measure P on $(\Omega, \mathcal{F})$ such that if

$$A_1, A_2 \in \mathcal{F}, \mu(A_1), \mu(A_2) > 0 \;\; \frac{\mu(A_1)}{\mu(A_2)} = \frac{P(A_1)}{P(A_2)}?$$

2.7 If P_1 and P_2 are two probability measures, then show that $P = \alpha P_1 + (1 - \alpha) P_2$, $0 < \alpha < 1$ is also a probability measure.

2.8 Let $(\Omega, \mathcal{F}, P)$ be a probability space with $\mathcal{F} = \mathcal{B}(\mathcal{G})$, where $\mathcal{G}$ is a field of subsets of Ω. Then show that for every $A \in \mathcal{F}$ and all $\varepsilon > 0$, there exists a set $B_\varepsilon \in \mathcal{G}$ such that $P(A \Delta B_\varepsilon) < \varepsilon$. Hence any $A \in \mathcal{F}$ can be approximated by a $B_\varepsilon \in \mathcal{G}$ in the sense that $|\, P(A) - P(B_\varepsilon)\,| < \varepsilon$.

2.9 Let $(\Omega, \mathcal{F}, P)$ be a probability space and $\mathcal{N} = \{N \mid N \subset \Omega$, there exists $A \in \mathcal{F}$, such that $N \subset A, P(A) = 0\}$.

Define $\bar{\mathcal{F}} = \{B \cup N \mid B \in \mathcal{F}, N \in \mathcal{N}\}$ and $\bar{P}$ on $\bar{\mathcal{F}}$ by putting $\bar{P}\,(B \cup N) = P(B)$. Then show that

(a) $\bar{\mathcal{F}}$ is a σ-field coinciding with the minimal σ-field containing $\mathcal{F}$ and $\mathcal{N}$.

---

* and ** indicate relatively difficult problems.

(b) the set function $\bar{P}$ is a probability measure.

(c) $(\Omega, \bar{\mathcal{F}}, \bar{P})$ is a complete probability measure.

2.10 A non-symmetric coin is tossed until a head appears. Give a space Ω, a σ-field $\mathcal{F}$ and a probability measure P on $\mathcal{F}$ to describe the probability space associated with the experiment.

2.11 Let $(\Omega, \mathcal{F}, P)$ be a probability space with $\Omega = (\omega_1, \omega_2, \omega_3, \omega_4)$, $\mathcal{F}$ = class of all subsets of Ω and $P(\{\omega_i\}) = 1/10$, $i = 1, 2, 3, 4$. Define the r.v. X as

$$X(\omega_i) = \begin{cases} 1 & \text{if } i = 1, 2 \\ 0 & \text{if } i = 3, 4 \end{cases}$$

Determine the probability space induced by X.

2.12 Let $\{A_n, n \geq 1\}$ be a sequence of subsets of a space Ω and $\{I_{A_n}, n \geq 1\}$ be the corresponding sequence of indicator functions. Let

$$s_1 = \sum_{j=1}^{n} I_{A_j}, s_2 = \sum_{1 \leq j_1 \leq j_2 \leq n} I_{A_{j_1} A_{j_2}}, \ldots, s_n = \sum_{1 \leq j_1 \leq j_2 \leq n} I_{A_{j_1} \ldots A_{j_n}},$$

B_m = the set of points belonging to exactly $m(1 \leq m \leq n)$ of $A_1, \ldots, A_n$ and

$$T_k = \sum_{1 \leq j_1 < \cdots < j_k \leq n} \Phi(A_{j_1} \ldots A_{j_k}), \quad \text{for } k = 1, 2, \ldots, n,$$

where Φ is a finite non-negative additive set function on an algebra $\mathcal{A}$ such that $A_j \in \mathcal{A}$. Show that

(a) $I_{B_m} = s_m - \binom{m+1}{m} s_{m+1} + \binom{m+2}{m} s_{m+2} - \ldots + (-1)^{n-m} \binom{n}{m} s_n$

(b) $F(B_m) = T_m - \binom{m+1}{m} T_{m+1} + \binom{m+2}{m} T_{m+2} - \ldots + (-1)^{n-m} \binom{n}{m} T_n$

(c) if $A_1, A_2, \ldots, A_n$ are interchangeable events with

$p_j = P(A_1, \ldots, A_j)$, $1 \leq j \leq n$, $p_0 = 1$

then

$$P(B_m) = \binom{n}{m} \sum_{j=0}^{n-m} (-1)^j \binom{n-m}{j} p_{m+j}$$

2.13 Let $A_1, \ldots, A_N$ be N events in relation to an experiment E. Define

$$S_r = \sum_{1 \leq i_1 \ldots i_r \leq N} P(A_{i_1} \ldots A_{i_r}).$$

Let $P_{[m]}$ and $P_{(m)}$ denote respectively the probabilities of "exactly m" and "at least m" events out of N events to occur. Then show that

(a) $S_r = \sum_{k=r}^{N} \binom{k}{r} P_{[k]} = \sum_{k=r}^{N} \binom{k-1}{r-1} P_{(k)}$

(b) $\sum_{k=r}^{N} \binom{k-1}{r-1} P_{(m)} = \binom{m}{r} P_{(m)} \leq S_r \leq P_{(m)} \left\{ \sum_{k=m}^{N} \binom{k-1}{r-1} \right\} + \sum_{k=r}^{m-1} \binom{k-1}{r-1}$

(c) Gumbel's inequality: $P_{(N)} \geq \dfrac{S_r - \binom{N-1}{r}}{\binom{N-1}{r-1}}$

(d) Frechet's inequality: $P_{(N)} \leq \dfrac{S_r}{\binom{N}{r}}$

2.14 Find the probability generating functions of
(a) $P(X \leq n)$, (b) $P(X < n)$, (c) $P(X \geq n)$, (d) $(X > n + 1)$, (e) $P(X = 2n)$ in terms of the generating function $P(s)$ of X.

2.15 Let U_n be the probability of the combination SF occurring for the first time at trial numbers $(n - 1)$ and n. Find the generating function, mean and variance of U_n.

2.16 If $X_1, X_2, \ldots$ are i.i.d. geometric r.vs with probability distribution, $P(X = k) = p(1 - p)^k$, $0 < p < 1$, and N is an independent Poisson r.v. with parameter λ, find the probability generating function of $\sum_{i=1}^{N} X_i$.

2.17 If $\{X_n, n \geq 0\}$ is a sequence of i.i.d geometric r.vs. with parameter p and N, independent of $\{X_n, n \geq 1\}$ has a geometric distribution with parameter p_0, then show that $\sum_{i=1}^{N} X_i$ also has a geometric distribution.

2.18 Let $\{X_n\}$ be a sequence of r.vs. Show that for every $B \in \mathcal{B}\,(X_1, X_2, \ldots)$ and every $\varepsilon > 0$ there exists a finite set of integers $\{n_1, \ldots, n_k\}$ and an event $B_\varepsilon \in \mathcal{B}\,(X_{n_1}, \ldots, X_{n_k})$ such that $P(B \,\Delta\, B_\varepsilon) < \varepsilon$.

3

Distribution Functions

3.1 DISTRIBUTION FUNCTION (UNIVARIATE)

Definition 3.1 If X is a r.v. its d.f. F_X is defined by

$$F_X(x) = P[X \le x] \text{ for all } x \in (-\infty, \infty)$$

Theorem 3.1 If X is a r.v. then its d.f. F_X has the following properties.

(a) F_X is non-decreasing i.e. if $-\infty < x' \le x < \infty$, then $F_X(x') \le F_X(x)$,

(b) $\lim\limits_{x \to \infty} F_X(x) = 1$ and $\lim\limits_{x \to -\infty} F_X(x) = 0$, and

(c) F_x is right continuous, i.e. $\lim\limits_{h \to 0+} F_X(x + h) = F_X(x)$ for all real x.

Proof.

(a) Since $[X \le x'] \subset [X \le x]$ by the monotone property of probability measure $F_X(x') \le F_X(x)$.

(b) By continuity theorem of probability measure

$$[X \le -x] \to \phi \text{ as } x \to \infty \Rightarrow F_X(x) \to 0, \text{ as } x \to -\infty$$

$$[X \le x] \to \Omega \text{ as } x \to \infty \Rightarrow F_X(x) \to 1, \text{ as } x \to \infty.$$

(c) Let $h \ge 0$. Then $F_X(x + h) - F_X(x) = P[x < X \le x + h]$

But $[x < X \le x + h] \to \phi$ as $h \to 0 + (h \downarrow 0$ and $h > 0)$

So, $P[x < X \le x + h] \to 0$ as $h \downarrow 0$ i.e. $F_X(x + h) - F_X(x) \to 0$ as $h \downarrow 0$.

Note that if $F_X^*(x) = P[X < x]$, *then it is left continuous.*

Theorem 3.2 (*Converse of Theorem 3.1 also known as the Fundamental Theorem of Probability*). If F is a function defined on $(-\infty, \infty)$ which satisfies properties (a), (b) and (c) of Theorem 3.1, then there exists a probability space $(\Omega, \mathscr{B}, P)$ and a r.v. defined on it such that $F_X(x) = P[X \le x] = F(x)$ for all real x.

38

The proof is deferred till the section on Measure Extension (Section 14.1, Chapter 14).

Definition 3.2   A distribution function F is said to be *absolutely continuous,* if there exists a non-negative Borel measurable function f (f if measurable with respect to Borel σ-field) over $(-\infty, \infty)$ such that $F(x) = \int_{-\infty}^{x} f(t)\ dt$ for all real x

or alternatively there exists a non-negative integrable function f such that for every $x \le x'$,

$$F(x) - F(x') = \int_{x'}^{x} f(t)\, dt.$$

NOTE.   The function f is called a *density* of F and F has a derivative equal to F almost everywhere. In this context it is worthwhile to quote a definition from Real Analysis.

Definition 3.3 $f : [a, b] \to (-\infty, \infty)$ is absolutely continuous if for each $\varepsilon > 0$ there exists a $\delta > 0$, such that for all positive integers n and families $(a_1, b_1), \ldots, (a_n, b_n)$ of non-overlapping open subintervals of $[a, b]$ with $\sum_{1}^{n} |b_i - a_i| < \delta$ we have, $\sum_{i}$ $|f(b_i) - f(a_i)| \le \varepsilon$. It is known from real analysis that every absolutely continuous function is almost everywhere differentiable and is equal to the indefinite integral of its derivative.

$$P[x' < X \le x] = [F(x) - F(x')] \to P[X = x] \text{ as } x' \uparrow x.$$

Since

F is right continuous at x, $F(x +) = F(x)$, i.e.

$$\lim_{h \downarrow 0} F(x + h) = F(x +) = F(x).$$

Hence

F will be continuous at x iff $P[X = x] = 0$.

Definition 3.4   A d.f. F is continuous at x if $P(X = x) = 0$. But this does not imply F is differentiable at x or its derivative $f(x)$ (say) exists. If $P[X = x] > 0$, then we say that F has a jump of magnitude $P[X = x]$ at x.

EXAMPLE 3.1   Let $X = I_A$, i.e. $X(\omega) = I_A(\omega)$ with $P(A) = p > 0$. Then

$$F_X(x) = \begin{cases} 0 & \text{if } x < 0 \\ (1 - p) & \text{if } 0 \le x < 1 \\ 1 & \text{if } x \ge 0 \end{cases}$$

since x takes 1 with probability p and takes 0 with $(1 - p)$. If a d.f. F has a jump of magnitude 1 at c then F is called *degenerate*. That is X is a degenerate r.v. with d.f. F and $P[X = c] = 1$.

If F takes only finite number of values, then they are the jumps, i.e. F is a step function. Probability of having jumps at $x = 0, 1, \ldots$ of geometric random variable is given by $(1 - \theta)\theta^k (0 < \theta < 1)$, i.e. $P[X = k] = p_k = (1 - \theta)\theta^k$, $k = 0, 1, 2, \ldots$ This is an example of a d.f. with a countably infinite number of jumps.

The *Lebesgue-Stieltjes measure μ_F induced by a distribution function F* is the measure defined by $\mu_F(a, b) = F(b) - F(a)$ for interval $(a, b]$ and for other Borel sets on the real line by the method of measure extension (see Section 14.2).

Definition 3.5   A distribution function F is said to be (continuous) *singular* if it is continuous and if there exists a Borel set S on the real line of Lebesgue measure zero and $\mu_F(S) = 1$, where μ_F is the Lebesgue-Stieltjes measure induced by F.

Theorem 3.3   Every distribution function F can be written as convex combination of a discrete, a singular continuous and an absolute continuous functions (i.e. $F = \alpha_1 F_1 + \alpha_2 F_2 + \alpha_3 F_3$, $\alpha_i \geq 0$ and $\sum\limits_i \alpha_i = 1$). Such a decomposition is unique. Proof of Theorem 3.3 depends on the following lemma.

Lemma (*Jordan decomposition*).   Every distribution function has at most countable number of discontinuity points and determines two distribution functions F_c and F_d such that F_c is continuous and F_d is a step function and $F = F_c + F_d$.

Proof.   Consider the interval $(l, l + 1)(l = 0, \pm 1, \pm 2, \pm 3 \ldots)$. Let $x_1 < x_2 < x_3 < \ldots < x_n$ be n points of discontinuity of $F(x)$ in $(l, l + 1)$ with jumps of magnitude exceeding $1/m$ $(m = 1, 2, \ldots)$ $F(l) \leq F(x_1 - 0) < F(x_1) \leq F(x_2 - 0) < F(x_2) < \ldots < F(x_n - 0) < F(x_n) \leq F(l + 1)$. Let $p_k = F(x_k) - F(x_k - 0)$. Therefore,

$$\sum\limits_1^n p_k = \sum\limits_{k=1}^n (F(x_k) - F(x_k - 0)) \leq F(l+1) - F(l).$$

Thus the number of discontinuity points of F in $(l, l + 1)$ of magnitude exceeding $1/m$ cannot be larger than $m[F(l + 1) - F(l)] \leq m$. Hence as $m \to \infty$, the number of discontinuities in $(l, l + 1)$ is countable. Since the number of such intervals is countable the first part follows. The proof of second part and also that of Theorem 3.3 is given in Appendix II.

3.2   MULTIVARIATE DISTRIBUTION FUNCTION

For a k-dimensional random vector $\mathbf{X} = (X_1, \ldots, X_k)$, the multivariate distribution function or k-dimensional joint distribution is defined as

$$F_{\mathbf{X}}(x_1, \ldots, x_k) = P(X_1 \leq x_1, \ldots, X_k \leq x_k), \qquad -\infty < x_i < \infty. \ 1 \leq i \leq k$$

$$= P[\bigcap\limits_{i=1}^k (X_i \leq x_i)]$$

It is to be noted that we can get $k - 1$ or lower dimensional marginal joint distributions from k-dimensional joint distribution by taking limits $x_k \to \infty$, $x_{k-1} \to \infty$, etc., e.g.

$$\lim\limits_{x_k \to \infty} F_{X_1 \ldots X_k}(x_1, \ldots, x_k) = F_{X_1, X_2 \ldots X_{k-1}}(x_1, \ldots, x_{k-1}), \text{ the } (k-1)$$

dimensional marginal distribution. Henceforth we shall use the notation $F(x_1, ..., x_k)$ for a k-dimensional distribution function. Let $(\mathbf{a}, \mathbf{b}] = \{(x_1, ..., x_k) \mid a_i < x_i \le b_i, 1 \le i \le k\}$ be the k-dimensional rectangle or cell in R^k and $\Delta_{i,k}$ denote the set of $\binom{k}{i}$ k-tuples $(x_1, ..., x_k)$, where each x_j is either a_j or b_j and such that exactly i of the x_j's are a_j. Then $\Delta = \bigcup_{i=0}^{k} \Delta_{i,\,k}$ is the set of 2^k vertices of the cell $(\mathbf{a}, \mathbf{b}]$ in R^k. Let δ be an arbitrary vertex. Like one-dimensional d.f. a multivariate distribution function is characterized by the following theorem which is stated below without proof.

Theorem 3.4   (a) $\displaystyle\lim_{\substack{\min x_i \to \infty \\ 1 \le i \le k}} F(x_1, ..., x_k) = 1$

(b) for each i, $1 \le i \le k$, $\displaystyle\lim_{x_i \to -\infty} F(x_1, ..., x_k) = 0$

(c) $F(x_1, ..., x_k)$ is continuous from above in each argument

(d) $\displaystyle P[\bigcap_{1}^{k}(a_i < X_i \le b)] = m_F(\mathbf{a}, \mathbf{b}] = \sum_{i=0}^{k}(-1)^i \sum_{\delta \in \Delta_{i,k}} F(d) \ge 0$

$$= \Sigma(-1)^{k-\ell} F(b_{i_1}, ..., b_{i_l}, a_{j_1}, ..., a_{j_{k-\ell}})$$

where the summation is over all permutations of $i_1, ..., i_l, j_1, ..., j_{k-l}$. $l = 0\ 1, ..., k$ where $\mu F(\mathbf{a}, \mathbf{b}]$ is the Lebesgue-Stieltjes measure of the cell $(\mathbf{a}, \mathbf{b}]$. In two dimension,

$$\mu_F(\mathbf{a}, \mathbf{b}] = F(b_1, b_2) - F(a_1, b_2) - F(b_1, a_2) + F(a_1, a_2) \ge 0$$

The proof is very similar in principle to Theorem 3.1 but is more complicated and hence we omit it.

NOTE.   Condition (d) is very important. There are functions of several variables satisfying the first three properties (a), (b) and (c) but not (d). The following example will demonstrate this.

EXAMPLE 3.2   Show that the univariate marginals of the bivariate function

$$F(x, y) = \begin{cases} 1 - e^{-2x} & \text{if } x > 0, y \ge 1 \\ 1/2(1 - e^{-x} & \text{if } x > 0, 0 \le y < 1 \\ 0 & \text{otherwise} \end{cases}$$

are proper distribution functions but itself is not a bivariate distribution function.

Answer. $F(x, \infty) = 1 - e^{-2x}$, $x > 0$ is the univariable exponential distribution function.

$$F(\infty, y) = \begin{cases} 0 & \text{if } y < 0 \\ 1/2 & \text{if } 0 \le y < 1 \text{ is a d.f.} \\ 1 & \text{if } y \ge 1 \end{cases}$$

But μ_F (log 4, 1/2), (log 6, 2)] = F(log 6, 2) − F(log 6, 1/2) − F(log 4, 2) + F(log 4, 1/2) = − 1/44 < 0 shows that $F(x, y)$ cannot be a bivariate d.f. Note that using difference operator condition (d) can be rewritten as $\Delta_{h_1}^{(1)} \Delta_{h_2}^{(2)} \ldots \Delta_{h_k}^{(k)} F(x_1, \ldots, x_k)$ ≥ 0 for all $h_k \ge 0$ for every real number x_k, where $\Delta_h^{(j)} F(x_1, \ldots, x_k) = F(x_1, \ldots, x_{j+h}, \ldots, x_k) - F(x_1, \ldots, x_k)$.

3.3   STIELTJES INTEGRAL

Throughout this section F is a distribution function.

3.3.1   Riemann-Stieltjes Integral

Let $g(x)$ be a continuous function in the interval $a \le x \le b$. By inserting points $x_1 < x_2 < \ldots < x_n < x_{n+1} = b$ in this interval, the interval is subdivided into $n + 1$ sub-intervals. In each of these we arbitrarily select points $\xi_1, \ldots, \xi_{n+1}$ and form the sum

$$S = \sum_{i=1}^{n+1} g(\xi_i)[F(x_i) - F(x_{i-1})]$$

(This is a weighted sum of $\{g(\xi_i)\}_1^{n+1}$.)

The limit of S is called the *Riemann-Stieltjes integral* (R-S integral) and is given by

$$\lim_{n \to \infty} S = \lim_{n \to \infty} \Sigma g(\xi_i)[F(x_i) - F(x_{i-1})]$$

and is denoted by $\int_a^b g(x)\, dF(x)$. In particular, if $F(x) = x$, then this is called simply

Riemann integral of $g(x)$ and is denoted by $\int_a^b g(x)\, dx$. For a bounded measurable function $g(x)$, (not necessarily continuous), take an arbitrary partition of $[a, b]$ by intervals $(x_{i-1}, x_i]$ such that $\max_i (x_i - x_{i-1}) \to 0$ as $n \to \infty$. Let

$$M_i = \sup_{x \in (x_{i-1}, x_i]} g(x), \text{ the upper bound of } g(x) \text{ in } (x_{i-1}, x_i]$$

and

$$m_i = \inf_{x \in (x_{i-1}, x_i]} g(x), \text{ the lower bound of } g(x) \text{ in } (x_{i-1}, x_i].$$

Then

$$Z = \sum_{1}^{n} M_i[F(x_i) - F(x_{i-1})]$$

is the so-called upper Darboux-sum and

$$z = \sum_{1}^{n} m_i[F(x_i) - F(x_{i-1})]$$

is the lower Darboux-sum.

We say that Riemann-Stieltjes integral of $g(x)$ with respect to $F(x)$ exists if $\lim_{n \to \infty}$ $z = \lim_{n \to \infty} Z$ independently of partitions and in that case the common limiting value is denoted by $\int_a^b g(x)\,dF(x)$ and in short we write R-S integral.

NOTES.

1. Like the Riemann integral the R-S integral exists if $g(x)$ is continuous on $[a, b]$.

2. Riemann-Stieltjes integral exists even in the more general case, $g(x)$ is bounded in $[a, b]$ and has atmost a finite number of discontinuity points $\{r_i\}$ provided that $F(x)$ is continuous in every r_i.

3. If g is a monotonic function and $F(x)$ is continuous in $[a, b]$, then also the R-S integral exists.

The *Lebesgue-Stieltjes (L-S)* integral of $g(x)$ over a Borel set S with respect to a probability measure is defined as

$$\lim_{n \to \infty} Z = \lim_{n \to \infty} z \text{ (independently of chosen partition of } S) = \int_S g(x)\,dP$$

$$= \int_S g(x)\,dF(x)$$

where

$$Z = \sum_{i=1}^{n} M_i P(S_i), \quad z = \sum_{i=1}^{n} m_i P(S_i), \quad \{S_i\}_{i=1}^{n}$$

is a partition of S and M_i and m_i are upper and lower bounds of $g(x)$ in S_i. Here $g(x)$ is bounded for all x belonging to the given set S.

NOTES.

1. If g is continuous on $[a, b]$, and L-S integral $\int_{[a,b]} g\,dF$ becomes a Riemann-Stieltjes integral.

2. If $F(x) = x$, then the Lebesgue-Stieltjes integral becomes Lebesgue integral and the Riemann-Stieltjes integral becomes the Riemann integral.

3. In case $F(x)$ is continuous everywhere in $[a, b]$ and has a continuous derivative $F'(x)$ except at most in a finite number of points we have for every $(x_{j-1}, x_i]$ not containing any of the exceptional points $F(x_i) - F(x_{i-1}) = (x_{i-1} - x_i) F'(\xi_i)$ where ξ_i is a point in $(x_{i-1}, x_i]$. In this case the R-S integral $\int_a^b g(x)\, dF(x) = \int_a^b g(x)\, F'(x)\, dx$ becomes the Riemann integral of $g(x)F'(x)$.

4. If $F(x)$ is a step-function with discountinuities $p_1, p_2, \ldots$ at the points $x_1, x_2, \ldots$ the Stieltjes integral coincides with the sum $\sum_i p_i g(x_i)$ which is a finite sum or an absolutely convergent infinite series according as the set of points of discontinuity is finite or infinite.

3.3.2   Extension of Riemann-Stieltjes Integral

All those definitions extend themselves to the case of a complex valued function $g(x)$ as also to infinite intervals subject to the condition that $g(x)$ is integrable over $[a, b]$ with respect to $F(x)$. We have then

$$\lim_{n \to \infty} \sum_{i=1}^{n} g(s_i)(F(x_i) - F(x_{i-1})) = \int_{-\infty}^{\infty} g(x)\, dF(x)$$

whenever $\max_i (x_i - x_{i-1}) \to 0$ as $n \to \infty$ and $a \to -\infty$ and $b \to \infty$.

We shall state some important properties of Stieltjes integrals in most of the cases without proof. Proofs can be found in standard textbooks (like Rudin, Principle of Real Analysis).

Simple properties of Riemann-Stieltjes integral.   Let g, g_1 and g_2 be continuous in $[a, b]$ and F, F_1 and F_2 be d.fs. Then

(i) $\displaystyle \int_a^b [g_1(x) + g_2(x)]\, dF(x) = \int_a^b g_1(x)\, dF(x) + \int_a^b g_2(x)\, dF(x)$

(ii) $\displaystyle \int_a^b [g(x)\, d(F_1(x) + F_2(x)) = \int_a^b g(x)\, dF_1(x) + \int_a^b g(x)\, dF_2(x)$

(iii) $\displaystyle \int_a^b g(x)\, dF(x) = \int_a^c g(x)\, dF(x) + \int_c^b g(x)\, dF(x), a < c < b$

(iv) If g is R-S integrable and $m \le g \le M$, and also f is continuous on $[m, M]$, then $h(x) = \phi(g(x))$ is R-S integrable on $[a, b]$.

The mean value theorem.   If g is continuous on $[a, b]$, then

$$\int_a^b g(x)\, dF(x) = g(\xi)\, (F(b) - F(a)), a \le \xi \le b.$$

Integration by parts. $\displaystyle \int_a^b g(x)\, dF(x) = g(b)F(b) - g(a)F(a) - \int_a^b F(x)\, dg(x),$
provided

1. $g(x)$, $F(x)$ are both non-decreasing and continuous on $[a, b]$ except at most on a finite number of discontinuity points which are all interior points of $[a, b]$.

2. No point in (a, b) is a discontinuity point for both g and F.

Proof. Choosing the subintervals so that no x_i is a discountinuity point of either of the functions g and F, we have

$$g(b)F(b) - g(a)F(a) = \sum_{i=1}^{n} (g(x_i)(F(x_i) - g(x_{i-1})F(x_{i-1}))$$

$$= \sum_{1}^{n} g(x_i)(F(x_i) - F(x_{i-1}))$$

$$+ \sum_{1}^{n} F(x_{i-1}) (g(x_i) - g(x_{i-1})) \text{ (since } x_n = b, x_0 = a)$$

The two sums are included between the lower and upper Darboux sums corresponding to the integrals $\int_a^b g \, dF$ and $\int_a^b F \, dg$ respectively. Passing to the limit we thus obtain the formula of partial integration

$$\int_a^b d(gF) = \int_a^b g \, dF + \int_a^b F \, dg.$$

Dominated convergence theorem. If $\lim_{n \to \infty} g_n(x) = g(x)$ exists and $|g_n(x)| \le G(x)$ for all $n \ge 1$ and for $x \in [a, b]$, where $G(x)$ is integrable with respect to F in $[a, b]$, then $g(x)$ is integrable with respect to F in $[a, b]$ and

$$\lim_{n \to \infty} \int_a^b g_n(x) \, dF(x) = \int_a^b g(x) \, dF(x)$$

NOTE. The generalization to infinite range is straightforward.

Helly's theorem. If (i) non-decreasing (in x) sequence of functions $\{F_n(x)\}$ converges to $F(x)$, (ii) $g(x)$ is everywhere continuous and (iii) a, b are continuity points of $F(x)$, then

$$\lim_{n \to \infty} \int_a^b g(x) \, dF_n(x) = \int_a^b g(x) \, dF(x)$$

Proof. As each $F_n(x)$ is non-decreasing, so is $F(x)$. For all $n \ge 1$, $F_n(x + h) - F_n(x) \ge 0$ if $h > 0$.
Taking

$$\lim_{n \to \infty} , F(x + h) - F(x) \ge 0, \text{ if } h > 0.$$

Now

$$\int_a^b g(x) \, dF(x) = \sum_{i=1}^{k-1} \int_{x_i}^{x_{i+1}} g(x) \, dF(x)$$

where

$$x_k = b, \ x_0 = a, \ a = x_0 < x_1 < \ldots < x_k = b \text{ are the continuity points of } F.$$

$$\int_a^b g(x)\,dF(x) = \Sigma\left(\int_{x_i}^{x_{i+1}} (g(x) - g(x_i))dF(x) + \int_{x_i}^{x_{i+1}} g(x_i)\,dF(x)\right)$$

$$= \Sigma\int_{x_i}^{x_{i+1}} (g(x) - g(x_i))dF(x) + \sum_{i=0}^{k-1} g(x_i)(F(x_{i+1}) - F(x_i))$$

$$\leq \theta_1 \varepsilon / 3 + \Sigma g(x_i)(F(x_{i+1}) - F(x_i)) \tag{1}$$

Since g is continuous,

$$|g(x) - g(x_i)| < \frac{\varepsilon/3}{F(b) - F(a)} \text{ for } x_i \leq x \leq x_{i+1} \text{ and where } |\theta_1| \geq 1. \text{ Similarly,}$$

$$\int_a^b g(x)\,dF_n(x) \leq \theta_2 \varepsilon / 3 + \Sigma g(x_i)(F_n(x_{i+1}) - F_n(x_i)) \tag{2}$$

Since

$$F_n(x) \to F(x) \text{ at continuity points of } F,$$

$$F_n(x_{i+1}) - F(x_{i+1}) < \frac{\varepsilon}{6 \sum_1 g(x_i)} \text{ for all } i \text{ and large } n \text{ and hence the absolute}$$

difference of (1) and (2) is less than ε for large n.

Helly-Bray theorem.   If

(a) $g(x)$ is continuous,

(b) $F_n(x) \to F(x)$ in every continuity point of $F(x)$ and

(c) for any $\varepsilon > 0$, we can find A such that

$$\int_{-\infty}^{-A} |g(x)|\,dF_n(x) + \int_A^\infty |g(x)|\,dF_n(x) < \varepsilon \text{ for all } n = 1, 2, \ldots$$

Then

$$\lim_{n \to \infty} \int_{-\infty}^\infty g(x)\,dF_n(x) + \int_{-\infty}^\infty g(x)\,dF(x)$$

Proof.   We shall first prove for $F_n(x) \to F(x)$ for all x. Taking limit $n \to \infty$, from (c) it follows that

$$\int_{-\infty}^A |g(x)|\,dF(x) + \int_A^\infty |g(x)|\,dF(x) < \varepsilon \tag{1}$$

From the previous result,

$$\lim_{n \to \infty} \int_a^b g(x)\,dF_n(x) = \int_a^b g(x)\,dF(x) \tag{2}$$

and from condition (c) it follows that for continuity points $b > a$, $g(x)$ is integrable over $(-\infty, \infty)$ with respect to $F(x)$. Now if in (2) we take $a = -A$, $b = A$, then

$$\lim_{n \to \infty} \int_{-A}^{A} g(x)\, dF_n(x) = \int_{-A}^{A} g(x)\, dF(x) \tag{3}$$

Now

$$\int_{-\infty}^{\infty} g(x)\, dF_n(x) - \int_{-\infty}^{\infty} g(x)\, dF(x) = \left(\int_{-\infty}^{-A} + \int_{-A}^{A} + \int_{A}^{\infty} \right) g(x)\, dF_n(x)$$

$$- \left(\int_{-\infty}^{-A} + \int_{-A}^{A} + \int_{A}^{\infty} \right) g(x)\, dF(x)$$

$$\lim_{n \to \infty} \left| \int_{-\infty}^{\infty} g(x)\, dF_n(x) - \int_{-\infty}^{\infty} g(x)\, dF(x) \right| \le \lim_{n \to \infty} \left[\int_{-A}^{A} | g(x)\, dF_n(x) \right.$$

$$\left. + \int_{A}^{\infty} | g(x) |\, dF_n(x) \right] + \lim_{n \to \infty} \left| \int_{-A}^{A} (g(x)\, dF_n(x) - g(x)\, dF(x)) \right|$$

$$+ \int_{-\infty}^{-A} | g(x) |\, dF(x) + \int_{A}^{\infty} | g(x) |\, dF(x)$$

$$< \varepsilon + \varepsilon + \varepsilon = 3\varepsilon \text{ if } n \text{ and } A \text{ are both large.}$$

So we have proved the result if $F_n(x) \to F(x)$ at all x.

Now we can choose only continuity points of $F(x)$, $x_1, x_2, \ldots$, since between two discontinuity points we can choose a continuity point and for monotonic F_n the points of discontinuities are at most countable. Our result is true, if $F_n(x) \to F(x)$ at all continuity points of F.

Differentiation under the integral Sign. If for continuity points x of the function g and for a fixed value of t, (i) the partial derivatives $\delta g(x, t)/\delta t$ exists, and (ii) $| \delta g(x, t)/\delta t | \le G(x)$ which is integrable over $(-\infty, \infty)$, then

$$\delta/\delta t \int_{-\infty}^{\infty} g(x, t)\, dF(x) = \int_{-\infty}^{\infty} \delta g(x, t)/\delta t\, dF(x)$$

Change of order of integration (Fubini theorem). If

(i) $g(x, t)$ is continuous in t in every interval (a, b) for all values of x in $(-\infty, \infty)$.

(ii) $| g(x, t) | \le G_1(x)$ for all $t \in (a, b)$,

(iii) $\int_{-\infty}^{\infty} |g(x, t)|\, dt \le G_2(x)$, where $G_1(x)$ and $G_2(x)$ are integrable over $(-\infty, \infty)$, then

$$\int_{-\infty}^{\infty}\int_{-\infty}^{\infty} [g(x,t)\,dF(x)]\,dt = \int_{-\infty}^{\infty}\left[\int_{-\infty}^{\infty} g(x,t)dt\right]dF(x).$$

3.3.3   Stieltjes Integral for Function of p Variables

For a p-dimensional point $\mathbf{x} = (x_1, \ldots, x_p)$ of the d.f. $F(x_1, \ldots, x_p)$, $I_i = \prod_{j=1}^{p} [x_{ij} - h_{ij}, x_{ij} + h_{ij}]$ is a partition of the rectangle $I = [\mathbf{a}, \mathbf{b}]$, where $\mathbf{x}_i = (x_{i1}, \ldots, x_{ip})$. Then the Stieltjes integral of $g(x)$ with respect to $F(x_1, \ldots, x_p)$ is defined as $\lim_{n\to\infty} \Sigma_j\ g(s^j)$

$$\Delta_p F(x_1, \ldots, x_p) = \int g(x)\,dF(x_1, \ldots, x_p),$$ where $s^j = (s_1^j, \ldots, s_p^j)$ is an arbitrary point of I_i and Δp is the pth order difference of F at $(x_1, \ldots, x_p)$ and $\max_{i,j} h_{ij} \to 0$ as

$n \to \infty$. Let us close the section with a *Lebesgue integrable* function which is not Riemann integrable. Let

$$f(x) = 1 \text{ if } x \text{ is rational}$$
$$= 0 \text{ if } x \text{ is irrational}$$

Then it is not Riemann integrable but it is Lebesgue integrable (with the value zero), since the upper and lower Riemann integrals are

$$\overline{R}\int_0^1 f(x)\,dx = 1 \text{ and } \underline{R}\int_0^1 f(x)\,dx = 0.$$

In fact, if we divide the interval into two disjoint parts S_i and S_r containing respectively the irrational and rational numbers, since Lebesgue measure of S_i is zero and S_r is 1, upper and lower Darboux sums corresponding to this division are equal to 0. Hence lower integral and upper integral are both zero. So

$$\int_0^1 f(x)\,dx = 0.$$

PROBLEMS AND COMPLEMENTS

3.1   Show that equivalent random variables are identically distributed but the converse is not true.

3.2   Let F_1 and F_2 be the d.fs. of two r.vs. Will $F = \beta F_1 + (1 - \beta)F_2$ and $F^* = F_1^{\beta_1} F_2^{(1-\beta)}$, $0 < \beta < 1$ be d.fs.?

3.3   Example that marginal distributions do not determine a joint distribution: Let f_1, f_2, f_3 be three p.d.fs. with corresponding d.fs. F_1, F_2, F_3. Define for $|\alpha| < 1$,

$$f_\alpha(x_1, x_2, x_3) = f(x_1)f(x_2)f(x_3)\ \{1 + \alpha[2F_1(x_1) - 1][2F_2(x_2) - 1][2F_3(x_3) - 1]\}$$

Show that f_α is a p.d.f. for each $\alpha \in [-1, 1]$ and the collection of joint densities $\{f_\alpha, -1 \le \alpha \le 1\}$ has same marginal densities f_1, f_2, f_3.

3.4 Show that the continuity property of one-dimensional distributions, i.e. $P[\omega : X(\omega) = x] = 0$ for each x in some non-degenerate interval (finite or infinitie) may fail in the multi-dimensional case.

3.5 Show that if $X \stackrel{d}{=} Y$ then $g(X) \stackrel{d}{=} g(g(Y)$ for any one-to-one measurable function g but $XZ \stackrel{d}{=} YZ$ for any r.v. Z is false.

3.6 Show that the Lévy-Prokhorov metric between two d.fs. F and G defined by

$$d(F, G) = \inf_{h \ge 0} \{h : F(x - h) - h \le G(x) \le F(x + h) + h\}, \text{ for all } x \in \mathbb{R} \text{ is}$$

a genuine distance function.

3.7 Give examples of sequences of distributions which (i) does not converge, (ii) converges but not to a d.f.

3.8 Give examples of r.vs. which converge in distribution but not in densities or mass functions.

3.9 Give an example of a p.d.f. whose m.g.f. does not exist but moments of all order do.

4

Measurable Functions

4.1 ALGEBRA OF MEASURABLE FUNCTIONS

Since not all sets are measurable, it is of great importance to know that sets which arise naturally in certain construction of functions are measurable. If f is a real-valued function defined on a bounded interval, we can talk about Riemann integral of f, at least if f is piece-wise continuous. Probability theory and other branches of mathematics require much more general integration process, one that applies to functions from an arbitrary set to the extended reals, provided that certain measurability conditions are satisfied. Probability considerations may again be used to motivate the concept of measurable functions, i.e. random variables. Suppose that $(\Omega, \mathcal{B}, P)$ is a probability space, and that f is a function from Ω to $\overline{R}$. Thus if the outcome of the experiment corresponds to the point $\omega \in \Omega$, we may compute the number $f(\omega)$. Suppose that we are interested in the probability that $a \leq f(\omega) \leq b$, in other words, we wish to compute $P\{\omega : f(\omega) \in A\}$, where $A = [a, b]$. For this to be possible, the set $\{\omega : f(\omega) \in A\} = f^{-1}(A)$ must be in the σ-field $\mathcal{B}$. If $f^{-1}(A) \in \mathcal{B}$ for each interval A, then f is a "measurable function", in other words, probabilities of events involving f can be computed. Now let us formally define a measurable function and give a theorem which will state that a measurable function can be defined in several convenient ways. Let $(X, \mathcal{B})$ be a measurable space and $(X, \mathcal{B}) \xrightarrow{f} (\overline{R}, \mathcal{F})$ be an extended real valued function.

Definition 4.1 An extended real valued function f on X is said to be measurable if for each real number a the set $\{x : f(x) < a\} \in \mathcal{B}$.

Recall that an almost surely finite measurable function on a probability space is called a random variable.

Theorem 4.1 The following statements are equivalent:

    (a) f is measurable, i.e. $\{x : f(x) \leq a\} \in \mathcal{B}$ for each real number a.

    (b) $\{x : f(x) \leq a\} \in \mathcal{B}$ for each real number a.

    (c) $\{x : f(x) > a\} \in \mathcal{B}$ for each real number a.

    (d) $\{x : f(x) \geq a\} \in \mathcal{B}$ for each real number a.

50

(e) $\{x : a < f(x) < b\} \in \mathcal{B}$ for each real number a and b.

(f) $f^{-1}(G) \in \mathcal{B}$ for each open set G of R.

(g) $f^{-1}(F) \in \mathcal{B}$ for each closed set F of R.

(h) $f^{-1}(E) \in \mathcal{B}$ for each Borel set E of R.

Proof. We shall prove (a) $\Rightarrow$ (h).

Let $\mathcal{D} = \{E : f^{-1}(E) \in \mathcal{B}\}$, E is a Borel set of R. We shall prove that $\mathcal{D}$ is a σ-algebra of subsets of R. If $E_i \in \mathcal{D}$, then $f^{-1}(E_i) \in \mathcal{B}$ for all $i \geq 1$. Now

$$f^{-1}(\bigcup_i^{\infty} E_i) = \bigcup_i^{\infty} f^{-1}(E_i) \in \mathcal{B}, \text{ since } \mathcal{B} \text{ is a } \sigma\text{-algebra.}$$

If $E \in \mathcal{D}$, then $f^{-1}(E) \in \mathcal{B}$ and $f^{-1}(E^C) = (f^{-1}(E))^C \in \mathcal{B}$.
Similarly for intersections. Hence $\mathcal{D}$ is a σ-algebra.

$$\{x : f(x) < a\} = f^{-1}[-\infty, a) \in \mathcal{B} \text{ implies } [-\infty, a) \in \mathcal{D} \text{ for every real } a.$$

Therefore $\mathcal{D} \supset$ Borel σ-field of R. Hence $f^{-1}(E) \in \mathcal{B}$ for all Borel sets E of R.

We have (c) $\Leftrightarrow$ (b), since $\{x : f(x) \leq a\} = X - \{x : f(x) > a\} \in \mathcal{B}$ and vice versa.

(a) $\Leftrightarrow$ (d), since they involve sets which are complements of each other.

(c) $\Leftrightarrow$ (d), since $\{x : f(x) \geq a\} = \bigcup_{n=1}^{\infty} \{x : f(x) > a - 1/n\} \in \mathcal{B}$ and the countable union of measurable sets is measurable. (c) or (d) $\Rightarrow$ (e), since

$$\{x : a < f(x) < b\} = \{x : f(x) > a\} - \{x : f(x) \geq b\}$$

$$= \{x : f(x) > a\} - \bigcap_{n=1}^{\infty} \{x : f(x) > b - 1/n\} \in \mathcal{B}$$

(f) $\Leftrightarrow$ (g), since $G^C = F$ and $F^C = G$.

(f) $\Leftrightarrow$ (g), since any open set in R can be written as countable union of open intervals and obviously and open interval is an open set.

It follows from Theorem 4.1 that if we restrict ourselves to measurable functions, the most important sets connected with them are measurable. It also follows from (f) or (g) in Theorem 4.1 that a continuous function is measurable (note that f defined on R is continuous iff $f^{-1}(G)$ is open for every open set G in R and an open set is a Borel set). Of course each step function is measurable. If f is a measurable function and E is a measurable subset of the domain of f, then the function obtained by restricting f to E is also measurable. The following theorem tells us that certain operations performed on measurable functions lead again to a measurable function.

Theorem 4.2 If c is a constant and the functions f and g are measurable then so are the functions $f + c$, cf, $f + g$, $f - g$, $f \cdot g$, max (f, g) and f/g (assuming these are well defined, in other words, $f(x) + g(x)$ is never of the form $+\infty - \infty$ and $f(x)/g(x)$ is never of the form ∞/∞ or $a/0$). Moreover if $\{f_n\}$ is a sequence of measurable functions, then $\sup_n f_n$, $\inf_n f_n$, $\overline{\lim}_{n\to\infty} f_n$, $\underline{\lim}_{n\to\infty} f_n$ are all measurable functions. Also $\lim_{n\to\infty} f_n$ is a measurable function if it exists.

Proof. $\{x : f(x) + c < a\} = \{ x : f(x) < a - c\} \in \mathscr{B}$ if f is a measurable function. Hence $m(x) = f(x) + c$ is a measurable function. If $c \neq 0$,

$$\{x : f(x) \cdot c < a\} = \{x : f(x) < a/c\} \in \mathscr{B}.$$

The case $c = 0$ is trivial. Let $h(x) = f(x) + g(x)$. Then $\{x : f(x) + g(x) < a\} = \underset{\{r\}}{\cup}$ $\{x : f(x) < r\} \cup \{x : g(x) < a - r\}$, where $\{r\}$ is the set of all rationals. Since the rationals are countable, this set is measurable. Therefore, $h = (f + g)$ is measurable. Since

$$fg = \frac{1}{2}((f + g)^2 - f^2 - g^2),$$

fg is a measurable function if we can show that f^2 or g^2 is a measurable function. It is enough to show that either f^2 or g^2 is a measurable function. Now

$$\{f : f^2(x) > a\} = \{x : f(x) > \sqrt{a}\} \cup \{x : f(x) < -\sqrt{a}\} \text{ for } a \geq 0.$$

NOTE. $\{x : f^2(x) > a\} = X$ if $a > 0$ and $X \in \mathscr{B}$.

Hence f^2 is a measurable function, and so fg is measurable if f and g are both measurable. Again,

$$\{x : \max (f, g) < a\} = \{x : f(x) < a\} \cap \{x : g(x) < a\}$$

$$\{x : \inf_n f_n(x) < a\} = \bigcup_{n=1} \{x : f_n(x) < a\}.$$

Also note that

$$\sup_n f_n(x) = \inf_n (-f_n(x)).$$

Therefore, f_n is measurable implies $-f_n$ is measurable, and $\inf_n (-f_n) = \sup_n (f_n)$ are all measurable functions. Now,

$$\overline{\lim_{n \to \infty}} f_n(x) = \inf_{n \geq 1} \sup_{m \geq n} f_m(x) \; and \; \underline{\lim_{n \to \infty}} f_n(x) = \sup_{n \geq 1} \inf_{m \geq n} f_m(x)$$

and hence they are also measurable functions. If $\lim_{n \to \infty} f_n$ exists then

$$\overline{\lim_{n \to \infty}} f_n = \underline{\lim_{n \to \infty}} f_n = \lim_{n \to \infty} f_n.$$

Hence they are all measurable functions and $\lim_{n \to \infty} f_n$ is a measurable function.

Let
$$(X, \mathscr{B}) \xrightarrow{f} (\overline{R}, \mathscr{F}).$$

Definition 4.2   A property is said to hold almost everywhere (a.e.) or almost surely (a.s) if the set of points where it fails to hold is a set of measure zero (probability zero).

Definition 4.3   A measure μ is called complete if subsets of sets of μ measure zero are measurable.

Theorem 4.3   If f is a measurable function and $f = g$ a.e., then g is measurable if μ is a complete measure.

Proof.   Let E be the set $\{x : f(x) \neq g(x)\}$ such that $\mu(E) = 0$.
Now $\{x : g(x) > a\} = \{x : f(x) > a\} \cup \{x \in E : g(x) > a\} - \{x \in E : g(x) \geq a\}$.
Since f is measurable the set $\{x : f(x) > a\}$ is measurable.

$$\{x \in E : g(x) > a\} \subset E \text{ and } \{x \in E : g(x) \leq a\} \subset E \text{ and } \mu(E) = 0$$

imply that these sets are measurable. Therefore $\{x : g(x) > a\}$ is measurable and that implies g is a measurable function.

Definition 4.4   An extended real-valued function ϕ on $(X, \mathcal{B}, \mu)$ is called simple if it is $\mathcal{B}$-measurable and assumes only a finite number of values say $\alpha_1, \ldots, \alpha_n$, i.e.

$$\phi(x) = \sum_1^n \alpha_i I_{A_i}(x).$$

NOTE.   The sum, product and difference of two simple functions are simple. This representation of ϕ is not unique.

Alternative definition.   An extended real-valued function ϕ on $(X, \mathcal{B}, \mu)$ is called simple if it can be expressed as

$$\phi(x) = \sum_1^n \alpha_i I_{A_i}(x).$$

where A_i are disjoint measurable sets with $\bigcup_1^n A_i = X$ and α_i are distinct non-zero constants. This form is called the canonical form of a simple function. Here A_i's are given by $A_i = \{x : f(x) = \alpha_i\}$.

Theorem 4.4   Every extended real-valued measurable function f is the limit of a sequence $\{f_n\}$ of simple finite valued functions. If f is non-negative then each function may be taken to be non-negative and the sequence $\{f_n\}$ may be assumed to be non-decreasing.

Proof.   Let f be a measurable function. Then

$$f_n(x) = -nI_{[f(x)<-n]}(x) + \sum_{i=-n \cdot 2^n}^{n \cdot 2^n} \frac{i-1}{2^n} I_{\left[\frac{i-1}{2^n} \leq f(x) < \frac{i}{2^n}\right]}(x) + nI_{[f(x) \geq n]}(x)$$

Note that

$$f(x) = \pm \infty \Rightarrow f_n(x) = \pm n \to \pm \infty \text{ and } |f(x)| < \infty \Rightarrow \text{ for each } n \text{ there exists an}$$
i such that

$$\frac{i-1}{2^n} \leq f(x) < i/2^n \text{ with } -n \cdot 2^n \leq i \leq n \cdot 2^n$$

and this implies $f_n(x) = (i-1)/2^n$. Hence

$|f_n(x) - f(x)| \leq 1/2^n \to 0$ as $n \to \infty$. Suppose $f(x) \geq 0$. For every $n = 1, 2, 3, \ldots$ and for every $x \in X$, we write

$$f_n(x) = \begin{cases} (i-1)/2^n & \text{if } (i-1)/2^n \leq f(x) < i/2^n, i = 1, \ldots, n2^n \\ n & \text{if } f(x) \geq n. \end{cases}$$

Clearly, $f_n \geq 0$ and simple and also non-decreasing, i.e. $f_{n+1}(x) \geq f_n(x)$. If $f(x) < \infty$, then for some n, $0 \leq f_n(x) - f(x) \leq 1/2^n$. If $f(x) = \infty$ then $f_n(x) = n$ for all integer n. Hence $f_n(x) \to f(x)$ for all $x \in X$.

4.2 ALMOST EVERYWHERE CONVERGENCE

Definition 4.2(a)   A sequence $\{f_n\}$ of extended real-valued function is said to converge almost everywhere (a.e.) or almost surely (a.s.) on the measurable space $(X, \mathscr{B}, \mu)$ $((\Omega, \mathscr{B}, P))$ to a limit function f if the following conditions hold:

(a) There exists $E \in \mathscr{B}$ with $\mu(E) = 0$ $(P(E) = 0)$.

(b) If $x \in E^C$ and $\varepsilon > 0$, then an integer $n_0 = n_0(x, \varepsilon)$ can be found with the property that

$$\left. \begin{array}{ll} f_n(x) < -1/\varepsilon & \text{if } f(x) = -\infty \\ |f_n(x) - f(x)| < \varepsilon & \text{if } |f(x)| < \infty \\ f_n(x) > 1/\varepsilon & \text{if } f(x) = \infty \end{array} \right\} \text{ whenever } n \geq n_0$$

(In case of almost sure convergence first and third inequalities are meaningless.)

NOTE.   (a) If $n_0(x, \varepsilon) = n(\varepsilon)$ is independent of x then the convergence is said to be uniform convergence a.e.

(b) if $f_{n+\gamma} - f_n \to 0$ on a set A uniformly in γ (or $f_m - f_n \to 0$ on A as $m, n \to \infty$), we say that the function f_n mutually (Cauchy) converges on A. According to Cauchy criterion, if functions $\{f_n\}$ are finite then $f_n \to f$ finite iff $f_m - f_n \to 0$ as $m, n \to \infty$ or equivalently $f_{n+\gamma} - f_n \to 0$ as $n \to \infty$ uniformly in γ. In this connection we shall state a theorem without a formal proof.

Theorem 4.5   A sequence $\{f_n\}$ of a.e. finite measurable functions converges a.e. to a finite measurable function f iff the sequence mutually converges a.e. (a.s.). (Note that such a sequence is called fundamental a.e. Since for fixed x, a.e. convergence is pointwise convergence, the proof follows easily from completeness of real line and the following discussions.)

If f_n and f are real-valued functions (i.e. finite) then the set of convergence can be written as

$$\{x : f_n \to f\} = \bigcap_{\varepsilon > 0} \bigcup_{n=1}^{\infty} \bigcap_{\gamma=1}^{\infty} \{x : |f_{n+\gamma}(x) - f(x)| < \varepsilon\}$$

$$= \bigcap_{k=1}^{\infty} \bigcup_{n=1}^{\infty} \bigcap_{\gamma=1}^{\infty} \{x : |f_{n+\gamma}(x) - f(x)| < 1/k\} \text{ is}$$

$\mathcal{B}$-measurable, where $\varepsilon = 1/k > 0$. Similarly the set of mutual convergence is given by

$$\{x : f_{n+\gamma} - f_n \to 0\} = \bigcap_{k=1}^{\infty} \bigcup_{n=1}^{\infty} \bigcap_{\gamma=0}^{\infty} \{x : | f_{n+\gamma}(x) - f_n(x) | < 1/k\}.$$

They are obtained by countable set of operations and hence the above set is $\mathcal{B}$-measurable. Let f_n and f are a.e. finite measurable functions. (In a probability space $(\Omega, \mathcal{B}, P)$ instead of f_n and f the random variables X_n and X are used.) Therefore

(a) $f_n \to f$ a.e. iff for all $\varepsilon > 0$, $\mu[\bigcap_{n=1}^{\infty} \bigcup_{\gamma=1}^{\infty} \{| f_{n+\gamma} - f_n | \geq \varepsilon\}] = 0$

(b) $(f_{n+\gamma} - f_n)$ iff for every $\varepsilon > 0$, $\mu[\bigcap_{n=1}^{\infty} \bigcup_{\gamma=1}^{\infty} \{| f_{n+\gamma} - f | \geq \varepsilon\}] = 0$

(c) If μ is finite then, $f_n \to f$ a.e., iff for every $\varepsilon > 0$,

$\mu(\bigcup_{\gamma} \{| f_{n+\gamma} - f | \geq \varepsilon\}) \to 0$ as $n \to \infty$.

(d) If μ is finite then, $f_{n+\gamma} - f_n \to 0$ a.e., as $n \to \infty$ and all $\gamma = 0, 1, 2, \ldots$ iff

for all $\varepsilon > 0$, $\mu (\bigcup_{\gamma} \{|f_{n+\gamma} - f_n | \geq \varepsilon\}) \to 0$ as $n \to \infty$. (c) follows from (a) by continuity theorem of measure. Similarly (d) follows from (b).

We state without proof the following result (known as Egoroff's theorem) which establishes an interesting and useful connection between convergence and uniform convergence.

Theorem 4.6 (Egoroff). If E is a measurable set of finite measure, and if $\{f_n\}$ is a sequence of *a.e. finite valued* measurable functions which converges *a.e. on E* to a finite valued measurable function f, then, for every $\varepsilon > 0$, there exists a *measurable subset F on E* such that $\mu(F) < \varepsilon$ and such that the sequence $\{f_n\}$ converges to f uniformly on E-F.

4.3   CONVERGENCE IN MEASURE

Definition 4.5   A sequence $\{f_n\}$ of measurable functions is said to converge to f in measure, if, given $\varepsilon > 0$ there exists positive integer N such that for all $n \geq N$ we have

$$\mu\{x : | f_n(x) - f(x) | \geq \varepsilon\} < \varepsilon \text{ and we write } f_n \xrightarrow{\mu} f.$$

In a probability space we similarly define convergence in probability of $\{X_n\}$ to X if

$$P\{| X_n - X | > \varepsilon\} \to 0 \text{ as } n \to \infty \text{ and we write } X_n \xrightarrow{P} X.$$

Theorem 4.7   If $f_n \to f$ a.e. then $f_n \xrightarrow{\mu} f$ if μ is finite (If $X_n \to X$ a.s. then $X_n \xrightarrow{P} X$ for random variables, since probability measure is finite.)

Proof.   Let

$$E_{n,\varepsilon} = \bigcap_{i=n}^{\infty} \{x : |f_i(x) - f(x)| < \varepsilon\}$$

is a non-decreasing sequence of sets for fixed $\varepsilon > 0$. Now the set of convergence is given by

$$C = \{x : f_n \to f\} = \bigcap_{k=1}^{\infty} \bigcup_{n=1}^{\infty} E_{n,1/k}, \text{ taking } \varepsilon = 1/k$$

Now $\mu(C^C) = 0$ (since $f_n \to f$ a.e. implies that the measure of the set where it fails is zero). Therefore,

$$\mu = (\bigcup_{k=1}^{\infty} \bigcup_{n=1}^{\infty} E_{n,1/k})^C) = 0$$

i.e. $$m(\bigcap_{n=1}^{\infty} E_{n,1/k}^C) = 0 \text{ for each } k = 1, 2, 3, \ldots$$

But $E_{n,1/k}^C \downarrow$ in n for fixed k. This implies (μ being finite)

$$\mu(E_{n,1/k}^C) \to 0 \text{ as } n \to \infty \text{ for all } k \geq 1 \text{ and } \mu(E_{n,\varepsilon}^C) \to 0 \text{ as } n \to \infty \text{ for all}$$

$\varepsilon > 0$. Therefore

$$\mu\{x : |f_n(x) - f(x)| \geq \varepsilon\} \leq \mu\{\bigcup_{i=n}^{\infty} \{x : |f_i(x) - f(x)| \geq \varepsilon\}]$$

$$= \mu(E_{n,\varepsilon}^C) \to 0 \text{ as } n \to \infty \text{ for every } \varepsilon > 0. \text{ Hence } f_n \xrightarrow{\mu} f.$$

Theorem 4.8   Let $\{f_n\}$ be a sequence of measurable functions converges in measure to f. Then there exists a subsequence $\{f_{n_k}\}$ which converges to f a.e.

Proof.   Given γ there is an integer n_γ such that for all $n \geq n_\gamma$ we have

$$\mu\{x : |f_n(x) - f(x)| \geq 2^{-\gamma}\} < 2^{-\gamma}, \text{ choosing } \varepsilon = 2^{-\gamma}. \text{ Let}$$

$$E_\gamma = \{x : |f_{n_\gamma}(x) - f(x)| < 2^{-\gamma}\}.$$

Then if $x \notin \bigcup_{\gamma=k}^{\infty} E_\gamma$ we have $|f_{n_\gamma}(x) - f(x)| < 2^{-\gamma}$ for all $\gamma \geq k$ and so $f_{n_\gamma}(x) \to f(x)$ pointwise as $\gamma \to \infty$.

Hence $f_{n_\gamma}(x) \to f(x)$ for any $x \notin A = \bigcap_{k=1}^{\infty} \bigcup_{\gamma=k}^{\infty} E_\gamma$, the set of divergence. But $\mu(A)$

$$\leq \mu(\bigcup_{\gamma=k}^{\infty} E_\gamma) \leq \sum_{\gamma=k}^{\infty} \mu(E_\gamma) < \sum_{\gamma=k}^{\infty} 2^{-\gamma} = 2^{-k+1} \text{ for all } k \geq 1. \text{ Hence } \mu(A) = 0.$$

Definition 4.6   A sequence $\{f_n\}$ of a.e. finite valued measurable function is fundamental in measure (i.e. Cauchy sequence in measure) if for every $\varepsilon > 0$.

$$\mu(x : |f_n(x) - f_m(x)| \geq \varepsilon) \to 0 \text{ as } m, n \to \infty.$$

Similarly a sequence of random variables is fundamental in probability (i.e. Cauchy sequence in probability) if for all $\varepsilon > 0$.

$$P(\omega : \mid X_n(\omega) - X_m(\omega) \mid \geq \omega\} \to 0 \text{ as } m, n \to \infty$$

Theorem 4.9   If a sequence $\{f_n\}$ of measurable functions converges in measure to f, then $\{f_n\}$ is fundamental in measure. If also $\{f_n\}$ converges in measure to g, then $f = g$ a.e.

Proof.   The first assertion of the theorem follows from the relation

$$\{x : \mid f_n(x) - f_m(x) \mid \geq \varepsilon\} \subset \{x : \mid f_n(x) - f(x) \mid$$

$$\geq \varepsilon/2\} \cup \{x : \mid f_m(x) - f(x) \mid \geq \varepsilon/2\}.$$

Hence

$$\mu\{x : \mid f_n(x) - f_m(x) \mid \geq \varepsilon\} \leq \mu\{x : \mid f_n(x) - f(x) \mid \geq \varepsilon/2$$

$$+ \mu\{x : \mid f_m(x) - f(x) \mid \geq \varepsilon/2\}$$

Similarly for the second assertion,

$$\mu\{x : \mid f(x) - g(x) \mid \geq \varepsilon\} \leq \mu\{x : \mid f_n(x) - f(x) \mid \geq \varepsilon/2\}$$

$$+ \mu\{x : \mid f_n(x) - g(x) \mid \geq \varepsilon/2\} = \delta/2 + \delta/2 = \delta \text{ if } n \geq n_0.$$

Since $\delta > 0$ can be arbitrarily small,

$$\mu\{x : \mid f(x) - g(x) \mid \geq \varepsilon\} = 0 \text{ for any } \varepsilon > 0.$$

Since $\varepsilon > 0$ is arbitrary, $f = g$ a.e.

Theorem 4.10   If $\{f_n\}$ is a sequence of measurable functions which is fundmental in measure, then there exists a subsequence $\{f_{n_k}\}$ which converges a.e. to a finite measurable function f, and hence $f_n \overset{\mu}{\to} f$.

Proof.   For every integer $k \geq 1$ let $n(k)$ be an integer such that for $n \geq n(k)$ and $\gamma = 1, 2, \ldots$, $\mu\{x : \mid f_{n+\gamma} - f_n \mid > 2^{-k}\} < 2^{-k}$. Let $n_1 = n(1)$, $n_2 = \max (n_1 + 1, n(2))$, $n_3 = \max (n_2 + 1, n(3)), \ldots$ so that $n_1 < n_2 < \ldots$ and the sequence $\{f_{n_k}\}$ is indeed an infinite subsequence of $\{f_n\}$. Let

$$f_k^* = f_{n_k},$$

$$A_k = \{x : \mid f_{k+1}^* - f_k^* \mid \geq 2^{-k}\}, \text{ and } B_n = \bigcup_{k \geq n} A_n.$$

Then $\mu(A_k) < 2^{-k}$ and hence

$$\mu(B_n) \leq \sum_{k=n}^{\infty} \mu(A_k) < 2^{-n+1}.$$

Thus for given $\varepsilon > 0$ and n large enough, so that $2^{-n+1} < \varepsilon$ for all $\gamma \geq 1$ we have on B_n^C

$$\mid f_{n+\gamma}^* - f_n^* \mid \leq \sum_{k=n}^{\infty} \mid f_{k+1}^* - f_k^* \mid < \sum_{k=n}^{\infty} 2^{-k} = 2^{-n+1} < \varepsilon$$

i.e. $B_n^C \subset \bigcap\limits_{\gamma=1}^{\infty} \{x : |f_{n+\gamma}^* - f_n^*| < \varepsilon\}$ i.e. $B_n \supset \bigcup\limits_{\gamma=1}^{\infty} \{x : |f_{n+\gamma}^* - f_n^*| \geq \varepsilon\}$

Hence

$$\mu[\bigcap\limits_{n=1}^{\infty}\bigcup\limits_{\gamma=1}^{\infty} \{x : |f_{n+\gamma}^* - f_n^*| \geq \varepsilon\}] \leq \mu[\bigcup\limits_{\gamma=1}^{\infty} \{x : |f_{n+\gamma}^* - f_n^*| \geq \varepsilon\}] \leq \mu(B_n) < 2^{-n+1} \to 0 \quad \text{as}$$

$n \to \infty$. f_n^* converges uniformly outside B_n for all $n \geq k$. So $\{f_k^*\}$ is almost uniformly fundamental. Therefore, $f_{n+\gamma}^* - f_n^* \to 0$ a.e. Thus by Theorem 4.5 there is an a.e. finite measurable function f such that $f_n^* \to f$ a.e. where $f(x) = \lim\limits_{k \to \infty} f_{n_k}(x)$ for every x for which the limit exists. To prove $f_n \overset{\mu}{\to} f$ (finite measurable), if f_n is fundamental in measure, please note that we already proved $|f_n^* - f| \leq \varepsilon$ a.e. on B_n^C. Now

$$B_n^C - N = B_n^C \cap N^C \subset [|f_{n+\gamma}^* - f| \leq \varepsilon] \Rightarrow [|f_{n+\gamma}^* - f| > \varepsilon] \subset B_n \cup N$$

for all $\gamma \geq 0$.

Therefore, $\mu\{|f_n^* - f| > \varepsilon\} \leq \mu(N) + \mu(B_n)$, where N is a subset of X such that $\mu(N) = 0$.

$$\mu\{|f_n^* - f| > \varepsilon\} < 2^{-n+1} \to 0 \text{ as } n \to \infty \text{ (just proved), i.e.}$$

$$\mu(x : |f_{n_k}(x) - f(x)| > \varepsilon) \to 0 \text{ as } n \to \infty \text{ for every } \varepsilon > 0 \tag{1}$$

Hence

$$\mu\{x : |f_n(x) - f(x)| > \varepsilon\} \leq \mu\{x : |f_n(x) - f_{n_k}(x)| > \varepsilon/2\}$$
$$+ \mu\{x : |f_{n_k}(x) - f(x)| \geq \varepsilon/2\}.$$

First part $\to 0$, since $\{f_n\}$ is Cauchy in measure and second part $\to 0$ as $n \to \infty$ due to equation (1). Therefore, $f_n \overset{\mu}{\to} f$.

PROBLEMS AND COMPLEMENTS

4.1   Let f and g be extended real-valued Borel measurable functions on $(\Omega, \mathscr{F})$ and for $A \in \mathscr{F}$ define

$$h(x) = \begin{cases} f(x) & \text{if } x \in A \\ g(x) & \text{if } x \in A^C \end{cases}$$

Show that h is Borel measurable.

4.2   Let $f = g$ a.e. where f is a continuous function. Show that ess sup $f =$ ess sup $g = \sup f$.

4.3   Let f be a continuous function and g be a measurable function. Show that the composite function $f \circ g$ is measurable.

4.4 Show that $\sup\limits_{\alpha} \{f_\alpha, \; \alpha \in T\}$ is not necessarily measurable even if each f_α is.

4.5 Let $x \in [0, 1]$ have the expansion to the base k, i.e. $x = 0. \, x_1 \, x_2 \, \dots \, x_k \, \dots$ for some integer k, the non-terminating expansion being used in cases of ambiguity. Show that $f_n(x) = x_n$ is a measurable function of x for each n.

4.6 Let f be a bounded measurable function on $\mathbb{R}^1$. Then show that for each positive integer n there are simple functions g_n and h_n such that

(a) $g_n(x) \leq f(x) \leq h_n(x)$ for all $x \in \mathbb{R}^1$
(b) $\mid g_n(x) - h_n(x) \mid \; \leq 1/n$ for all $x \in \mathbb{R}^1$.

Hence prove that a bounded measurable function can be approximated uniformly by a sequence of simple functions.

4.7 Let $f(x) = \begin{cases} x \sin\left(\dfrac{1}{x}\right) & \text{if } x > 0 \\ 0 & \text{if } x = 0 \end{cases}$

Find the Lebesgue measure of the set $\{x : f(x) \geq 0\}$

4.8 Let $E_1, E_2, \dots, E_n$ be measurable subsets of $[0, 1]$. If each point of $[0, 1]$ belongs to at least three of these sets, show that at least one of the sets has Lebesgue measure $\geq 3/n$.

4.9 Let f be a Lebesgue measurable function not almost everywhere infinite. Show that there exists a set of positive Lebesgue measure on which f is bounded.

4.10 (a) If $F(x)$ is a Lebesgue measurable function for all $x \in [a, b]$ and $f(x) = F'(x)$ exists for every $x \in [a, b]$, then prove that f is also Lebesgue measurable.

(b) If f is Riemann integrable on $[a, b]$ and $F(x) = \displaystyle\int_a^x f(t)\, dt$, prove that $F'(x) = f(x)$ a.e. on $[a, b]$.

5

Integration Theory and Expectation

5.1 DEFINITION AND PROPERTIES

Let $(X, \mathscr{B}, \mu)$ be a measure space and let $\phi(X)$ be a simple function i.e. $\phi(\cdot) = \sum_{i=1}^{r} a_i I_{A_i}(\cdot)$, where A_i's are disjoint sets in $\mathscr{B}$. We define,

$$\int \phi\, d\mu = \sum_{1}^{r} a_i \mu(A_i)$$

as long $+\infty$ and $-\infty$ do not both appear in the sum; if they do, we say that the integral does not exist. Strictly speaking, it must be verified that if ϕ has a different representation say, $\phi(\cdot) = \sum_{j=1}^{s} b_j I_{B_j}(\cdot)$, then

$$\sum_{1}^{r} a_i \mu(A_i) = \sum_{1}^{s} b_j \mu(B_j)$$

Proof. Note that ϕ can be written as

$$\sum_{1}^{r}\sum_{1}^{s} C_{ij} I_{A_i \cap B_j}, \text{ where } A_i = \bigcup_{j=1}^{\infty}(A_i B_j),$$

$$B_j = \bigcup_{1}^{r}(A_i B_j), \text{ and } A_i \cap B_j \text{ are disjoint.}$$

In fact,

$$A_i = \{x : \phi(x) = a_i\}, \ B_j = \{x : \phi(x) = b_j\} \text{ and } C_{ij} = \begin{cases} a_i & \text{on } A_i \\ b_j & \text{on } B_j \end{cases}$$

Then from the definition of integral of a simple function

$$\int \phi\, d\mu = \sum_{1}^{r}\sum_{1}^{s} C_{ij}\, \mu(A_i \cap B_j) = \sum_{1}^{r} a_i \sum_{1}^{s} \mu(A_i \cap B_j) = \sum_{i=1}^{r} a_i \mu(A_i)$$

and similarly
$$\int \phi \, d\mu = \sum_{1}^{s} b_j \mu(B_j).$$

Let f be a non-negative extended real-valued measurable function on the measure space $(X, \mathscr{B}, \mu)$. Then define

$$\int f \, d\mu = \sup_{\phi} \left\{ \int \phi \, d\mu : \phi \text{ simple and } 0 \le \phi \le f \right\}.$$

NOTE. (a) This agrees with the previous definition if f is simple. Furthermore, we may if we like to restrict f to be finite valued. Our definition will hold good for expectation also if X is a non-negative r.v. *Note that in a probability space* $(\Omega, \mathscr{B}, P)$ instead of the integral sign $\int$ the expectation sign E is used and for a r.v. X which takes only finitely many real values we write

$$E(X) = \int x \, dP = \sum_{1}^{r} x_i P[X = x_i]$$

(b) The integral of a non-negative Borel (Baire, i.e. continuous functions and limits of continuous functions) measurable function always exists, it may be $+ \infty$.

Finally, if f is an arbitrary Borel measurable function, let

$$f^+ = \max \, (f, 0) \text{ and } f^- = \max \, (-f, 0) \text{ i.e.}$$

$$f^+(x) = \begin{cases} f(x) & \text{if } f(x) \ge 0 \\ 0 & \text{if } f(x) < 0 \end{cases}$$

$$f^-(x) = \begin{cases} -f(x) & \text{if } f(x) \ge 0 \\ 0 & \text{if } f(x) > 0 \end{cases}$$

f^+ is called the positive part of f, f^- is the negative part of f. Therefore $|f| = f^+ + f^-$, $f = f^+ - f^-$ and f^+ and f^- are Borel measurable functions.

Definition 5.1   (1) $\displaystyle\int f \, d\mu = \int f^+ \, d\mu - f^- \, d\mu$ if this is not of the form $+ \infty - \infty$, if it is, we say that the integral does not exist. The function f is said to be integrable iff $\displaystyle\int |f| \, d\mu < \infty$, i.e. if $\displaystyle\int f^+ \, d\mu$ and $\displaystyle\int f^- \, d\mu$ are both finite.

(2) If $A \in \mathscr{B}$ we define $\displaystyle\int_A f \, d\mu = \int_A f \cdot I_A \, d\mu$

NOTE. If f is a bounded continuous function from R to R and μ is Lebesgue measure on R, $\displaystyle\int_A f \, d\mu$ agrees with the Riemann integral. However, the integral of f with respect to Lebesgue measure exists for many functions that are not Riemann integrable.

$$e.g.\ f(x) = \begin{cases} 0 & \text{if } x \text{ is irrational} \\ 1 & \text{if } x \text{ is rational} \end{cases}$$

Then the upper Riemann integral $\overline{\int_a^b} f(X)\ dx = b - a$ and lower Riemann integral

$\underline{\int_a^b} f(X)\ dx = 0$, where $\mu = m$, the Lebesgue measure and $\int_{[a,b]} f\ dm = 0$. For a

partition $a = x_0 < x_1 \ldots < x_n = b$, upper Darboux sum:

$$\sum_i \max_{x_{i-1} < x \le x_i} f(x)(x_i - x_{i-1}) = b - a$$

and lower Darboux sum:

$$\sum_i \max_{x_{i-1} < x \le x_i} f(x)(x_i - x_{i-1}) = 0$$

but
$$\int_{[a,b]} f\ dm = \begin{cases} 0 \times m & \text{(set of irrationals)} = 0 \\ (b-a) \times m & \text{(set of rationals)} = 0 \end{cases}$$

Theorem 5.1   Let f be a Borel measurable function from X to R.

(a) If $\int f\ d\mu$ exists and $c \in R$, then $\int cf\ d\mu$ exists and is equal to $c \int f\ d\mu$.

(b) If $g(x) \ge f(x)$ a.e. then $\int g\ d\mu \ge \int f\ d\mu$ in the sense that if $\int f\ d\mu$ exists

and $> -\infty$, then $\int g\ d\mu$ exists and $\int g\ d\mu \ge \int f\ d\mu$; if $\int g\ d\mu$ exists and

$< \infty$, then $\int f\ d\mu$ exists and $\int f\ d\mu \le \int g\ d\mu$. Thus if both integrals exist,

$\int g\ d\mu \ge \int f\ d\mu$ whether or not the integral are finite.

(c) If $\int f\ d\mu$ exists, then $\left| \int f\ d\mu \right| \le \int |f|\ d\mu$.

(d) If $f \ge 0$ a.e. and $B \in \mathcal{B}$, then
$$\int_B f\ d\mu = \sup_\phi \left[\int_B \phi\ d\mu : 0 \le \phi \le f, \phi \text{ is simple} \right].$$

(e) If $\int f\ d\mu$ exists, so does $\int_A f\ d\mu$ for each $A \in \mathcal{B}$; if $-\infty < \int f\ d\mu <$

∞ then $-\infty < \int_A f\ d\mu < \infty$ for each $A \in \mathcal{B}$.

Proof. (a) It is immediate that this holds when f is simple. If $f \geq 0$ and $c > 0$, then

$$\int cf\, d\mu = \sup_{\phi} \left\{ \int \phi\, d\mu, 0 \leq \phi \leq cf, \phi \text{ is simple} \right\}.$$

$$= c \left\{ \sup_{\psi} \int \psi\, d\mu, 0 \leq \psi \leq f, \psi \text{ is simple} \right\} \text{ where } \psi = \phi/c$$

$$= c \int f\, d\mu.$$

In general, if $f = f^+ - f^-$ and $c > 0$, then $(cf)^+ = cf^+$ and $(cf)- = cf^-$, $cf = c(f^+ - f^-)$.
Hence

$$\int cf\, d\mu = c \int f^+\, d\mu - c \int f^-\, d\mu \text{ (by what we have just proved)}$$

$$= c \left[\int f^+ d\mu - \int f^-\, d\mu \right] = c \int f\, d\mu.$$

If $c < 0$ and general f (i.e. a measurable function and not necessarily non-negative) then $(cf)^+ = -cf^-$ and $(cf)^- = -cf^+$, so that

$$\int cf\, d\mu = -c \int f^- d\mu + c \int f^+\, d\mu = c \left[\int f^+\, d\mu - \int f^- d\mu \right] = c \int f\, d\mu.$$

(b) If g and f are non-negative and $0 \leq \phi \leq f$, ϕ is simple, then

$$0 \leq \phi \leq g \text{ and } g(x) \geq f(x) \text{ a.e.}$$

Hence

$$\int f\, d\mu \leq \int g\, d\mu \text{ (by definition)}.$$

In general, $f \leq g$ a.e. $\Rightarrow f^+ \leq g^+$ and $f^- \geq g^-$ a.e. If $\int f\, d\mu > -\infty$,

we have $\int g^-\, d\mu \leq \int f^-\, d\mu < \infty$; hence $\int g\, d\mu$ exists and equals

$$\int g^+\, d\mu - \int g^-\, d\mu \geq \int f^+\, d\mu - \int f^-\, d\mu = \int f\, d\mu.$$

The case $\int g\, d\mu < \infty$ is similar, and the last part of (b) is trivial now.

(c) $-|f| \leq f \leq |f|$; so by (a) and (b)

$$-\int |f|\, d\mu \leq \int f\, d\mu \leq \int |f|\, d\mu; \text{ hence the result } \left| \int f\, d\mu \right| \leq \int |f|\, d\mu.$$

Note that f is measurable $\Rightarrow f^+$ and f^- are measurable and hence $|f| = f^+ + f^-$ is also measurable.
(d) and (e) are left as exercises.

Since we are taking supremum over a large collection of simple functions, it is not apparent from the definition that $\int (f + g)\, d\mu = \int f\, d\mu + \int g\, d\mu$. Consequently, we begin our treatment of the integral by first establishing the convergence theorems. This then enables us to determine the value of $\int f\, d\mu$ by taking the limit of $\int \phi_n\, d\mu$ for any increasing sequence $\{\phi_n\}$ of simple functions which converges to f.

5.2   CONVERGENCE THEOREMS FOR INTEGRATIONS AND EXPECTATIONS

Theorem 5.2 (*Fatou's lemma*).   Let $\{f_n\}$ be a sequence of non-negative measurable functions which converges a.e. on a set $E \in \mathcal{B}$ to a measurable function f. Then

$$\int_E f\, d\mu \leq \underline{\lim} \int_E f_n\, d\mu$$

(it may be easier to remember $\int \lim_{n\to\infty} \inf f_n d\mu \leq \lim_{n\to\infty} \inf \int f_n\, d\mu$).

Here the equality is not always true, but the inequality is. For example, let

$$g(x) = \begin{cases} 0 & \text{if } 0 \leq x \leq 1/2 \\ 1 & \text{if } 1/2 < x \leq 1 \end{cases}$$

Define

$$f_{2k}(x) = g(x),\ 0 \leq x \leq 1$$

$$f_{2k+1}(x) = g(1 - x),\ 0 \leq x \leq 1$$

Then

$$\lim_{n\to\infty} \inf f_n(x) = \lim_{n\to\infty} \inf_{k \geq n} f_k(x) = \lim (0) = 0.$$

$$\int_0^1 f_{2k}(x)\, dx = \int_0^{1/2} 0\, dx + \int_{1/2}^1 1\, dx = 1/2 \text{ and}$$

$$\int f_{2k+1}(x)\, dx = \int_0^{1/2} 1\, dx + \int_{1/2}^1 0\, dx = 1/2.$$

Hence

$$\lim_{n\to\infty} \int_0^1 f_n(x)\, dx = \frac{1}{2}, \text{ but } \int_0^1 \lim_{n\to\infty} \inf f(x)\, dx = 0.$$

Proof (Fatou's lemma).   Given $\lim_{n\to\infty} f_n(x) = f(x)$ a.e. and $f_n(x) \geq 0$ a.e. $\Rightarrow f(x) \geq 0$ a.e. Without loss of generality we may assume that

$$f_n(x) \to f(x) \text{ for each } x \in E \text{ (since } \mu(E^C) = 0.$$

From the definition of $\int f \, d\mu$ it suffices to show that if ϕ is any non-negative simple function with $\phi \leq f$, then

$$\int_E \phi \, d\mu \leq \underline{\lim} \int_E f_n \, d\mu$$

Case 1. If $\int_E \phi \, d\mu = \infty$, then, there is a measurable set $A \subset E$, with $\mu(A) = \infty$ and $\phi > a > 0$ on A.

Set $A_n = \{x \in E : f_k(x) > a \text{ for all } k \geq n\} = \bigcap_{k \geq n} \{x \in E : f_k(x) > a\}$.

Then $\{A_n\}$ is an increasing sequence of measurable sets whose union contains A, since $\phi \leq \lim f_n = f$.

Thus $\lim_{n \to \infty} \mu(A_n) = \mu(\overset{\infty}{\underset{1}{\cup}} A_n)$ (by continuity theorem of measure).

Now $A = \{x \in E : \phi > a > 0\} \subset \{x : f > a\} \subset \lim A_n$.
Therefore $\lim \mu(A_n) = \mu(\lim A_n) \geq \mu(A) = \infty$.

Since $\int_E f_n \, d\mu \geq a\mu(A_n)$, we have

$$\underline{\lim} \int_E f_n \, d\mu = \infty = \int_E \phi \, d\mu \leq \int_E f \, d\mu \text{ (since } \phi \leq f \text{ a.e.)}$$

Therefore

$$\underline{\lim} \int_E f_n \, d\mu = \infty = \int_E f \, d\mu.$$

Case 2. If $\int_E \phi \, d\mu < \infty$, then there is a measurable set $A \subset E$ with $\mu(A) < \infty$; in fact $A = \{x \in E \mid f(x) > 0\}$ such that $\phi = 0$ on $E - A$. Let $M = \max_{x \in E} \phi(x)$. For $\varepsilon > 0$, set $A_n = \{x \in E : f_k(x) > (1 - \varepsilon)\phi(x) \text{ for all } k \geq n\}^{\uparrow}$ and

$$\overset{\infty}{\underset{1}{\cup}} A_n \supset A. \text{ So } \{A - A_n\} \downarrow \text{ and } \underset{n}{\cap} (A - A_n) = \phi \text{ (null set).}$$

Therefore,

$$\lim \mu(A - A_n) = \mu(\lim (A - A_n)) = \mu(\phi) = 0$$

(since $\mu(A) < \infty$). So we can find an n such that $\mu(A - A_k) < \varepsilon$ for all $k \geq n$ and $\varepsilon > 0$ is arbitrarily small.

Thus for $k \geq n$,

$$\int_E f_k \, d\mu \geq \int_{A_k} f_k \, d\mu \geq (1 - \varepsilon) \int_{A_k} \phi \, d\mu \geq (1 - \varepsilon) \left\{ \int_E \phi \, d\mu \right.$$

$$\left. - \int_{A - A_k} \phi \, d\mu \right\} \geq (1 - \varepsilon) \int_E \phi \, d\mu - \int_{A - A_k} \phi \, d\mu$$

and
$$\int_{A-A_k} \phi \, d\mu < M \int_{A-A_k} d\mu = M \cdot \mu(A - A_k) < \varepsilon M.$$

Hence

$$\int_E f_k \, d\mu \geq \int_E \phi \, d\mu - \varepsilon \left(\int_E \phi \, d\mu + M \right).$$

Now the last inequality is true for all $k \geq n$ and since

$$\underline{\lim} \int_E f_n \, d\mu = \lim_{n \to \infty} \inf_{k \geq n} \int_E f_k \, d\mu \quad \text{we have}$$

$$\underline{\lim} \int_E f_n \, d\mu \geq \int_E \phi \, d\mu - \varepsilon \left(\int_E \phi \, d\mu + M \right).$$

Since $\varepsilon > 0$ is arbitrary and $\left(\int_E \phi \, d\mu + M \right) < \infty,$

$$\underline{\lim} \int_E f_n \, d\mu \geq \int_E \phi \, d\mu.$$

As an easy corollary to Fatou's lemma we can easily deduce (taking μ to be a probability measure P):

Theorem 5.3 (*Fatou's theorem for random variables*). Let $\{X_n\}$, Y and Z be random variables on a probability space. Then

   (a) If $Y \leq X_n$ and Y is integrable, then $E\,(\underline{\lim}\ X_n) \leq \underline{\lim}\ E(X_n)$

   (b) If $X_n \leq Z$ and Z is integrable, the $E\,(\overline{\lim}\ X_n) \geq \overline{\lim}\ E(X_n)$

   (c) If $Y \leq X_n \leq Z$; Y, Z are integrable, and $X_n \overset{a.s.}{\to} X$, the $\lim\limits_{n \to \infty} E(X_n) = E(X_n)$.

   (d) is a special case of dominated convergence theorem.

Theorem 5.4 (*Monotone convergence theorem, Lebesgue*). Let $\{f_n\}$ be an increasing sequence of non-negative measurable functions which converges a.e. to a measurable function f. Then

$$\lim_{n \to \infty} \int f_n \, d\mu = \int f \, d\mu.$$

In case of r.vs. we have:

$$\text{if } 0 \leq X_n \uparrow X, \text{ then } E(X_n) \to E(X).$$

Proof. Now $0 \leq f_n \leq f$ for all $n \geq 1$.

Since

$$f_n \leq f, \text{ we have } \int f_n \, d\mu \leq \int f \, d\mu. \text{ Thus by Fatou's lemma,}$$

$$\int f \, d\mu \leq \underline{\lim} \int f_n \, d\mu \leq \overline{\lim} \int f_n \, d\mu \leq \int f \, d\mu$$

Hence

$$\lim \int f_n \, d\mu = \int f \, d\mu$$

Before stating the usual dominated convergence theorem let us state a more general version of this theorem.

Extended Lebesgue dominated convergence theorem.   If $\{f_n\}$ is a sequence of measurable functions, $f_n \overset{\mu}{\to} f$ and $|f_n(x)| \leq g(x)$ where g is integrable $\left(i.e. \int |g| \, d\mu < \infty\right)$ then

$$\lim_{n \to \infty} \int f_n \, d\mu = \int f \, d\mu$$

The proof of this is to be attempted as an exercise (see Problem 5.11).

Theorem 5.5 *(Dominated convergence theorem).*   Let $g \geq 0$ be integrable over E and suppose that $\{f_n\}$ is a sequence of measurable functions such that on E, $|f_n(x)| \leq g(x)$ a.e. and $f_n(x) \to f(x)$ a.e. on E. Then

$$\int_E f_n(x) \, d\mu \to \int_E f(x) \, d\mu.$$

Proof.   Define $h_n = g + f_n \geq 0$ a.e., $u_n = g - f_n \geq 0$ a.e. Applying Fatou's lemma on h_n and u_n we get,

$$\int_E g \, d\mu + \int_E f \, d\mu = \int_E (g + f) \, d\mu \leq \underline{\lim} \int_E (g + f_n) \, d\mu \quad \text{(Note: } g + f \geq 0 \text{ a.e.}$$

$$= \int_E g \, d\mu + \underline{\lim} \int f_n \, d\mu \tag{1}$$

[Note that if f and g are non-negative measurable functions and a and b are non-negative constants then $\int (af + bg) \, d\mu = a \int f \, d\mu + b \int g \, d\mu$ is used in the proof which follows from Theorem 5.1(a) and (d).]

Now $g - f \geq 0$ a.e.

$$\int_E g \, d\mu - \int_E f \, d\mu = \int_E (g - f) d\mu \leq \underline{\lim} \int_E (g - f_n) d\mu = \int_E g \, d\mu - \overline{\lim} \int_E f_n \, d\mu \tag{2}$$

since $\int_E g \, d\mu < \infty$ and hence $\int_E f \, d\mu < \infty$ (since $|f_n| \leq g$ a.e.)

By (1), $$\int_E f \, d\mu \leq \underline{\lim} \int_E f_n \, d\mu \leq \int f \, d\mu < \infty \tag{3}$$

By (2),
$$\int_E f \, d\mu \geq \overline{\lim} \int_E f_n \, d\mu \tag{4}$$

So by (3) and (4) we get $\overline{\lim} \int_E f \, d\mu \leq \int_E f_n \, d\mu \leq \int_E f \, d\mu \leq \underline{\lim} \int_E f_n \, d\mu$.

But $\underline{\lim} \int_E f_n \, d\mu \leq \overline{\lim} \int_E f_n \, d\mu$ and this implies that the limit $\int_E f_n \, d\mu$ exists

and equals $\int f \, d\mu$.

Proposition 5.1   If f and g are measurable functions and a and b are constants then $\int (af + bg) \, d\mu = a \int f \, d\mu + b \int g \, d\mu$. If $af + bg$ is well defined, $\int f \, d\mu$ and $\int g \, d\mu$ are well defined.

If $f \geq 0$ a.e. then $\int f \, d\mu \geq 0$ with equality only if $f = 0$ a.e. In particular if f and g are integrable, so is $f + g$.

The proof is to be attempted as an exercise (see Problem 5.2).

Corollary (to monotone convergence theorem).   Let $\{f_n\}$ be a sequence of non-negative measurable functions. Then

$$\int \left(\sum_1^\infty f_n \right) d\mu = \sum_{n=1}^\infty \int f_n \, d\mu.$$

In case of non-negative r.vs. $E\left(\sum_1^\infty X_n \right) = \sum_1^\infty E(X_n)$.

Proof.   By previous proposition:

$$\int \left(\sum_1^k f_i \right) d\mu = \sum_1^k \int f_i \, d\mu$$

Let $g_k = \sum_1^k f_i$, then $0 \leq g_k \uparrow \sum_1^\infty f_n$, by monotone convergence theorem and hence

$$\int \sum_1^\infty (f_n) \, d\mu = \lim_{k \to \infty} \sum_1^k \int f_n \, d\mu = \sum_1^\infty \int f_n \, d\mu$$

5.3   MOMENTS AND INEQUALITIES

EXAMPLE 5.1   Let X be a Cauchy r.v. with p.d.f. $f(x) = 1/(1 + x^2)$; $-\infty < x < \infty$.
Then

$$E(X^+) = \int_0^\infty \frac{x \, dx}{\pi(1 + x^2)} = \infty = E(X^-) = \frac{1}{\pi} \int_{-\infty}^0 \frac{x \, dx}{1 + x^2}$$

Hence $E(X)$ does not exist but the Cauchy principle value is

$$\int_{-\infty}^{\infty} \frac{x\,dx}{\pi(1+x^2)} = \frac{1}{\pi} \lim_{A \to \infty} \int_{-A}^{A} \frac{x\,dx}{1+x^2} = 0$$

This example shows that Lebesgue integral may not exist, while Cauchy principal value may exist.

EXAMPLE 5.2   Let X take the values $0, \pm 1, \pm 2, \ldots$ with

$$P(X = k) = P(X = -k) = \frac{3}{\pi^2 k^2}, \quad k = 1, 2, \ldots \text{ and } P(X = 0) = 0.$$

But $EX^+ = \sum_{k=1}^{\infty} \frac{3k}{\pi^2 k^2} = 3/\pi^2 \cdot \sum_1^{\infty} \frac{1}{k} = \infty = EX^-$. So $E(X)$ does not exist.

If X takes values $1, 2, 3, \ldots$ with $P(X = k) = 6/(\pi^2 k^2)$, then $EX^+ = \infty$, $EX^- = 0$, and hence $E(X)$ exists and is equal to ∞.

5.3.1   Kolmogorov's Definition of Expectation

Let (Ω, β, P) be a probability space and X be a measurable function.

From historical point of view, it is worthwhile to give the definition of expectation due to Kolmogorov (1928).

Definition 5.2   (1) If $P(X = \infty) > 0$ and $X \geq 0$ define $E(X) = \infty$. If $P(X < \infty) = 1$ and $X \geq 0$ define $E(X) = \lim_{n \to \infty} \sum_{k=1}^{\infty} k/2^n \, P[k/2^n < X \leq (k+1)/2^n]$.

(2) In general, we say $E(X)$ exists, if either $EX^+ < \infty$ or $EX^- < \infty$ (may be both and define $E(X) = EX^+ - EX^-$. Kolmogorov's definition is classical and based on the so called elementary functions. In this connection recall the definitions of simple and elementary functions.

$$\Psi_n(\omega) = \sum_1^k a_i \, I_{A_i}(\omega) \text{ is a simple function, where } \Omega = \bigcup_1^k A_i.$$

A_i's are measurable sets and

$$\phi_n(\omega) = \sum_{k=1}^{\infty} a_k I_{A_k}(\omega)$$

is an elementary function, where

$$\Omega = \bigcup_1^{\infty} A_k, \, A_k\text{'s are measurable and } -\infty \leq a_k \leq \infty \text{ for all } k \geq 1.$$

Now we discuss some inequalities which are very important in L_p spaces and in many other areas.

5.3.2   Holder's and Minkowski's Inequalities

Holder's inequality.   If p and q are non-negative extended real numbers such that $1/p + 1/q = 1$ and if $f \in L_p$ and $g \in L_q$, then

$$fg \in L_1 \text{ and } \|f\|_1 = \int |fg| \, d\mu \le \|f\|_p \, \|g\|_q \tag{1}$$

where $\|f\|_p = \left(\int |f|^p \, d\mu \right)^{1/p}$. Equality holds in (1) iff for some non-zero constants α and β we have

$$\alpha \, |f|^p = \beta \, |g|^q \text{ a.e.}$$

Special case $p = q = 2$ gives Cauchy-Schwartz inequality

$$\int |fg| \, d\mu \le \left(\int f^2 \, d\mu \right)^{1/2} \left(\int g^2 \, d\mu \right)^{1/2}$$

Minkowski's inequality.   If f and g are in L_p then so is $f + g$ and $\|f + g\|^p \le \|f\|_p + \|g\|_p$, i.e.

$$\left[\int |f + g|^p \, d\mu \right]^{1/p} \le \left[\int |f|^p \, d\mu \right]^{1/p} + \left[\int |g|^p \, d\mu \right]^{1/p}$$

(*Note.* This is the triangle inequality in L_p space. Consequently $\| \cdot \|_p$ is a norm and hence is a distance function.)

Proof of Holder's inequality.   If $p = 1$, $q = \infty$, the proof is obvious, since

$$\int |fg| \, d\mu \le \|f\|_q \, \|g\|_\infty, \text{ where } \|g\|_\infty = \sup_\omega |g(\omega)|.$$

Assume $p > 1$, and consequently $1 < q < \infty$.

Let us first assume $\|f\|_p = \|g\|_q = 1$.

The general proof of Holder's inequality depends on:

A lemma in analysis. Let α and β be non-negative real numbers and suppose $0 < \lambda < 1$. The $\alpha^\lambda \beta^{1-\lambda} \le \lambda \alpha + (1 - \lambda)\beta$ with equality if $\alpha = \beta$.

Proof.   Consider the function ϕ defined for non-negative real numbers by

$$\phi(t) = (1 - \lambda) + \lambda t - t^\lambda$$

Then $\phi'(t) = \lambda(1 - t^{\lambda-1})$, sine $(\lambda - 1) < 0$ we have $\phi'(t) < 0$ for $t < 1$ and $\phi'(t) > 0$ for $t > 1$.

Thus for $t \ne 1$, we have $\phi(t) > \phi(1) = 0$.

Hence $1 - \lambda + \lambda t \ge t^\lambda$ and equality iff $t = 1$.

If $\beta \ne 0$, put $t = \alpha/\beta$ then $\alpha^\lambda \beta^{1-\lambda} \le \lambda \alpha + (1 - \lambda)\beta$.

If $\beta = 0$, put $t = 1$ then $\alpha = \beta$ and equality is achieved.

Proof of Holder's inequality (continued).

Put $\alpha = |f(x)|^p$ and $\beta = |g(x)|^q$, $\lambda = 1/p$ and $1 - \lambda = 1/q$

Then

$$|f(x)g(x)| \le \lambda \, |f(x)|^p + (1 - \lambda) \, |g(x)|^q \tag{2}$$

Integrating both sides of (2), we get

$$\int |fg| \, d\mu \le \lambda \int |f|^p \, d\mu + (1 - \lambda) \int |g|^q \, d\mu = \lambda + (1 - \lambda) = 1 \tag{3}$$

Holder's inequality is trivial if $\| f \|_p = 0$ or $\| g \|_q = 0$ (i.e. $f = 0$ a.e. or $g = 0$ a.e.).

Then $f/\| f \|_p$ and $g/\| g \|_q$ both have norm 1, i.e. without loss of generality $\| f \|_p = 1$ and $\| g \|_q = 1$. Substituting it in (3) we get

$$(\| f \|_p \| g \|_q)^{-1} \int | fg | \, d\mu = \int \frac{f}{\| f \|_p} \cdot \frac{|g|}{\| g \|_q} \, d\mu \le 1$$

i.e. $\int | fg | \, d\mu \le \| f \|_p \| g \|_q$. Equality in (2) holds only if $| f(x) |^p = | g(x) |^q$ and equality in (3) can hold only if it holds a.e. in (2). Hence equality holds only if $\| g \|_q | f |^p = \| f \|^p | g |^q$ a.e.

Proof of Minkowski's inequality.   The case $p = 1$ follows from $\int | f + g | \, d\mu$ $\le \int | f | \, d\mu + \int | g | \, d\mu$. Define $M = $ essential sup (ess sup) $f(\omega) = \inf \{ M : \mu[\omega. : f(\omega) > M] = 0 \} = \| f \|_\infty$. The case $p = \infty$ follows from ess sup $| f + g | \le$ ess sup $| f | + $ ess sup $| g |$ i.e.

$$\| f + g \|_\infty \le \| f \|_\infty + \| g \|_\infty.$$

Now we can assume $\qquad\qquad 1 < p < \infty$.

Integrating both sides, $f + g$ is in L_p if $f \in L_p$ and $g \in L_p$.

Also $\qquad \int | f + g |^p \, d\mu \le \int | f + g |^{p-1} | f | \, d\mu + \int | f + g |^{p-1} | g | \, d\mu.$

By Holder's inequality $\int | f + g |^{p-1} | g | \, d\mu \le \| g \|_p \| (f + g)^{p-1} \|_q$

and $\int | f + g |^{p-1} | f | \, d\mu \le \| f \|_p \| (| f + g |^{p-1}) \|_q$. Now $p = (p - 1)q$ and

$\| f + g \|_p^p \le (\| f \|_p + \| g \|_p) (\| f + g \|_p)^{p/q}$ imply $\| f + g \|_p \le \| f \|_p + \| g \|_p$.

PROBLEMS AND COMPLEMENTS

5.1 Let $(\Omega, \mathscr{F}, P)$ be a probability space and $\{A_n\}$ a sequence of events in $\mathscr{F}$. If for some $A \in \mathscr{F}$, $P(A_n \Delta A) \to 0$ as $n \to \infty$ then show that $P(A_n) \to P(A)$.

5.2 If f and g are measurable integrable functions then

$$\int (f + g) \, d\mu = \int f \, d\mu + \int g \, d\mu.$$

5.3 Let f be a fixed measurable function and suppose that $\int f \, d\mu$ is defined. Then the set function

$$v(E) = \int_E f \, d\mu$$

is σ-additive.

5.4 Let f be a non-negative integrable function. Show that $\forall \; \varepsilon > 0 \; \exists \; \delta > 0$ s.t.
$\forall \; E$ with $\mu(E) < \delta, \int_E f \, d\mu < \varepsilon$.

5.5 (a) If f is a lebesgue integrable function then show that f is finite a.e.

  (b) If $\int f \, d\mu = 0$ and $f \geq 0$ a.e., show that $f = 0$ a.e.

5.6 (a) If f is a Lebesgue integrable function such that $\int_F f \, d\mu = 0$ for every measurable set F, then show that $f = 0$ a.e.

  (b) If f and g are measurable with $\int_E f \, d\mu \geq \int_E g \, d\mu$ for every measurable set F, then show that $f \geq g$ a.e.

5.7 If f is an integrable function, then the set

$$N(f) = \{x : f(x) \neq 0\} \text{ has } \sigma\text{-finite measure.}$$

5.8 If f and g are non-negative and $\int_A f \, d\mu = \int_A g \, d\mu$ for all $A \in \mathscr{F}$ and if μ is σ-finite, then show that $f = g$ a.e.

5.9 If f and g are integrable and $\int_A f \, d\mu = \int_A g \, d\mu$ for all $A \in \mathscr{P}$, where $\mathscr{P}$ is a π-system generating $\mathscr{F}$ and X is finite or countable union of $\mathscr{P}$-sets, then show that $f = g$ a.e.

5.10 Let $f(x) = 0$ for x rational and $f(x) = n$ for x irrational, $0 \leq x \leq 1$, where n is the number of zeros immediately after the decimal point in the representation of x on the decimal scale. Show that f is Lebesgue measurable and find

$$\int_0^1 f(x) \, dx.$$

5.11 (Beppo Lévi's theorem): Suppose that f_n are integrable and $\sup_n \int |f_n| \, d\mu < \infty$. Show that if $f_n \uparrow f$, then f is integrable and $\int f_n \, d\mu \to \int f \, d\mu$.

5.12 (The Weierstrass M-test for series): If $|x_{nm}| < M_m, \; \sum_m M_m < \infty$ and $\lim_{n \to \infty} x_{nm} = x_m$ for each $m \geq 1$, then show by the dominated convergence theorem that

$$\lim_{n \to \infty} \sum_m x_{nm} = \sum_m x_m.$$

5.13 (Scheffe's theorem): Suppose $v_n(A) = \int_A f_n \, d\mu$ and $v(A) = \int_A f \, d\mu$ for densities f_n and f. If $v_n(\Omega) = v(\Omega) < \infty$ for all $n \geq 1$ and if $f_n \to f$, then show that

$$\sup_{A \in \mathcal{F}} |v_n(A) - v(A)| = \frac{1}{2} \int |f_n - f| \, d\mu \to 0$$

5.14 For $n \geq 1$, let

$$f_n(x) = \begin{cases} 2n^2 & \text{if } \dfrac{1}{2n} \leq x \leq \dfrac{1}{n} \\ 0 & \text{if } x \in \left(0, \dfrac{1}{2n}\right) \cup \left(\dfrac{1}{n}, 1\right). \end{cases}$$

Obtain $\displaystyle\int_0^1 [\lim_{n \to \infty} f_n(x)] \, dx$ and $\displaystyle\lim_{n \to \infty} \int f_n(x) \, dx$ and show that Fatou's lemma holds but the dominated convergence theorem does not.

5.15 Show that

$$\lim_{n \to \infty} \int_R f_n(x) \, dx \neq \int_R \lim_{n \to \infty} f_n(x) \, dx$$

where
$$f_n(x) = \begin{cases} -n^2 & \text{for } x \in \left(0, \dfrac{1}{n}\right) \\ 0, & \text{otherwise} \end{cases}$$

Explain why this does not contradict Fatou's lemma.

5.16 If $\mu(X) < \infty$ and $f \in L_2$ on X, then prove that $f \in L_1$ on X. Give a counter example to show that this is not true if $\mu(X) = \infty$.

5.17 Let f and g be non-negative measurable functions. Then for any k and m such that $k + m = km$, $0 < k < 1$ or $k < 0$,

$$\int fg \, d\mu \geq \left(\int f^k \, d\mu\right)^{1/k} \left(\int g^m \, d\mu\right)^{1/m}.$$

5.18 Let $0 < p < 1$ and $f \geq 0$, $g \geq 0$, $f, g \in L^p(\mu)$. Show that

$$\| f + g \|_p \geq \| f \|_p + \| g \|_p.$$

5.19 (Cantelli's first moment inequality): Let X be a r.v. with $E(X) = 0$ and var $(X) = \sigma^2$. Then

$$P(X > x) \leq \frac{\sigma^2}{\sigma^2 + x^2} \quad \text{if } x > 0$$

$$\geq \frac{\sigma^2}{\sigma^2 + x^2} \quad \text{if } x < 0$$

5.20 (Cantelli's second moment inequality): Let X be a r.v. with $E(X) = 0$, $E(X^2) = \sigma^2$ and $E(X^4) = \mu_4 < \infty$. Prove that

$$P(|X| \geq k\sigma) \leq \begin{cases} 1 & \text{if } k^2 < 1 \\ 1/k^2 & \text{if } 1 \leq k^2 < \dfrac{\mu_4}{\sigma^4} \\ \dfrac{\mu_4 - \sigma^4}{m_4 + k^4\sigma^4 - 2k^2\sigma^4} & \text{if } k^2 \geq \dfrac{\mu_4}{\sigma^4} \end{cases}$$

5.21 (Basic inequality): Let X be a r.v. and g a non-negative Borel function.

(a) If g is even and non-decreasing on $[0, \infty]$, then

$$\frac{E\,g(X) - g(a)}{ess \, \sup g(X)} \leq P[|X| \geq a] \leq \frac{E\,g(X)}{g(a)},\, a > 0.$$

(b) If g is non-decreasing on $(-\infty, \infty)$, then for all real a,

$$\frac{E\,g(X) - g(a)}{ess \, \sup g(X)} \leq P(X \geq a) \leq \frac{E\,g(X)}{g(a)}$$

(c) $\dfrac{E|X|^r - a^r}{ess \, \sup g|X|^r} \leq P[|X| \geq a] \leq \dfrac{E|X|^r}{a^r}$ for $r > 0, a > 0$.

NOTE: Essential supremum of a function f is defined as

$$ess \, \sup |f| = \inf \{a : P[|f| \leq a] = 1\}.$$

5.22 Let $(\Omega, \mathcal{B}, \mu)$ be a measure space and f a non-negative measurable function with $M = ess \, \sup f > 0$. Then show that if $0 < \mu(\Omega) < \infty$, $\lim\limits_{n \to \infty} \dfrac{\int f^{n+1} d\mu}{\int f^n d\mu} = M$.

5.23 Let X be a r.v. such that $E|X|^k < \infty$ for some $k > 0$. Then show that $n^k\, P(|X| > n) \to 0$ as $n \to \infty$. Give a counter example to show that the converse does not hold.

5.24 Let X be a r.v. with $n^\alpha\, P(|X| > n) \to 0$ as $n \to \infty$, $\alpha > 0$. Show that $E|X|^\beta < \infty$ for $0 < \beta < \alpha$.

5.25 Let X be a r.v. satisfying

$$\frac{P(|X| > \alpha k)}{P(|X| > k)} \to 0 \text{ as } k \to \infty \text{ for all } \alpha > 1.$$

Show that X possesses moments of all orders.

5.26 Let $\{X_n\}_1^\infty$ be a sequence of r.vs. on $(\Omega, \mathcal{F}, P)$. Then show that $\sum\limits_{n=1}^{\infty} X_n$ converges absolutely a.s. if $\sum\limits_{n=1}^{\infty} E|X_n| < \infty$.

6

Types of Convergence and Limit Theorems

6.1 SLUTSKY'S THEOREM AND INTERRELATIONS BETWEEN DIFFERENT TYPES OF CONVERGENCE

In this section all the random variables are defined on a common probability space.

Theorem 6.1 (*Slutsky, 1925*).   Let $X_n \xrightarrow{d} X, Y_n \xrightarrow{P} c$ (real constant), then

(a) $X_n + Y_n \xrightarrow{d} X + c$ (i.e. the d.f. of $X_n + Y_n$ tends to $F(x - c)$ at each continuity point of F).

(b) $X_n Y_n \xrightarrow{d} cX$ (i.e. the d.f. of $X_n Y_n$ tends to $F(x/c)$ at each continuity point of F).

(c) $X_n/Y_n \xrightarrow{d} X/c$, $c \neq 0$, (i.e. the d.f. $X_n Y_n$ tends to $F(cx)$ at each continuity point of F).

[Note that $F(x) = P(X \leq x)$ implies $F(x - c) = P(X \leq x - c) = P(X + c \leq x)$.]

Proof.   (a) Choose and fix x such that $(x - c)$ is a continuity point of F. Let $\varepsilon > 0$ be such that $x - c + \varepsilon$ and $x - c - \varepsilon$ are also continuity points of F. Then

$$F_{X_n + Y_n}(x) = P[X_n + Y_n \leq x] \leq P[X_n + Y_n \leq x, |Y_n - c| < \varepsilon]$$

$$+ P[|Y_n - c| \geq \varepsilon] \text{ (since } P(A) = P(AB) + P(AB^C) \leq P(AB) + P(B^C))$$

$$\leq P[X_n \leq x - (c - \varepsilon)] + P[|Y_n - c| \geq \varepsilon]$$

$\overline{\lim} \, F_{X_n + Y_n}(x) \leq \overline{\lim} \, P(X_n \leq (x - c + \varepsilon)) + \overline{\lim} \, P(|Y_n - c| < \varepsilon)$. Since $Y_n \xrightarrow{P} c$ the last term $\to 0$ and hence

$$\overline{\lim} \, F_{X_n + Y_n}(x) \leq F(x - c + \varepsilon) \text{ for } \varepsilon > 0 \tag{1}$$

75

Now

$$P(X_n + Y_n > x) \leq P(X_n + Y_n > x, \, | \, Y_n - c \, | < \varepsilon) + P(| \, Y_n - c \, | \geq \varepsilon)$$

$$\leq P(X_n + c + \varepsilon > x) + P(| \, Y_n - c \, | \geq \varepsilon)$$

Similarly, $P(X_n \leq x - c - \varepsilon) \leq P(X_n + Y_n \leq x) + P(|Y_n - c \, | \geq \varepsilon)$

and hence $$F(x - c - \varepsilon) \leq \lim F_{X_n + Y_n}(x) \tag{2}$$

Since $x - c$ is a continuity point of F and $\varepsilon > 0$ may be taken to be arbitrarily small, (1) and (2) yield

$$\lim_{n \to \infty} F_{X_n + Y_n}(x) = F(x - c).$$

(b) Let cx, $(c + \varepsilon)x$, $(c - \varepsilon)x$ be all continuity points of F. Then

$$P(X_n/Y_n \leq x) \leq P(X_n/Y_n \leq x, \, | \, Y_n - c \, | < \varepsilon) + P(| \, Y_n - c \, | \geq \varepsilon)$$

$$\leq P(X_n \leq (c + \varepsilon) \, x) + P(| \, Y_n - c \, | \geq \varepsilon)$$

Therefore,

$$\overline{\lim} \, P(X_n / Y_n \leq x) \leq \overline{\lim} \, P[X_n \leq (c + \varepsilon)x]$$

$$+ \lim P(| \, Y_n - c | \geq \varepsilon) = F((c + \varepsilon)x) \tag{3}$$

Similarly,

$$P(X_n \leq (c - \varepsilon)x) \leq P(X_n/Y_n \leq x) + P(|Y_n - c \, | \geq \varepsilon).$$

Therefore,

$$\underline{\lim} \, P(X_n \leq (c - \varepsilon)x) = F((c - \varepsilon)x) \leq \underline{\lim} \, P(X_n/Y_n \leq x) \tag{4}$$

Equations (3) and (4) imply (b)

(c) Let x/c, $x/(c - \varepsilon)$, $x/(c + \varepsilon)$ be continuity points of F. Then

$$F_{X_n Y_n}(x) = P(X_n Y_n \leq x) \leq P(X_n Y_n \leq x, |Y_n - c| \leq \varepsilon)$$

$$+ P(| \, Y_n - c \, | \geq \varepsilon) \leq P(X_n \leq x/(c - \varepsilon)) + P(| \, Y_n - c \, | \geq \varepsilon)$$

$$\overline{\lim} \, P(X_n Y_n \leq x) \leq \overline{\lim} \, P(X_n \leq x/(c - \varepsilon)) + \overline{\lim} \, P(| \, Y_n - c | \geq \varepsilon)$$

$$= F(x/(c - \varepsilon)) \tag{5}$$

Similarly, $P(X_n Y_n \leq x) + P(| \, Y_n - c \, | \geq \varepsilon) \geq P(X_n \leq x/(c + \varepsilon))$

$$\Rightarrow \underline{\lim} \, P(X_n Y_n \leq x) + \lim P(| \, Y_n - c \, | \geq \varepsilon) \geq \underline{\lim} \, p(X_n \leq x/(c + \varepsilon))$$

$$\Rightarrow \underline{\lim} \, P(X_n Y_n \leq x) \geq F(x/(c + \varepsilon))$$

Therefore,

$$\lim_{n \to \infty} F_{X_n Y_n}(x) = F(x/c)$$

The following theorem is very useful in large sample theory of Statistics and whose proof will be given after Theorem 6.3.

Theorem 6.2   Let $X_1, X_2, \ldots$ be a sequence of k-dimensional r.vs. and $g: R^k \to$

R^m is a measurable mapping which is continuous over a Borel set $B \subset R^k$ for which $P(X \in B) = 1$ and if $X_n \overset{P}{\to} X$, then $g(X_n) \overset{P}{\to} g(X)$.

Theorem 6.3

(a) $X_n \overset{P}{\to} X \Rightarrow X_n \overset{P}{\to} X$

(b) If X is a constant c a.s., then $X_n \overset{d}{\to} c \Rightarrow X_n \overset{d}{\to} c$

(c) Let $p_j^{(n)} = P(X_n = j)$, $p_j = P(X = j)$ where X_n and X are integer valued r.vs. Then $p_j^{(n)} \to p_j$ for all $j \geq 1$ iff $X_n \overset{d}{\to} X$.

Proof (a) First we prove that $X_n \overset{P}{\to} X \Rightarrow X_n \overset{d}{\to} X$.

Now $X_n \overset{P}{\to} X \Rightarrow P(|X_n - X| \geq \varepsilon) \to 0$ as $n \to \infty$ for all $\varepsilon > 0$. Let $x \in c(F)$ where F is the d.f. of X and $\varepsilon > 0$ be arbitrary. Then $P(X_n \leq x) - P(X \leq x + \varepsilon) \leq P(X_n \leq x, X > x + \varepsilon)$

$$(\text{here } P(A) - P(B) \leq P(AB^C) \text{ is used})$$

$\leq P(X > X_n + \varepsilon) \leq P(|X_n - X| \geq \varepsilon) \to 0$ as $n \to \infty$ for all $\varepsilon > 0$.

$$\Rightarrow \overline{\lim} F_n(x) - F(x + \varepsilon) \leq 0 \Rightarrow \overline{\lim} F_n(x) \leq F(x + \varepsilon)$$

Again $P(X \leq x - \varepsilon) - P(X_n \leq x) \leq P(X_n \geq x, X \leq x - \varepsilon) < P(|X_n - X| \geq \varepsilon) \to 0$

$$\text{as } n \to \infty \text{ for all } \varepsilon > 0$$

Hence

$$\underline{\lim} F_n(x) \geq F(x - \varepsilon).$$

Now

$$x \in c(F) \Rightarrow \lim_{\varepsilon \to 0} F(x - \varepsilon) = \lim_{\varepsilon \to 0} F(x + \varepsilon) = F(x).$$

Therefore, $\overline{\lim} F_n(x) \leq F(x) \leq \underline{\lim} F_n(x)$. But $\underline{\lim} F_n(x) \leq \overline{\lim} F_n(x)$ is always true. Hence

$$\lim F_n(x) = F(x) \text{ if } x \in c(F) \text{ and } X_n \overset{d}{\to} X \text{ is proved.}$$

(b) Since X is a constant c a.s., we can without loss of generality write $X(\omega) = c$ for all ω's.

$$F(x) = \begin{cases} 0 & \text{if } x < c \\ 1 & \text{if } x \geq c \end{cases}$$

So c is the only discontinuity point of $F(x)$.

$P(|X_n - X| \geq \varepsilon) = P(|X_n - c| \geq \varepsilon) = P(X_n \geq c + \varepsilon) + P(X_n \leq c - \varepsilon)$

$= 1 - F_n(c + \varepsilon) + F_n(c - \varepsilon)$, since $c \pm \varepsilon \in c(F)$

$\Rightarrow 1 - F(c + \varepsilon) + F(c - \varepsilon) = 0$ (since $F(c + \varepsilon) = 1$ and $F(c - \varepsilon) = 0$)

Therefore, $X_n \overset{P}{\to} c$. [proved]

Proof of Theorem 6.2 Given $\mathbf{X}_n \overset{P}{\to} \mathbf{X}$ we have to prove $g(\mathbf{X}_n) \overset{P}{\to} g(\mathbf{X})$.

Let $\{g(\mathbf{X}_{n_j})\}$ be any subsequence of $\{g(\mathbf{X}_n)\}$. We are only to prove that there is a subsequence $\{g(\mathbf{X}_{r_j})\}$ of $\{g(\mathbf{X}_{n_j})\}$ which converges a.s. to $g(\mathbf{X})$.
Since

$$\mathbf{X}_n \xrightarrow{P} \mathbf{X}, \text{ there is a subsequence } \{\mathbf{X}_{r_j}\} \text{ of } \mathbf{X}_{n_j}\} \text{ such that } \mathbf{X}_{r_j} \to \mathbf{X}$$
a.s.

Let $A = [\mathbf{X}_{r_j} \to \mathbf{X}] \cap [\mathbf{X} \in B]$. Then $P(A) = 1$ and $\omega \in A$ implies that

$\mathbf{X}_{r_j}(\omega) \to \mathbf{X}(\omega) \in B$. Since g is continuous on B, then

$$g(\mathbf{X}_{r_j}) \to g(\mathbf{X}(\omega)) \text{ as } j \to \infty \text{ for all } \omega \in A;$$

i.e. $\qquad\qquad g(\mathbf{X}_{r_j}) \to g(\mathbf{X})$ a.s. Hence $g(X_n) \xrightarrow{P} g(\mathbf{X})$.

Proposition 6.1   Let $X_n \xrightarrow{P} X$ and $Y_n \xrightarrow{P} Y$. Then (a) $X_n + Y_n \xrightarrow{P} X + Y$ and (b) $X_nY_n \xrightarrow{P} XY$.

Proof.

(a) $P(|\,X_n + Y_n - (X + Y)\,| \geq \varepsilon) \leq P(|\,X_n - X\,| \geq \varepsilon/2) + P(|\,Y_n - Y\,| \geq \varepsilon/2$
$\Rightarrow P(|\,X_n + Y_n - (X + Y)| \geq \varepsilon) \to 0$ (as first and second terms $\to 0$ as $n \to \infty$).

(b) First assume that $X = 0$ a.s.
Let $\varepsilon > 0$. Choose k_0 such that $P(|\,Y\,| \geq k_0) < \eta > 0$. Since Y is a r.v., it is finite a.s.

Since

$$[|\,X_nY_n\,| \geq \varepsilon] \subset [|\,X_n\,| \geq \varepsilon/(k_0 + 1)] \cup [|\,Y_n\,| \geq k_0 + 1],$$

$$\overline{\lim}\, P(|\,X_nY_n\,| \geq \varepsilon) \leq \overline{\lim}\, P(|\,Y_n\,| \geq k_0 + 1) + \overline{\lim}\, P(|\,X_n\,| \geq \varepsilon/(k_0 + 1))$$

$$= \overline{\lim}\, P(|\,Y_n\,| \geq k_0 + 1)(\text{since } X_n \xrightarrow{P} 0)$$

Now,

$$P(|\,Y_n\,| \geq k_0 + 1) \leq P(|\,Y_n - Y\,| \geq 1) + P(|\,Y\,| \geq k_0)$$

Therefore,

$$\overline{\lim}\, P(|\,X_nY_n\,| \geq \varepsilon) \to 0 \text{ as } n \to \infty$$

and $\qquad\qquad \overline{\lim} > \underline{\lim}$ so that $\underline{\lim}\, P(|X_nY_n| \geq \varepsilon) \to 0$ as $n \to \infty$

Hence $\qquad\qquad \lim_{n \to \infty} P(|\,X_nY_n\,| \geq \varepsilon) = 0$ for all $\varepsilon > 0$.

Therefore $\qquad\qquad\qquad X_nY_n \xrightarrow{P} 0.$

In general,

$$X_nY_n - XY = (X_n - X)(Y_n - Y) + X(Y_n - Y) + Y(X_n - X)$$

and all the terms converge in probability to zero by previous argument. So

$$X_nY_n \xrightarrow{P} XY.$$

NOTE.   Proposition 6.1 follows from Theorem 6.2 for proper choice of g but independent proof is given because of its importance.

6.1.1   Convergence in L_p

1. A measurable function f is said to be in $L_p(0 < p < \infty)$ if

$$\int |f|^p \, d\mu < \infty$$

(If $p = 2$, such a sequence of measurable functions forms a Hilbert Space.)

2. A sequence of measurable functions $\{f_n\}$ *converges in* L_p if

$$\int |f_n - f|^p \, d\mu \to 0 \text{ as } n \to \infty.$$

A sequence of r.vs. $\{X_n\}$ converges in pth *mean to* X if

$$E\,|\,X_n - X\,|^p \to 0 \text{ as } n \to \infty.$$

Theorem 6.4   If $f_1, \ldots, f_n, \ldots \in L_p$, $(0 < p < \infty)$, then

$$f_n \xrightarrow{L_p} f \Rightarrow f_n \xrightarrow{\mu} f.$$

The proof of the theorem depends on the following well known Lemma.

Lemma 6.1 (*Markov-Chebychev's inequality*).   Let f be a non-negative extended real-valued measurable function on $(X, \mathscr{B}, \mu)$. If $0 < p < \infty$ and $0 < \varepsilon < \infty$ then

$$\mu[x : f(x) \geq \varepsilon] \leq \frac{1}{\varepsilon^p} \int f^p \, d\mu$$

Proof. $\displaystyle \int f^p d\mu \geq \int_{[x:f(x)\geq\varepsilon]} f^p d\mu = \int f^p \cdot I_{[f\geq\varepsilon]} d\mu \geq \varepsilon^p \, \mu[x : f(x) \geq \varepsilon]$

Hence $$\mu[f \geq \varepsilon] \leq \varepsilon^{-p} \int f^p d\mu.$$

In case of r.vs. $E(X - \mu)^2 \geq E\{|\,X - \mu\,|^2 \, I_{[|X-\mu|\geq\varepsilon]}\} \geq \varepsilon^2 P[|\,X - m\,| \geq \varepsilon]$

Hence $$P[|\,X - \mu\,| \geq \varepsilon \leq \varepsilon^{-2} E(X - \mu)^2.$$

Proof of Theorem 6.4

$$\mu[x : |\,f_n(x) - f(x)\,| \geq \varepsilon] \leq \frac{1}{\varepsilon^p} \int |f_n - f|^p \, d\mu \to 0 \text{ as } n \to \infty.$$

Corollary.   If $\{X_n\}$ be a sequence of r.vs. defined on a probability space $(\Omega, \mathscr{F},$ $P)$ with $E(|\,X_n\,|^p) < \infty$ for all n, then $X_n \xrightarrow{L_p} X \Rightarrow X_n \xrightarrow{P} X(0 < p < \infty).$

6.2 INTERRELATIONS BETWEEN DIFFERENT TYPES OF CONVERGENCE AND COUNTER EXAMPLES

The following diagram gives the relations between different types of convergence in probability theory.

Convergence a.s. $\Rightarrow$ convergence in probability $\Rightarrow$ convergence in law (in distribution)

Convergence in pth mean (the case $p = 2$, is known as quadratic mean convergence)

Example of convergence in law but not in probability.   Here,

$$\Omega = [0, 1], \; X_n(\omega) = \begin{cases} 1 & \text{if } \omega \in \left[0, \dfrac{1}{2}\right) \\[2mm] 0 & \text{if } \omega \in \left[\dfrac{1}{2}, 1\right] \end{cases} \text{and } X(\omega) = \begin{cases} 0 & \text{if } \omega \in \left[\dfrac{1}{2}, 1\right] \\[2mm] 1 & \text{if } \omega \in \left[0, \dfrac{1}{2}\right) \end{cases}$$

on the probability space $(\Omega, \mathcal{B}, m)$, m is the Lebesgue measure on the unit interval. Then

$$F_{X_n}(x) = F_X(x) \text{ but } |X_n - X| = 1$$

with probability one for $n = 1, 2, \ldots$

Example of convergence in probability but not a.s. Let $Y_1, \ldots, Y_n$ be i.i.d. r.vs. Then

$$X_n = \frac{Y_n}{n} \xrightarrow{P} 0,$$

Since

$P[|Y_1| > n\varepsilon] \to 0$ as $n \to \infty$ for every $\varepsilon > 0$.

But $X_n \to 0$ a.s. iff $E(|Y_1|) < \infty$ follows from Kolmogorov's strong law for i.i.d. r.vs. and will be proved during its proof.

Take for example $Y_1, Y_2, \ldots$ are i.i.d. cauchy r.vs. then

$$X_n \xrightarrow{P} 0 \text{ but } X_n \nrightarrow 0 \text{ a.s.}$$

Example of convergence in probability but not in pth Mean

$$X_n = \begin{cases} n & \text{with probability } 1/\log n \\ 0 & \text{with probability } 1 - 1/\log n \end{cases}$$

Then
$X_n \xrightarrow{P} 0$. For any $p > 0$,

$$E|X_n|^p = \frac{n^p}{\log n} \to \infty.$$

But $P[|X_n| < 1/2] = 1 - 1/\log n \to 1$ as $n \to \infty$. Hence for example $\varepsilon = 1/2$ and $X_n \xrightarrow{P} 0$.

Example of convergence a.s. ***but not in*** L_p. As before let $(\Omega, \mathscr{B}, m)$ be the probability space with m the Lebesgue measure on $\Omega = [0, 1]$ and

$$X_n(\omega) = \begin{cases} e^n & \text{if } 0 \le \omega \le 1/n^2 \\ 0 & \text{otherwise} \end{cases}$$

Then

$$E|X_n|^p = \frac{e^{np}}{n^2} \to \infty,\ p > 0$$

By Borel-Cantelli lemma $X_n \to 0$ a.s., since

$$\sum_{n \ge 1} P(|X_n| > \varepsilon) = \sum_{n \ge 1} 1/n^2 < \infty,\ \varepsilon > 0$$

***Example of convergence in p*th *Mean but not* a.s.**
Let

$$\Omega = [0, 1],\ X_{n,m}(\omega) = \begin{cases} 1 & \text{if } (m-1)/n \le \omega \le m/n \quad m = 1, 2, \ldots, n \text{ and } n = 1, 2, \ldots \\ 0 & \text{otherwise} \end{cases}$$

Then $E|X_{nm}|^p = 1^p P[\omega : (m-1)/n \le \omega \le m/n] = 1/n \to 0$ for any $0 < p < \infty$. Arranging the double sequence in a single sequence, the arranged sequence X_{11}, $X_{21}, X_{22}, X_{31}, X_{32}, X_{33}, \ldots$ converges to 0 in pth mean and hence in probability. For any $\omega \ne 0$ the sequence $\{X_{nm}(\omega)\}$ has infinitely many zeros and infinitely many ones. Thus the set on which X_{nm} converges has probability zero.

Hence X_{nm} does not converge almost surely.

If $\mu(X) = \infty$ ***then*** a.e. ***convergence does not imply convergence in measure.*** Let $(\Omega, \mathscr{B}, m)$ be the Lebesgue measure space with

$$\Omega = R = (-\infty, \infty),\ f_n(x) = \begin{cases} 1 & \text{if } n \le x \le n + 1 \\ 0 & \text{otherwise} \end{cases}$$

So $f_n = I_{A_n}, A_n = [n, n + 1]$
Clearly for $0 < \varepsilon < 1$, $\mu[x : |f_n(x)| > \varepsilon] = 1 \nrightarrow 0$ as $n \to \infty$. But Obviously $f_n(x) \to f(x) \equiv 0$, everywhere.

PROBLEMS AND COMPLEMENTS

6.1　Give an example to show that convergence in distribution does not imply convergence of moments.

6.2　A point U_1 is chosen on [0, 1]. Then a second point is chosen at random on $[0, U_1]$ and so forth. Show that $U_n^{1/n}$ converges in probability to e^{-1}.

***6.3**　Show that on the real line $F_n \Rightarrow F$ (weakly) iff $F_n(x) \to F(x)$ at each continuity point of F.

---

*Denotes difficult problem.

6.4 Show that $X_n \xrightarrow{P} 0$ iff $E\left(\dfrac{|X_n|}{1+|X_n|}\right) \to 0$ as $n \to \infty$.

6.5 Let $\{X_n\}$ be a sequence of pairwise uncorrelated r.vs. with $E(X_i) = \mu_i$, var $(X_i) = \sigma_i^2$, $i = 1, 2, \ldots$ If $\sum\limits_{i=1}^{n} \sigma_i^2 \to \infty$ as $n \to \infty$, then show that $\dfrac{1}{\sum\limits_{i=1}^{n} \sigma_i^2} (\sum\limits_{i=1}^{n} (X_i - \mu_i)) \xrightarrow{P} 0$ as $n \to \infty$.

6.6 Let $\{X_n\}$ be a sequence of r.vs. with $\overline{X}_n = \dfrac{1}{n} \sum\limits_{i=1}^{n} X_i$. Show that a necessary and sufficient condition for the sequence $\{X_n\}$ to satisfy the weak law of large numbers is $E\left(\dfrac{\overline{X}_n^2}{1+\overline{X}_n^2}\right) \to 0$ as $n \to \infty$.

6.7 Let $\{X_n\}$ be a sequence of r.vs. with $E(X_n) = 0$, $E(X_n^2) = 1$ for all $n \geq 1$ and cov $(X_n, X_m) = \rho$ if $|n - m| = 1$ and 0 otherwise, Let $S_n = \sum\limits_{i=1}^{n} X_i$. Show that $\dfrac{S_n}{n} \xrightarrow{P} 0$.

6.8 Let $\{X_n\}$ be a strictly decreasing sequence of positive r.vs. and suppose that $X_n \xrightarrow{P} 0$. Then $X_n \xrightarrow{a.s.} 0$.

6.9 Let $\{X_n, n \geq 0\}$ be a sequence of r.vs. on a countable probability space $(\Omega, S(\Omega)\ P)$ where $S(\Omega) = \{$class of all subsets of $\Omega\}$. Show that $X_n \xrightarrow{P} X_0 \Rightarrow X_n \to X_0$ a.s.

6.10 If $X_n \xrightarrow{a.s.} 0$, then show that $\dfrac{1}{n} \sum\limits_{k=1}^{n} X_k \xrightarrow{a.s.} 0$. Give a counter example to show that if $X_n \xrightarrow{P} 0$, $\dfrac{1}{n} \sum\limits_{i=1}^{n} X_n \xrightarrow{P} 0$.

6.11 Let $\{X_n\}$ be a sequence of i.i.d. r.vs. with common mean μ and finite fourth moments. Then prove that $S_n/n \xrightarrow{a.s.} \mu$, where $S_n = \sum\limits_{i=1}^{n} X_i$.

6.12 Let $\{X_n\}$ be a sequence of r.vs. with $P(X_n = 0)$ $1 - 1/\log n$ and $P(X_n = n) = 1/\log n$, $n \geq 2$. Investigate whether the sequence converges in probability and in the rth mean.

6.13 Show that

(a) if $X_n \xrightarrow{L_2} X$, then $\lim\limits_{n \to \infty} E(X_n) = E(X)$ and $\lim\limits_{n \to \infty} E(X_n^2) = E(X^2)$

(b) if $X_n \xrightarrow{L_p}$ then $\lim\limits_{n \to \infty} E\,|X_n|^k = E\,|X|^k$ for $k \leq p$.

6.14 (a) Let $X_n \xrightarrow{L_2} X$ and $Y_m \xrightarrow{L_2} Y$. Then show that

$$\lim_{n \to \infty} \lim_{m \to \infty} E(X_n Y_m) = E(XY).$$

(b) Show that $X_n \xrightarrow{L_2} X$ iff $\lim_{n \to \infty} \lim_{m \to \infty} E(X_n X_m)$ exists and is independent of the manner in which $n, m \to \infty$.

6.15 Let $\{Y_n\}$ be a sequence of independent r.vs. with $P(Y_n = \pm 1) = \dfrac{1}{2} n^{-1/4}$ and $P(Y_n = 0) = 1 - n^{-1/4}$, $n \geq 1$. Let $E_n = \{Y_i = 0$ for all i satisfying $n - \sqrt{n} \leq i < n\}$. Define a second sequence of r.vs. $\{X_n\}$ such that $X_1 = Y_1$ and for $n \geq 2$,

$$X_n = \begin{cases} Y_n & \text{if } E_n^c \text{ occurs} \\ +1 \text{ or } -1 \text{ with equal probability} & \text{if } E_n \text{ occurs} \end{cases}$$

Show that $X_n \xrightarrow{L_2} 0$ but not almost surely. Does $X_n \xrightarrow{L_p} 0$ for integer $p > 0$?

6.16 (a) Let $X_n \xrightarrow{d} X$ and for some $p > 0$, $\sup_n E|X_n|^p = M < \infty$. Then show that for each $r < p$, $\lim_{n \to \infty} E|X_n|^r = E|X|^r < \infty$.

(b) If r is a positive integer, then $|X_n|^r$ and $|X|^r$ above can be replaced by X_n^r and X^r.

(c) In particular, if $X_n \in L_r$ then $X_n \xrightarrow{d} X$ and $\{|X_n|^r\}$ uniformly integrable for $r \geq 1 \Rightarrow X_n \xrightarrow{L_p} X$ (for definition see Section 6.2).

7

Independence

7.1 DEFINITIONS AND DYNKIN'S THEOREM

The most important and fundamental concept in probability theory is the concept of independence whose analogue is absent in general measure theory.

Definition 7.1 (*Independent classes of events*). Let $(\Omega, \mathcal{F}, P)$ be a probability space and consider the classes of events $\mathcal{G}_t \subset \mathcal{F}$ for each $t \in T$, where T is any index set $\neq \phi$. Then $\mathcal{G}_t$, $t \in T$ are said to be independent if for each $m = 2, 3, \ldots,$ each choice of $t_1, t_2, \ldots, t_m \in T$ ($t_1, \ldots, t_m$ are distinct)

and

$$A_j \in \mathcal{G}_{t_j}, \text{ for } j = 1, 2, \ldots, m,$$

we have

$$P(A_1 A_2 \ldots A_m) = P(A_1) \ldots P(A_m).$$

If $\mathcal{G}_t$, $t \in T$ are independent then $\mathcal{H}_t$, $t \in T^*$ are independent for $T^* \subset T$ and $\mathcal{H}_t \subset \mathcal{G}_t$ for each $t \in T^*$.

Definition 7.2 Random variables $X_1, \ldots, X_n$ are independent if $F_{X_1 \ldots X_n} = F_{X_1} \ldots F_{X_n}$ where $F_{X_1 \ldots X_n}$ is the joint d.f. and F_{X_i} is the marginal d.f. of the r.v. X_i.

Definition 7.3 (*σ-field generated by* r.vs.). For a r.v. X the smallest σ-field containing all the sets of the form $[X < \lambda]$, $|\lambda| < \infty$ is called the Borel field generated by X and we shall denote this by $\mathcal{B}(X)$. For a family of r.vs. $\{X_t, t \in T\}$, the smallest σ-field containing all sets of the form $[X_t < \lambda_t]$, $t \in T$ $|\lambda_t| < \infty$ is called the σ-field generated by $\{X_t, t \in T\}$ and is denoted by $\mathcal{B}(X_t, t \in T) = \mathcal{B}(X_T)$.

Definition 7.4 $\mathcal{B}(X_1)$ and $\mathcal{B}(X_2)$ are said to be independent if for every $A_1 \in \mathcal{B}(X_1)$ and $A_2 \in \mathcal{B}(X_2)$, A_1 and A_2 are independent, i.e. $P(A_1 A_2) = P(A_1) P(A_2)$.

Definition 7.5 A family $\{X_t, t \in T\}$ of r.vs. is said to be independent if $X_{t_1}, \ldots X_{t_n}$ are independent for every finite $(t_1, \ldots, t_n) \subset T$.

Theorem 7.1 (*Equivalence of two definitions of independent r.vs.*).

(a) X_1 and X_2 are independent iff $\mathcal{B}(X_1)$ and $\mathcal{B}(X_2)$ are independent.

84

(b) Random variables X_t, $t \in T$ are independent iff the σ-fields $\mathcal{B}(X_t)$, $t \in T$ are independent.

Proof. *If part.* Let t_1, t_2, ..., $t_m \in T$, $-\infty < \lambda_1$, λ_2, ..., $\lambda_m < \infty$, then

$$(X_{t_1} \le \lambda_1) \in \mathcal{B}(X_{t_1}), ..., (X_{t_m} \le \lambda_{t_m}) \in \mathcal{B}(X_{t_m}). \text{ Hence}$$

$$F_{x_{t_1}...x_{t_m}} (\lambda_1, \lambda_2, ..., \lambda_m) = F_{x_{t_1}}(\lambda_1) \, F_{x_{t_2}}(\lambda_2) \, ... \, F_{x_{t_m}}(\lambda_m).$$

Only if part. Now we have to prove the converse part of equivalence theorem, i.e. we have to prove that if X and Y are independent according to Definition 7.2, then $\mathcal{B}(X)$ and $\mathcal{B}(Y)$ are independent.

Let $\qquad \mathcal{G} = \{B \mid P(AB) = P(A)P(B), B \in \mathcal{B}(Y) \text{ and } A \in \mathcal{D}(X)\}$ $\qquad\qquad$ (1)

where $\mathcal{D}(X) = \{[\omega : X(\omega) \le \lambda], -\infty < \lambda < \infty\} \subset \mathcal{B}(X)$. Note that $\mathcal{G}$ is a non-empty class, since $\mathcal{G} \supset \mathcal{D}(Y)$

Here we are given $\mathcal{D}(X)$ and $\mathcal{D}(Y)$ as independent. We have to prove $\mathcal{B}(X)$ and $\mathcal{B}(Y)$ are independent. By symmetry it is sufficient to prove that $\mathcal{G} = \mathcal{B}(Y)$. To prove this result the following definitions and lemmas due to E.B. Dynkin (1961) will be needed.

Definition 7.6 Let $\mathcal{G}$ be a non-empty class of subsets of Ω. We call $\mathcal{G}$ a π-system (multiplicative system) if A, $B \in \mathcal{G}$ implies $AB \in \mathcal{G}$ and a λ-system if (a) $\Omega \in \mathcal{G}$, (b) A, $B \in \mathcal{G}$ and $B \subset A$ implies $A - B \in \mathcal{G}$, and (c) $A_n \in \mathcal{G}$ and $A_n \uparrow A$ implies $A \in \mathcal{G}$. (Note that if $\mathcal{G}$ is a λ-system, then A, $B \in \mathcal{G}$ and $A \cap B = \phi$ $\Rightarrow A \cup B \in \mathcal{G}$.)

Lemma 7.1 (*Dynkin's π–λ theorem*). If a λ-system $\mathcal{G}$ covers a π-system $\mathcal{D}$, then $\mathcal{G} \supset \mathcal{B}(\mathcal{D})$, the σ-field generated by $\mathcal{D}$. In particular, if $\mathcal{D}$ is a λ-system, $\lambda(\mathcal{D}) = \mathcal{B}(\mathcal{D})$. (Note that the set theoretic intersection of λ-systems is a λ-system and hence there exists a unique minimal λ-system $\lambda(\mathcal{D})$ containing a given class $\mathcal{D}$ of events.)

Proof. First we shall show that a class of events which is both a π-system and a λ-system is a σ-field. Being a λ-system it contains Ω and is closed under complementation and being a π-system it is closed under finite intersection and therefore it is a field. Being a λ-system, it is closed under both finite disjoint union and under monotone limits and therefore it is closed under countable union. Hence it is a σ-field. Let $\mathcal{G}^*$ be the minimal λ-system and $\mathcal{G}^* \supset \mathcal{D}$. It is enough to show that $\mathcal{G}^*$ is a π-system, because now $\mathcal{G}^*$ is a σ-field (being simultaneously π and λ-system) containing $\mathcal{D}$, $\mathcal{G}^* \supset \mathcal{B}(\mathcal{D})$ which implies $\mathcal{G} \supset \mathcal{G}^* \supset \mathcal{B}(\mathcal{D})$.

Set $\mathcal{F}_1 = \{A : AB \in \mathcal{G}^* \text{ for all } B \in \mathcal{D}, A \subset \Omega\}$. Then $\mathcal{F}_1 \supset \mathcal{D}$. We shall prove that $\mathcal{F}_1$ is a λ-system. Note that for A_1, $A_2 \in \mathcal{F}_1$, A_1B and $A_2B \in \mathcal{G}^*$ whenever $A_2 \subset A_1$ and $B \in \mathcal{D}$. Now $\mathcal{G}^*$ being a λ-system, $(A_1 - A_2)B = A_1B - A_2B \in \mathcal{G}^*$ and this implies $(A_1 - A_2) \in \mathcal{F}_1$. If $A_n \uparrow A$ and $A_1 \in \mathcal{F}_1$, then $A_nB \in \mathcal{G}^*$ for $B \in \mathcal{D}$ and $A_nB \uparrow AB$. Since $\mathcal{G}^*$ is a λ-system, $AB \in \mathcal{G}^*$ and that implies $A \in \mathcal{F}_1$. Obviously $\Omega \in \mathcal{F}_1$. Therefore $\mathcal{F}_1$ is a λ-system and $\mathcal{F}_1 \supset \mathcal{G}^*$. Hence $AB \in \mathcal{G}^*$ whenever $B \in \mathcal{D}$ and $A \in \mathcal{G}^*$. Thus if $\mathcal{F}_2 = \{B : AB \in \mathcal{G}^* \text{ for all } A \in \mathcal{G}^* \text{ and } B \subset \Omega\}$ then $\mathcal{F}_2 \supset \mathcal{D}$. Similar to the proof of $\mathcal{F}_1$ being a λ-system, $\mathcal{F}_2$ can also be shown to be a λ-system. Therefore $\mathcal{F}_2 \supset \mathcal{G}^*$ and so $AB \in \mathcal{G}^*$ if both A and $B \in \mathcal{G}^*$. Hence $\mathcal{G}^*$ is a π-system.

Lemma 7.2   Let $\mathcal{D}$ be a π-system and $\mathcal{G}$ be a non-empty class of subsets of Ω. If $\mathcal{D}$ and $\mathcal{G}$ are independent, then $\mathcal{B}(\mathcal{D})$ and $\mathcal{G}$ are independent.

Proof.   For any $B \in \mathcal{G}$, define

$$\mathcal{D}^* = \{A : A \in \mathcal{B}(\mathcal{D}) \text{ and } P(AB) = P(A)\,P(B) \tag{2}$$

Then $\Omega \in \mathcal{D}^*$, $A_1 - A_2 \in \mathcal{D}^*$ if $A_1, A_2 \in \mathcal{D}^*$ and $A_1 \supset A_2$.

$$A \in \mathcal{D}^* \text{ if } A = \lim_{n \to \infty} A_n, \text{ when } A_n \in \mathcal{D}^* \text{ and } A_n \subset A_{n+1} \text{ for } n \geq 1.$$

Thus $\mathcal{D}^*$ is a λ-system which contains $\mathcal{D}$ by assumption. Consequently $\mathcal{D}^* \supset \mathcal{B}(\mathcal{D})$ by Lemma 7.1.

In other words $P(AB) = P(A)P(B)$ for all $B \in \mathcal{G}$ and every $A \in \mathcal{B}(\mathcal{D})$. The proof of the lemma is complete.

Proof of Theorem 7.1(a) (*Only if part continued*).   Now note that $\mathcal{D}(X)$ and $\mathcal{D}(Y)$ are π-systems, because intersections of semiclosed intervals are semiclosed. Like $\mathcal{D}^*$ in (2) it is easy to see the $\mathcal{G}$ defined by (1) is a λ-system containing $\mathcal{D}(Y)$. Then by Lemmas 7.1 and 7.2, Theorem 7.1(a) is proved.

To prove Theorem 7.1(b) with the conjunction of Lemmas 7.1 and 7.2 it is sufficient to prove the following.

Theorem 7.2   Let $\{\mathcal{G}_t, t \in T\}$ be a family of classes of events. Assume that each class of this family is a π-system. Then for σ-fields $\mathcal{B}(\mathcal{G}_t)$, $t \in T$ to be independent, the independence of classes $\mathcal{G}_t$, $t \in T$ is sufficient and also necessary.

Proof.   It is sufficient to consider the case where the family $\mathcal{G}_t$, $t \in T$ is finite. Fix $t \in T$ and denote by $\mathcal{G}$ the class of events A_t such that

$$P(\bigcap_{j \in T} A_j) = \prod_{j \in T} P(A_j) \text{ for all } A_j \in \mathcal{G}_j, \quad t \neq j.$$

It is sufficient to show that $\mathcal{G} \supset \mathcal{B}(\mathcal{G}_t)$. It is clear that $\mathcal{G} \supset \mathcal{G}_t$ while (1) $\Omega \in \mathcal{G}$, (2) $\mathcal{G} \ni B_1 \subset B_2$ implies $B_2 - B_1 \in \mathcal{G}$ and $\{B_n, n = 1, 2, \ldots\} \subset \mathcal{G}$. $B_i \cap B_j = \phi$ when $i \neq j$ implies $\bigcup_{n \geq 1} B_n \in \mathcal{G}$, i.e. $\mathcal{G}$ is a λ-system. Thus a λ-system $\mathcal{G}$ contains a π-system $\mathcal{G}_t$. By Lemmas (7.1) and (7.2) the theorem follows.

Corollary 7.1   If the r.vs. X_t, $t \in T$ are independent and T_1 and T_2 are non-empty disjoint subsets of T, then

$$\mathcal{B}(X_t, t \in T_1) \text{ and } \mathcal{B}(X_t, t \in T_2) \text{ are independent.}$$

Proof.   Let $\mathcal{D}_i = [X_{t_1}(i) \leq x, \ldots, X_{t_n}(i) \leq x]$, $i = 1, 2$, $t_j^{(i)} \in T_i$, $1 \leq j \leq n$. Now $\mathcal{D}_1$ and $\mathcal{D}_2$ are independent π-systems. By application of Lemma 7.2 twice,

$$\mathcal{B}(\mathcal{D}_1) = \mathcal{B}(X_t, t \in T_1) \text{ and } \mathcal{B}(\mathcal{D}_2) = \mathcal{B}(X_t, t \in T_2) \text{ are independent.}$$

7.2   BOREL ZERO ONE LAW

In case of independence we can get a converse of Borel-Cantelli lemma which is the so called *Borel Zero One Law*. This theorem plays a fundamental role in modern probability theory. It is instrumental in proving strong law of large numbers and many other limit theorems.

Theorem 7.3 (*Borel-Cantelli*).   Let $\{A_n\}_1^\infty$ be independent events. Then

$$P(A_n \cdot \text{i.o.}) = 0 \text{ iff } \sum_1^\infty P(A_n) < \infty$$

i.e.
$$P(\limsup A_n) = 1 \text{ iff } \sum_1^\infty P(A_n) = \infty \tag{1}$$

Proof.   Assume $\sum_1^\infty P(A_n) = \infty$. Then by De-Morgan's rule and continuity theorem of measure,

$$1 - P(A_n \cdot \text{i.o.}) = P(\bigcup_{n=1}^\infty \bigcap_{k=n}^\infty A_k^C) = \lim_{n \to \infty} P(\bigcap_{k=n}^\infty A_k^C)$$

$$= \lim_{n \to \infty} P(\lim_{n \to \infty} \bigcap_{k=n}^m A_k^C), \text{ again by continuity theorem}$$

$$= \lim_{n \to \infty} \lim_{m \to \infty} (\bigcap_{k=n}^m A_k^C) = \lim_{n \to \infty} \lim_{m \to \infty} \prod_{k=n}^m P(A_k^C)$$

$$\leq \lim_{n \to \infty} \exp(-\sum_{k=n}^\infty P(A_k)) = 0 \text{ (by (1))}$$

Therefore $1 - P(A_n \cdot \text{i.o.}) \leq 0$. Hence $P(A_n \cdot \text{i.o.}) = 1$. The converse part is proved in First Borel-Cantelli lemma.

Borel zero one law is not true if A_n's are not independent as can be seen from the following example.

EXAMPLE 7.1   Let $\Omega = \{1, 2, 3, \ldots\}$

$$P(\{k\}) = \frac{1}{k(k+1)}, k = 1, 2, \ldots$$

Now
$$A_n = \{n, n+1, \ldots\} \downarrow \phi. \text{ Then } P(A_n) = \frac{1}{n}.$$

Hence
$$\sum_{k=n}^\infty P(A_k) = \infty, \text{ but } P(\phi) = 0 \neq 1.$$

7.3   KOLMOGORV'S ZERO ONE LAW

Let $\{X_n, n \geq 1\}$ be a sequence of r.vs. on a probability space $(\Omega, \mathscr{F}, P)$. The tail σ-field of $\{X_n, n \geq 1\}$ is

$$\bigcap_{n=1}^{\infty} \mathscr{B}(X_m, m \geq n) = \lim_{n \to \infty} \mathscr{B}(X_n, m \geq n)$$

The sets belonging to the *tail σ-field* are called *tail events* and the functions which are measurable with respect to the tail σ-field are called *tail functions*. A typical example of a tail event of $\{X_n, n \geq 1\}$ is the set of convergence of $\{X_n\}$. Before stating Kolmogorov's zero one law let us state a useful lemma which illustrates some typical tail functions.

Lemma 7.3 Let $(\mathscr{X}, \mathscr{A})$ be a measurable space and $X_1, \ldots, X_n$ be $\mathscr{A}$-measurable. Then

(a) $-X_n$, $a + bX_n$, $|a| < \infty$, $0 < b < \infty$, $\sup_n X_n$, $\inf_n X_n$,

   $\lim \sup_{n \to \infty} X_n$ and $\lim \inf_{n \to \infty} X_n$ are $\mathscr{A}$-measurable

(b) If $X_j > -\infty$ for every j, then $S_n = \sum_1^n X_j$ is $\mathscr{A}$-measurable.

(c) If moreover X_n be r.vs. then $\lim \sup X_n$. $\lim \inf X_n$, $\lim \sup \dfrac{S_n}{n}$ and $\lim$

   $\inf \dfrac{S_n}{n}$ are tail functions.

Proof of (a) and (b) are already given in measurable function (See Theorem 4.2).

Proof of (c).   Let

$$Y = \lim \sup S_n/n = \lim_{j \leq k \to \infty} \sup_{n \geq k} S_n/n \text{ for } j = 1, 2, 3, \ldots$$

Then

$$Y = \lim_{j \leq k \to \infty} \sup_{n \geq k} \frac{(X_1 + \ldots + X_{j-1}) + (X_j + \ldots + X_n)}{n} \text{ for } j = 1, 2, \ldots, n$$

Since X_i's are r.vs. they are finite a.s. Therefore

$$\sum_{k=1}^{j-1} X_k/n \to 0 \text{ a.s. as } n \to \infty \text{ and this}$$

$$\Rightarrow Y = \lim_{j \leq k \to \infty} \sup_{n \geq k} \frac{X_j + \ldots + X_n}{n} \text{ for } j = 1, 2, 3, \ldots$$

Since $X_j, \ldots, X_n$ are $\mathscr{B}(X_j, \ldots, X_n)$-measurable for all $j = 1, 2, \ldots$, $(X_j + \ldots + X_n)/n$ is $\mathscr{B}(x_j, X_{j+1}, \ldots)$-measurable for all $j = 1, 2, 3, \ldots$ Therefore $\sup_{n \geq k} (X_j + \ldots + X_n)/n$ is $\mathscr{B}(X_j, X_{j+1}, \ldots)$-measurable for all $j = 1, 2, 3, \ldots$

$$\inf_{j \leq k} \sup_{n \geq k} S_n/n = \lim_{j \leq k} \sup_{n \geq k} \frac{(X_j + \ldots + X_n)}{n} \text{ is } B(X_j, X_{j+1}, \ldots) \text{ measurable} \quad \text{for all}$$

$j = 1, 2, 3, \ldots$ Hence

$\limsup\limits_{n\to\infty} S_n/n$ is $\mathcal{B}\,(X_j, X_{j+1}, \ldots)$-measurable for all $j = 1, 2, \ldots$

Therefore, $\limsup\limits_{n\to\infty} S_n/n$ is measurable with respect to $\bigcap\limits_{j=1}^{\infty} \mathcal{B}(X_j, X_{j+1}, \ldots)$ which is the tail σ-field of $\{X_n\}_1^{\infty}$. Therefore $\limsup S_n/n$ is a tail function. Similarly $\liminf S_n/n$ is a tail function.

Kolmogorov's Zero-one Law (1932)

"The tail σ-field of a sequence of independent r.vs. is equivalent to $\{\phi, \Omega\}$" i.e. the tail events of independent r.vs. $X_1, X_2, \ldots$ have either probability zero or probability one. Therefore the tail functions of independent r.vs. are constants (i.e. degenerate) a.s.

Proof. By Corollary 7.1 of Section 7.1, $\mathcal{B}(X_k, 1 \leq k \leq n)$ and $\mathcal{B}(X_m, m > n)$ are independent. So are $\mathcal{B}(X_k, 1 \leq k \leq n)$ and $\bigcap\limits_{n=0}^{\infty} \mathcal{B}(X_m, m > n) = \tau$ for every $n \geq 1$. Therefore $\alpha = \bigcup\limits_{n=1}^{\infty} \mathcal{B}(X_k, 1 \leq k \leq n)$ and τ (tail σ-field) are independent. (Note that unions of σ-fields may not be σ-field, but is a π-systems and a field and hence $\bigcup\limits_{n=1}^{\infty} \mathcal{B}(X_k, 1 \leq k \leq n)$ is a π-system).

By Dynkin's Lemma 7.2, $\mathcal{B}(\bigcup\limits_{n=1}^{\infty} \mathcal{B}(X_k, 1 \leq k \leq n))$ and τ are independent. But τ is a sub-family of $\mathcal{B}(X_1, X_2, \ldots) = \mathcal{B}(\alpha)$ and hence τ is a sub-family of $\mathcal{B}(\bigcup\limits_{n=1}^{\infty} \mathcal{B}(X_k, 1 \leq k \leq n))$. Therefore τ is independent of itself, i.e. if $A \in \tau$, then $P(AA) = P(A) \cdot P(A)$, i.e. $P(A) = [P(A)]^2$ and this implies $P(A) = 1$ or 0. If Y is τ-measurable, $[Y \leq y] \in \tau$ for each $y \in R$. Then $P(Y \leq y) = 0$ or 1. Since $P[Y \leq y] \uparrow$ in y, it is zero for $y < y_0$ (say) and one for $y \geq y_0$. Thus for some y_0, $P[Y = y_0] = 1$ and Y is a.s. a constant.

Exercise 7.1 Let $X_1, X_2, \ldots, X_n, \ldots$ be independent r.vs. Prove that

$$P[\sum_{n=1}^{\infty} X_n \text{ converges}] = 1 \text{ or } 0$$

(Note that the convergence of the series depends entirely on the tail of the series.)

7.4 LÉVY'S INEQUALITY AND LÉVY'S THEOREM

The following remarkable result is due to P. Lévy and it is considered as a touchstone in the theory of independence.

Theorem 7.4 (*Lévy, 1937*). Let X_n's be independent r.vs. Then $\sum\limits_{n=1}^{\infty} X_n$ converges a.s. iff $\sum\limits_{n=1}^{\infty} X_n$ converges in probability.

Lévy's theorem depends upon the following maximal inequality of Levy. Unlike Kolmogorov's inequality it does not depend on existence of any moments.

Lévy's inequality. Let X_1, X_2, ..., X_n be independent r.vs. and $S_k = \sum_1^k X_i$. Then for every $\varepsilon > 0$.

$\quad$ (a) $P[\max_{1 \le k \le n} (S_k - \mu(S_k - S_n)) \ge \varepsilon] \le 2P(S_n \ge \varepsilon)$

$\quad$ (b) $P[\max_{1 \le k \le n} |S_k - \mu(S_k - S_n)| \ge \varepsilon] \le 2P(|S_n| \ge \varepsilon)$

where $\mu(Y)$ is a median of Y.

Definitions and observations. (1) Let X be a r.v. Then there exists at least one finite number $\mu(X)$, called the median of X, such that $P[X \le \mu(X)] \ge 1/2 \le P[X \ge \mu(X)]$.

$\quad$ If X has d.f. F, then $F(\mu(X) -) \le 1/2 \le F(\mu(X))$. A median may not be unique.

(2) A r.v. X is said to be symmetric (about zero) if $P(-X < \lambda) = P(X > -\lambda)$, $-\infty < \lambda < \infty$, i.e. X and $-X$ has the same d.f.

$\quad$ (a) If X is symmetric around zero, then zero is a median of X.

$\quad$ (b) If X_1 and X_2 are independent and symmetric about zero, so is $X_1 + X_2$, because ch.f. of X_1, $\phi_1(t) = E(e^{itX_1})$ is real and that of X_2, $\phi_2(t) = E(e^{itX_2})$ is also real. Then ch.f. of $X_1 + X_2$ is $\phi(t) = \phi_1(t) \cdot \phi_2(t)$ which must be real. So $X_1 + X_2$ is symmetric.

$\quad$ (c) If X_1, ..., X_n are independent and symmetric r.vs. then Levy's inequality simplifies to
$P(\max_{1 \le k \le n} S_k \ge \varepsilon) \le 2P(S_n \ge \varepsilon)$ and
$P(\max_{1 \le k \le n} |S_k| \ge \varepsilon) \le 2P(|S_n| \ge \varepsilon)$ for every $\varepsilon > 0$.

NOTE. Trivially we have a lower bound

$$P(\max_{1 \le k \le n} S_k \ge \varepsilon) \ge P(S_n \ge \varepsilon).$$

Proof of Lévy's inequality. Set $S_0 = 0$ and define

$$T = \begin{cases} \text{1st } k, 1 \le k \le n \text{ such that } S_k - \mu(S_k - S_n) \ge \varepsilon \\ n + 1 \text{ otherwise} \end{cases}$$

Let $\qquad A_k = \{\mu(S_k - S_n) \ge S_k - S_n\}$, $1 \le k \le n$.
Then $P(A_k) \ge 1/2$. Since
$$\{T = k\} \in \mathcal{B}(X_1, ..., X_k), A_k \in \mathcal{B}(X_{k+1}, ..., X_n),$$
$$\{S_n \ge \varepsilon\} \supset \bigcup_{k=1}^{n} A_k \{T = k\}$$

and $\{T = k\}$ and A_k are independent,

$$P\{S_n \ge \varepsilon\} \ge \sum_{j=1}^{n} P\{A_k[T=k]\} = \sum_{k=1}^{n} P(A_k)P\{T=k\} \ge 1/2\, P\{1 \le T \le n\}$$
$$= 1/2 P[\lim_{1 \le k \le n} (S_k - \mu(S_k - S_n)) \ge \varepsilon]$$

Rewriting inequality (a) with X_k replaced by $-X_k$, $1 \le k \le n$, recalling that $\mu(-X)$ $= -\mu(X)$ and adding this to (a) we get inequality (b).

Proof of Lévy's theorem. Since almost sure convergence implies convergence in probability, the necessity part is obvious. So it is enough to prove that the partial sum S_n converges in probability implies S_n converges a.s. Let $S_{m,n} = S_n - S_m$. By Theorem 4.9, for any $0 < \varepsilon < 1/4$, there exists an integer $m_0 \ni n > m \ge m_0 \Rightarrow$ $P(|S_{m,n}| > \varepsilon) < \varepsilon$ and that implies $|\mu(S_{m,n})| \le \varepsilon$ for $n > m \ge m_0$. By Lévy's inequality, for $j > m \ge m_0$.

$$P[\max_{m < n \le j} |S_{m,n}| > 2\varepsilon] = P[\max_{m < n \le j} |S_{m,n}| > 2\varepsilon, \max_{m < n \le j} |\mu(S_{m,n})| \le \varepsilon]$$

$$\le P[\max_{m < n \le j} |S_{m,n} - \mu(S_{m,n} - S_{m,j})| > \varepsilon]$$

$$\le 2P[|S_{m,j}| > \varepsilon] \le 2\varepsilon.$$

Hence, letting $j \to \infty$, if $m \ge m_0$,

$$P[\sup_{n > m} |S_{m,n}| > \varepsilon] \le 2\varepsilon.$$

and hence $S_n \overset{a.s.}{\to} S$, say.

PROBLEMS AND COMPLEMENTS

7.1   Give a counter-example to show that the relation $P(ABC) = P(A)P(B)P(C)$ does not always imply the mutual independence of the events A, B, C.

7.2   Give an example to show that mutually independent events can form families which are strongly dependent.

7.3   (Feller and Chung): Let $\{A_n, n \ge 1\}$ and $\{B_n, n \ge 1\}$ be sequences of events on $(\Omega, \mathcal{F}, P)$ and set $A_0 = \phi$. Show that if either (i) B_n and $A_n A_{n-1}^C \ldots A_0^C$ are independent for all $n \ge 1$ or (ii) the classes $\{B_n\}$ and $\{A_n, A_n A_{n+1}^C, A_n A_{n+1}^C A_{n+2}^C, \ldots\}$ are independent for all $n \ge 1$, then

$$P[\bigcup_{n=1}^{\infty} (A_n B_n)] \ge P[\bigcup_{n=1}^{\infty} A_n] \inf_{n \ge 1} P(B_n).$$

7.4   Consider a probability space with $\Omega = \{1, 2, 3, \ldots\}$ and

$$P\{k\} = \frac{1}{C_r} \cdot \frac{1}{k^{1+r}} \text{ where } C_r = \sum_{k=1}^{\infty} \frac{1}{k^{1+r}}, r > 0.$$

Let $\{p_n\}_1^{\infty}$ be a sequence of prime numbers $p_1 = 2$, $p_2 = 3$, $p_3 = 5$, … and consider the events $A_j = \{k : k \text{ is divisible by } p_j\}$

(a) Prove that A_1, A_2, … are independent.

(b) Hence prove Euler's formula

$$C_r = \sum_{k=1}^{\infty} \frac{1}{k^{1+r}} = \prod_{j=1}^{\infty} \left(1 - \frac{1}{p_j^{1+r}}\right)^{-1}$$

7.5   Give examples of dependent r.vs. X and Y such that X^2 and Y^2 are independent.

7.6   Show that the independence of the sequence of events A_n is essential for the Borel zero-one law.

7.7   Let $(\Omega, \mathcal{F}, P)$ be a countable probability space and $\{A_n\}$ a sequence of independent events in $\mathcal{F}$. For $0 \leq x \leq 1$ define $\phi(x) = \min\,(x, 1-x)$. Show that $\displaystyle\sum_{n=1}^{\infty} \phi(P(A_n)) < \infty$.

7.8 F and G are two independent d.fs. Show that if (i) x and y are points of increase of F and G, then $x + y$ is a point of increase of $F * G$, and if (ii) x and y are atoms of F and G, then $x + y$ is an atom of $F * G$.

7.9   Let N and $X_1, X_2, \ldots$ be independent r.vs. where $P(N = n) = q^{n-1}p,\ n \geq 1$ and each X_k has exponential density $\alpha e^{-ax},\ x > 0,\ \alpha > 0$. Find the density of

$$S_N = \sum_{k=1}^{N} X_k.$$

7.10   (Etemadi): Let $\{X_n\}$ be a sequence of independent r.vs. Show that
$$P[\max_{1 \leq k \leq n} |S_k| \geq 4\alpha] \leq 4 \max_{1 \leq k \leq n} P[|S_k| \geq \alpha],\ \alpha > 0.$$

7.11   (Lévy's theorem): Let $\{X_n\}$ be a sequence of independent r.vs. Then the partial sum S_n converges a.s. if it converges in probability.

7.12   If X_k's are independent with mean 0 and variance 1, then show that for $\alpha \geq \sqrt{2}$,

$$P\left[\frac{M_n}{\sqrt{n}} \geq \alpha\right] \leq 2P\left[\frac{S_n}{\sqrt{n}} \geq \alpha - \sqrt{2}\right],$$

where $M_n = \max_{0 \leq k \leq n} S_k,\ S_0 = 0$ and $S_n = \displaystyle\sum_{k=1}^{n} X_k.$

7.13   (a)   Suppose P_1 and P_2 are two probability measures on $\sigma(\mathcal{P})$, where $\mathcal{P}$ is a π-system. If P_1 and P_2 agree on $\mathcal{P}$, then show that they agree on $\sigma(\mathcal{P})$.

      (b)   Show that the above result is true if μ_1 and μ_2 are measures on $\sigma(\mathcal{P})$ such that they are σ-finite on $\mathcal{P}$.

      (c)   Also show that the result is true if μ_1 and μ_2 are finite measures on $\sigma(\mathcal{P})$.

7.14   Prove the following

      (a)   A λ-system is closed under complementation.

      (b)   A λ-system closed under finite unions is a σ-field.

      (c)   A λ-system which is also a π-system is a σ-field.

      (d)   There exists a unique minimal λ-system (respectively π-system) containing a given class $\mathscr{C}$ of sets.

7.15 If $\{E_j, 1 \le j < \infty\}$ are independent events then show that

$$P(\bigcap_{j=1}^{\infty} E_j) = \prod_{j=1}^{\infty} P(E_j),$$

where the infinite product is defined to be the obvious limit; similarly show that

$$P(\bigcup_{j=1}^{\infty} E_j) = 1 - \prod_{j=1}^{\infty} (1 - P(E_j)).$$

If moreover $P(E_j) < 1$ for all $j \ge 1$, then show that

$$P(\bigcap_{j=1}^{\infty} E_j) = 1 \text{ implies } \prod_{j=1}^{\infty} P(E_j) = \infty.$$

7.16 Let $\{X_n\}$ be a sequence of independent r.vs. Then show that $X_n \to 0$ a.s. iff

$$\sum_{n=1}^{\infty} P\{|X_n| > \varepsilon\} < \infty \text{ for all } \varepsilon > 0.$$

NOTE. Hsu and Robbins (1947) introduced the concept of complete convergence of a sequence of r.vs. A sequence of r.vs. $\{X_n\}$ is said to converge completely to a constant μ if $\sum_{n=1}^{\infty} P\{|X_n - \mu| > \varepsilon\} < \infty$ for every $\varepsilon > 0$. The above exercise proves that in case of independent r.vs. almost sure convergence and complete convergence are equivalent. In general, complete convergence implies almost sure convergence but the converse is not true.

7.17 (Extension of Borel–Cantelli lemma). Let $A_1, A_2 \ldots$ be events in a given probability space such that $\sum_{n=1}^{\infty} P(A_n) = \infty$ and

$$\liminf_{n \to \infty} \frac{\sum_{j,k=1}^{n} P(A_j A_k)}{(\sum_{j=1}^{n} P(A_j))^2} = 1.$$

(a) Show that the lim inf condition above is satisfied if the A_n are pairwise independent and $\sum_{n=1}^{\infty} P(A_n) = \infty$.

(b) Show that $\varliminf P(|\sum_{k=1}^{n} I_{A_k} - P(A_n)| > \frac{1}{2} \sum_{k=1}^{n} P(A_k)) = 0$.

(c) Conclude from (b) that there exists a subsequence $\{n_j\}$ such that

$$\sum_{k=1}^{n} I_{A_k} \ge \sum_{k=1}^{n_j} P(A_k) \text{ a.s. for large } j.$$

(d) Show that $P(\varlimsup A_n) = 1$.

7.18 (Hewitt–Savage Zero-one law). An *event A* is said to be *symmetric* iff the occurrence or non-occurrence of *A* is not affected by a permutation of finitely many of the r.vs. X_i where $A = \{(X_1, X_2, \ldots) \in A'\}$, $(A \in \mathcal{F}_1 = \sigma(X_1, X_2, \ldots)$ and $A' \in \mathcal{B}(R)$. (Note that a tail event is a symmetric event but the converse is not true.) Let $\{X_i\}$ be a sequence of i.i.d. r.vs. If A is a symmetric event in $\mathcal{F}_1$, then show that $P(A) = 0$ or 1.

8

Law of Large Numbers and Associated Limit Theorems

8.1 WEAK LAW OF LARGE NUMBERS (WLLN)

Proposition 8.1   (a) Let $X_1, X_2, \ldots, X_n$, be i.i.d. random variables with $EX_i = \mu$ and var $(X_i) = \sigma^2 < \infty$.

Then
$$\frac{S_n}{n} = \frac{1}{n} \sum_{i=1}^{n} X_i \xrightarrow{P} \mu.$$

Proof.   By Chebychev's inequality, for $\varepsilon > 0$

$$P\left[\left|\frac{S_n}{n} - \mu\right| \geq \varepsilon\right] \leq \frac{1}{\varepsilon^2} \text{ var } (S_n/n) = \frac{1}{\varepsilon^2} \sigma^2/n \to 0 \text{ as } n \to \infty.$$

(b) *Khintchine's weak law.*   Let $X_1, \ldots, X_n$, be i.i.d. r.vs. with $E|X| < \infty$, then $S_n/n \xrightarrow{P} \mu = E(X)$.

Proof.   Proof is given in ch.f., Section 9.2 as an application of Theorem 9.6.

(c) Let $X_1, \ldots, X_n$ be independent r.vs. with $E(X_i) = $ for all $i \geq 1$ and $\frac{1}{n^2} \sum_{i=1}^{n} \sigma_i^2 \to 0$ as $n \to \infty$. The $S_n/n \xrightarrow{P} 0$.

Proof.   The proof follows from Chebychev's inequality and the fact that var $(S_n) = \sum_{i=1}^{n} \sigma_i^2$.

8.2 CONVERGENCE OF SUMS OF INDEPENDENT RANDOM VARIABLES

We have seen in Chapter 7 (Kolmogorov's zero-one law) that if $\{X_n\}$ is a sequence of independent r.vs., $\{X_n\}$ or ΣX_n either converges a.s. or does not converge a.s. Therefore, the first question dealt with sums of independent r.vs. is when such

95

sums converge almost surely. A partial answer is given next in the form of Khintchine–Kolmogorov convergenc theorem—popularly known as Kolmogorov's convergence theorem. Complete answer to the above problem is found in Kolmogorov's three series theorem proved later in this section.

Theorem 8.1 (*Kolmogorov convergence theorem*).   Let $\{X_n\}$ be independent r.vs. with $EX_n = 0$ and $\sigma_n^2 = E X_n^2 < \infty$, $n \geq 1$.

If $\sum\limits_{n=1}^{\infty} E X_n^2 < \infty$, then $\sum\limits_{n=1}^{\infty} X_n$ converges a.s.

First proof.   For $\varepsilon > 0$, $\varepsilon^2 \, P[|\sum\limits_{k=n}^{m} X_k | \geq \varepsilon] \leq \sum\limits_{k=n}^{m} E X_k^2$ (by Chebychev's inequality)

$$< \varepsilon^3 \text{ if } m, n \geq n_0 \text{ (because } \sum\limits_{k=1}^{\infty} EX_k^2 < \infty).$$

Therefore,

$$P[|\sum\limits_{k=n}^{m} X_k | \geq \varepsilon] \leq \varepsilon \text{ if } m, n \geq n_0$$

i.e. $P[| S_m - S_{n-1} | \geq \varepsilon] \leq \varepsilon$ if $m, n \geq n_0$

i.e. $\{S_n\}$ is a Cauchy sequence in probability.

Therefore, $S_n \overset{P}{\to} S$, say. Hence by Lévy's theorem $S_n \to S$ (say) a.s. i.e. $\sum\limits_{n=1}^{\infty} X_n$ converges a.s.

Definition 8.1   Two sequences of r.vs. $\{X_n\}$ and $\{Y_n\}$ are said to be tail equivalent (Loéve) if $\sum\limits_{n=1}^{\infty} P(X_n \neq Y_n) < \infty$.

Theorem 8.2 (*Kolmogorov's three-series theorem, 1928*).   Suppose that $\{X_n\}_{n=1}^{\infty}$ be a sequence of independent r.vs. Let

$$X_n' = \begin{cases} X_n & \text{if } |X_n| \leq 1 \\ 0 & \text{if } |X_n| > 1 \end{cases} \text{ for all } n \geq 1.$$

Then the series $\sum\limits_{n} X_n$ converges a.s. iff the following series converges

(a) $\sum\limits_{n=1}^{\infty} P(\omega : | X_n(\omega) | > 1) < \infty$;   (b) $\sum\limits_{n=1}^{\infty} E(X_n')$ converges; and

(c) $\sum\limits_{n=1}^{\infty} \sigma_{X_n'}^2 < \infty$.

Proof.   Suppose that the three conditions hold. Because of (c) by Kolmogorov–Khintchine theorem $\sum\limits_{n=1}^{\infty} (X_n' - EX_n')$ converges a.s. Then (b) implies $\sum\limits_{n=1}^{\infty} X_n'$ converges a.s. By (a) and Borel–Cantelli lemma $P(| X_n | > 1 \text{ i.o.}) = 0$. So $| X_n | \leq 1$ eventually a.s. Thus $X_n = X_n'$ eventually a.s. Thus $\sum\limits_{n=1}^{\infty} X_n$ converges a.s. Conversely if $\sum\limits_{n} X_n$ converges a.s. then $X_n \to 0$ a.s. Hence $P(| X_n | > 1 \text{ i.o.}) = 0$. This implies (a) by Borel-zero one law. Now $\{X_n\}_{n=1}^{\infty}$ and $\{X_n'\}_{n=1}^{\infty}$ are tail equivalent sequences. So it is clear that $\sum\limits_{n=1}^{\infty} X_n'$ converges a.s. when $\sum\limits_{n=1}^{\infty} X_n$ converges a.s. Now $\{X_n'\}_{n=1}^{\infty}$ is a sequence of uniformly bounded independent r.vs. Let $S_n' = \sum\limits_{j=1}^{\infty} X_j'$. Since $\sum\limits_{n=1}^{\infty} X_n'$ converges a.s.,

$$\lim_{n \to \infty} P(\sup_{m \geq n} |S_m' - S_n'| \geq \varepsilon) = 0 \tag{1}$$

By the lower bound of Kolmogorov's inequality (Section 8.3)

$$P(\sup_{m \geq n} |S_m' - S_n'| \geq \varepsilon) \geq 1 - \frac{(2 + \varepsilon)^2}{\sum\limits_{j=n}^{\infty} \sigma_{X_j'}^2}$$

Now if

$$\sum_{j=1}^{\infty} \sigma_{X_j'}^2 = \infty, \text{ then we gave}$$

$$P(\sup_{m \geq n} |S_m' - S_n'| \geq \varepsilon) = 1.$$

This contradicts the contention (1). So $\sum\limits_{j=1}^{\infty} \sigma_{X_j'}^2 < \infty$ proving (c). Then Khintchine–Kolmogorov's theorem implies $\sum\limits_{n}(X_n' - EX_n')$ converges a.s. Now since $\sum\limits_{j=1}^{\infty} X_n'$ converges a.s. we have $\sum\limits_{n=1}^{\infty} E(X_n')$ convergent proving (b).

8.3 STRONG LAW OF LARGE NUMBERS AND KOLMOGOROV INEQUALITIES

Definition 8.2   A sequence $\{X_n\}$ of L_1 r.vs. is said to obey the classical strong law of large numbers (SLLN) if $\dfrac{1}{n} \sum\limits_{i=1}^{n} (X_i - E\,X_i) \to 0$ a.s.

(i) Kolmogorov's strong law of large numbers for independent r.vs. Let $X_1, \ldots, X_n$ be independent r.vs. with $EX_n = 0$, $n \geq 1$ and $\sum\limits_{n=1}^{\infty} E(X_n^2)/n^2 < \infty$. Then $S_n/n \to 0$ a.s. as $n \to \infty$.

(ii) Kolmogorov's (1929) strong law for i.i.d. r.vs. (or special case of Birkhoff's Ergodic theorem (1932)). If

$$X_1, \ldots, X_n \text{ are i.i.d. r.vs. and } S_n/n = \frac{1}{n} \sum_{i=1}^{n} X_i,$$

then $S_n/n \to c < \infty$ a.s. iff $E(|X_1|) < \infty$ and $c = E(X_1)$. The second proof of Kolmogorov's convergence theorem and the proof of Kolmogorov's strong law of large numbers for independent r.vs. depend on an extremely useful tool in probability theory known as maximal inequality (Kolmogorov's inequality). Therefore, we state and prove Kolmogorov inequality first.

Theorem 8.3 (*Kolmogorov's (maximal) inequality*). Let $X_1, \ldots, X_n, \ldots$ be independent random variables and $E(X_i^2) < \infty$, $i \geq 1$. If $S_n = \sum\limits_{i=1}^{n} X_i$ and $\varepsilon > 0$, then

(a) $\varepsilon^2 P(\max\limits_{1 \leq k \leq n} |S_k - ES_k| \geq \varepsilon] \leq \sum\limits_{k=1}^{n} \sigma_k^2$ and if moreover $|X_k| \leq c < \infty$ a.s.,

then

(b) $1 - \dfrac{(\varepsilon + 2c)^2}{\sum\limits_{k=1}^{n} \sigma_k^2} \leq P(\max\limits_{1 \leq k \leq n} |S_k - ES_k| \geq \varepsilon)$

Proof.   We assume $E X_k = 0$, $k \geq 1$. Define a r.v. (stopping time) t by

$$t = \begin{cases} 1 \text{st } k, n \geq k \geq 1 \text{ such that } S_k^2 \geq \varepsilon^2 \text{ if there is such a } k. \\ n + 1 \text{ otherwise} \end{cases}$$

Then

$$[\max\limits_{k \leq n} |S_k| \geq \varepsilon] = [t \leq n] \tag{1}$$

and $[t = k] \in \mathcal{B}(X_1, \ldots, X_k)$. Hence

$$\int_{[t=k]} S_k (S_n - S_k)\, dP = E[S_k I_{[t=k]}(S_n - S_k)]$$

$$(\text{since } (S_n - S_k) \in \mathcal{B}(X_{k+1}, \ldots, X_n)).$$

$$= E[S_k I_{[t=k]}] E(S_n - S_k)$$

$$= 0 \ (\text{since } E(S_n - S_k) = \sum\limits_{i=k+1}^{n} E(X_i) = 0)$$

Therefore,

$$\int_{[t=k]} S_n^2\, dP = \int_{[t=k]} (S_k + (S_n - S_k))^2\, dP = \int_{[t=k]} \{S_k^2 + (S_n - S_k)^2$$

$$+ 2(S_n - S_k) S_k\}\, dP \geq \int_{[t=k]} S_k^2\, dP \geq \varepsilon^2 P(t = k) \tag{2}$$

Therefore,

$$\varepsilon^2 P[t \leq n] = \varepsilon^2 \sum_1^n P[t=k] \leq \sum_{k=1}^n \int_{[t=k]} S_n^2 \, dP$$

$$= \int_{[t=k]} S_n^2 \, dP \leq \int S_n^2 \, dP = E(S_n^2) \tag{3}$$

But

$$E\, S_n^2 = \sum_{k=1}^n EX_k^2 \ (\text{since } X_1, \ldots, X_n \text{ are independent})$$

$$= \sum_{k=1}^n \sigma_k^2 \ (\text{since } EX_k = 0 \text{ for all } k \geq 1)$$

So from (3) and (1), $\varepsilon^2 \ P[\max_{1 \leq k \leq n} S_k \geq \varepsilon] \leq \sum_{k=1}^n \sigma_k^2$.

To prove the lower bound of Kolmogorov's inequality, let $f_k = I_{[t > k]}$. Then $f_k \cdot S_k$ and X_{k+1} are independent for $k = 0, 1, \ldots, n - 1$. Now $[t > k] = [t \geq k]^c \in \mathscr{B}$ $(X_1, \ldots, X_k)$, since $[t \leq k] \in \mathscr{B}(X_1, \ldots, X_k)$. Therefore, $E(f_k S_k X_{k+1}) = E(f_k S_k)$ $E(X_{k+1}) = 0$ (since $EX_{k+1} = 0$). Now

$$E(S_k^2 f_{k-1}) = E(S_k^2 f_{k-1}^2) = E(S_{k-1} f_{k-1} + X_k f_{k-1})^2$$

$$= E(S_{k-1}^2 f_{k-1}) + E(X_k^2 f_{k-1}) \ (\text{since } E(S_{k-1} f_{k-1}^2 X_k) = 0)$$

$$= E(S_{k-1}^2 f_{k-1}) + E(X_k^2)E(f_{k-1}) = E(S_{k-1}^2 f_{k-1}) + E(X_k^2)P(t > k - 1) \tag{4}$$

Again,

$$E(S_k^2 f_{k-1}) = E(S_k^2 f_k) + E(S_k^2 I_{[t=k]}) \tag{5}$$

$$(\text{since } f_{k-1} = f_k + I_{[t=k]})$$

From (5) and (4),

$$E(S_{k-1}^2 f_{k-1}) + E(X_k^2)P(t > k - 1) = E\,(S_k^2 f_{k-1})$$

$$= E(S_k^2 f_k) + E(S_k^2 I_{[t=k]}).$$

Since $|X_k| \leq c$ for all k and $|X_k - EX_k| \leq 2c$,

$$E(S_{k-1}^2 f_{k-1}) + E(X_k^2)P(t \geq k) \leq E(S_k^2 f_k) + (\varepsilon + 2c)^2 P(t = k) \tag{6}$$

Summing over (6) for $k = 1$ to n and after cancellation, we get

$$\sum_{k=1}^n E(X_k^2)P(t \geq k) \leq E(S_n^2 f_n) + (\varepsilon + 2c)^2 \, P(t \leq n).$$

Now $S_n^2 f_n^2 \leq \varepsilon^2$ (by definition of t) and $P(t > n) < P(t \geq k)$ if $k \leq n$ imply

$$\sum_{k=1}^n E(X_k^2)P(t \geq n) \leq \varepsilon^2 E(f_n) + (\varepsilon + 2c)^2 \, P(t \leq n)$$

or

$$\sum_{k=1}^{n} E(X_k^2) P(t > n) \le \varepsilon^2 P(t > n) + (\varepsilon + 2c)^2 P(t \le n)$$

$$\le (\varepsilon + 2c)^2 P(t \le n) + (\varepsilon + 2c)^2 P(t > n)$$

$$= (\varepsilon + 2c)^2$$

Hence,

$$1 - P(t \le n) \le (\varepsilon + 2c)^2 / \sum_{k=1}^{n} E(X_k^2) = \frac{(\varepsilon + 2c)^2}{\sum_{k=1}^{n} \sigma_k^2}$$

$$\Rightarrow P(t \le n) \ge 1 - (\varepsilon + 2c)^2 / \sum_{1}^{n} \sigma_k^2 .$$

Second proof of Kolmogorov's convergence theorem (for independent r.vs.). Assume $E(X_k) = 0$, $k \ge 1$. For $\varepsilon > 0$, by Kolmogorov's inequality (a)

$$\varepsilon^2 P[\max_{n \le k \le N} | \sum_{m=n}^{k} X_m | \ge \varepsilon] \le \sum_{m=n}^{N} \sigma_m^2 \to 0 \text{ as } N \to \infty.$$

Hence $(S_k - S_n) \to 0$ a.s. as $n, k \to \infty$ and hence S_n converges a.s. Here we have used the fact that

$$[\max_{k \le n} |S_k| \ge \varepsilon] = \bigcup_{k \le n} [|S_k| \ge \varepsilon] \text{ and } S_n \to 0 \text{ a.s. iff } \lim_{n \to \infty}$$

$$P[\max_{k \le n} |S_k| \ge \varepsilon] = 0.$$

Let us state a lemma from real analysis which is indispensable in probability theory to prove strong law of large numbers and many other important results.

Lemma 8.1 (*Kronecker's lemma*).   For sequences $\{a_n\}$ and $\{b_n\}$ of real numbers and $\sum_{1}^{\infty} a_n$ convergent and $b_n \uparrow$, $1/b_n \sum_{k=1}^{n} b_k a_k \to 0$ as $n \to \infty$.

Proof.   Since $\sum_{1}^{\infty} a_n$ converges, $S_n = \sum_{1}^{n} a_k \to S$(say)

$$\frac{1}{b_n} \sum_{k=1}^{n} b_k a_k = \frac{1}{b_n} \sum_{k=1}^{n} b_k (S_k - S_{k-1}) = \frac{1}{b_n} (\sum_{1}^{n} b_k S_k - \sum_{1}^{n} b_k S_{k-1})$$

$$= \frac{1}{b_n} (\sum_{1}^{n} b_k S_k - \sum_{1}^{n-1} (b_{k+1} S_k)) \text{ (since } S_0 = 0)$$

$$= \frac{1}{b_n} (b_n S_n + \sum_{1}^{n-1} (b_k - b_{k+1}) S_k) = S_n + \frac{1}{b_n} \sum_{1}^{n-1} (b_k - b_{k+1}) S_k \to S - S = 0,$$

Since

$$\frac{1}{b_n}\sum_1^{n-1}(b_k-b_{k+1})(S_k-S)+\frac{S}{b_n}\sum_1^{n-1}(b_k-b_{k+1})=\frac{1}{b_n}\sum_1^{n-1}(b_k-b_{k+1})S_k.$$

Now,

$$\frac{S}{b_n}\sum_1^{n-1}(b_k-b_{k+1})=\frac{S}{b_n}(b_1-b_n)\rightarrow -S\text{ as }b_n\uparrow\infty$$

and

$$\frac{1}{b_n}\sum_1^{n-1}(b_k-b_{k-1})(S_k-S)\rightarrow 0\text{ as }n\rightarrow\infty$$

Since

$$\left|\frac{1}{b_n}\sum_1^{n-1}(b_k-b_{k+1})(S_k-S)\right|\leq\left|\frac{1}{b_n}\right|\sum_{k=1}^{n_0}|(b_k-b_{k+1})(S_k-S)|$$

$$+\frac{1}{b_n}\left|\sum_{n_0-1}^{n-1}(b_k-b_{k+1})(S_k-S)\right|\text{ for }n>n_0$$

$$\leq\varepsilon+\frac{b_{n_0+1}-b_n}{b_n}\varepsilon\text{ if }n>n_0$$

$$\leq 2\varepsilon\text{ if }n>n_0$$

Proof of Kolmogorov's SLLN for independent r.vs. Put

$$b_k=k,\ a_k=\frac{X_k}{k}$$

in Kronecker's lemma. Then by Kolmogorov's convergence theorem $\sum_{k=1}^{\infty}a_k$ converges almost surely.

if

$$\sum_{k=1}^{\infty}\operatorname{var}(a_k)=\sum_{k=1}^{\infty}\sigma_k^2/k^2<\infty.$$

We shall state and prove a moment inequality which will be needed to prove strong law of large numbers and later on will be useful on many occasions in this book.

Lemma 8.2 (*Moments lemma (Loéve)*). Let X be a r.v. and $q(t)=P[|X|>t]$ $=1-F(t)=\overline{F}(t)$. For every $r>0,\ x>0$ we have

$$x^r\sum_{n=1}^{\infty}q(n^{1/r}x)\leq E|X|^r\leq x^r+x^r\sum_{n=1}^{\infty}q(n^{1/r}x)$$

Proof.

$$E|X|^r=\int_0^{\infty}t^r\,dP[|X|\leq t]=-\int_0^{\infty}t^r\,dq(t)$$

$$=-\sum_{n=1}^{\infty}\int_{(n-1)^{1/r}.x}^{n^{1/r}x}t^r\,dq(t),\ x>0$$

Now,
$$-\int_{(n-1)^{1/r}\cdot x}^{n^{1/r}x} t^r\, dq(t) \le nx^r[q\{(n-1)^{1/r}x\} - q(n^{1/r}x)]$$

(by mean value theorem of integral)

and
$$-\int_{(n-1)^{1/r}\cdot x}^{n^{1/r}x} t^r\, dq(t) \ge (n-1)x^r[q((n-1)^{1/r}x) - q(n^{1/r}x)]$$

(note that $0 \le q(n) \downarrow$).

If $E\,|X|^r = \infty$, the proof is obvious, and if $E\,|X|^r < \infty$ then $x^r\, N\, q(N^{1/r}x) \to 0$ as $N \to \infty$ (since $E\,|X|^r < \infty \Rightarrow x^r\, NP\,[|X| \ge N_x{}^r] \to 0$ as $N \to \infty$ (see Prob. 5.23).

In fact
$$\infty > E\,|X|^r \ge E\,|X|^r\, I_{[|X|>xN^{1/r}]} \ge Nx^r P[|X| > N^{1/r}x]$$

$$= Nx^r q[N^{1/r}x]$$

If $E|X|^r < \infty$, by absolute continuity of integral $Nx^r q(N^{1/r}x) \to 0$ as $N \to \infty$. On the other hand,

$$E|X|^r \ge \sum_{n=1}^{N} (n-1)x^r[q((n-1)^{1/r}x) - q(n^{1/r}x)]$$

$$= \sum_{n=1}^{N} x^r[q(n^{1/r}x)] - (N-1)x^r q(N^{1/r}x).$$

Since $Nq(N^{1/r}x) \to 0$, the right hand side of the last inequality tends to

$$\sum_{n=1}^{\infty} x^r q(n^{1/r}x).$$

Now if $\sum_{1}^{\infty} q(n^{1/r}x) < \infty$, then $nq(n^{1/r}x) \to 0$ and

$$E|X|^r \le \lim_{N\to\infty} \sum_{n=1}^{N} nx^r\,[q((n-1)^{1/r}x) - q(n^{1/r}x)]$$

$$= \lim_{N\to\infty} x^r[1 + \sum_{1}^{N-1} q(n^{1/r}x) - Nq(N^{1/r}x)] = x^r[1 + \sum_{1}^{\infty} q(n^{1/r}x)].$$

Hence the proof is complete.

Theorem 8.4 (*Kolmogorov's strong law of large numbers for i.i.d. r.vs.*).   Let $\{X_n\}$ be a sequence of i.i.d. r.vs. Then

$$S_n/n \to C < \infty \text{ a.s. iff } E\,|X_1| < \infty \text{ and then } C = E(X_1).$$

Proof.   For the only if part let $A_n = [|X_n| \ge n]$; then

$$\sum_1^\infty P(A_n) \le E(|\,X_1\,|) \le 1 + \sum_1^\infty P(A_n) \tag{1}$$

(by moments lemma, where $x = 1$, $r = 1$).

Now $$P(A_n) = P(|\,X_n\,| \ge n) = P(|\,X_1\,| \ge n).$$

If $\quad S_n/n \xrightarrow{\text{a.s.}} C < \infty$, then $\dfrac{X_n}{n} = \dfrac{S_n}{n} - \dfrac{n-1}{n}\dfrac{S_{n-1}}{n-1} \to C - C = 0$ a.s.

Hence $P[|X_1/n| > 1/2 \text{ i.o.}] = 0$. By Borel 0–1 law $\displaystyle\sum_{n=1}^\infty P\left[\,|\,X_n\,| > \dfrac{n}{2}\right] < \infty$

i.e. $$\sum_1^\infty P\left(|\,X_1\,| > \dfrac{n}{2}\right) < \infty \text{ (r.vs. are i.i.d.)}$$

$$\Rightarrow \infty > \sum_n P\left[\,|\,X_1\,| > \dfrac{n}{2}\right] \ge \sum_n P(A_n) \Rightarrow \sum_n P(A_n) < \infty \tag{2}$$

So, from (1) $E\,(|\,X_1\,|) < \infty$.

Conversely, let $E\,|\,X_1\,| < \infty$ and $C = EX_1$. Define $X_k^* = X_k 1_{[|X_k| \le k]}$, $k = 1, 2, 3, \ldots$

and $S_n^* = X_1^* + X_2^* + X_3^* + \ldots + X_n^*$. Then X_k^*, $k = 1, 2, \ldots, n$ are independent

and $|\,X_k^*\,| \le k$. Now,

$$\sum_{k=1}^\infty P[X_k \ne X_k^*] = \sum_1^\infty P[|\,X_k\,| > k] \le \sum_1^\infty P(A_k) < \infty \text{ (by moments lemma from (1))}$$

Therefore $P[X_k \ne X_k^* \text{ i.o.}] = 0$ (by Borel–Cantelli lemma).

Hence S_n/n and S_n^*/n tends to the same limit a.s. if they converge at all, i.e.

$$(S_n - S_n^*)/n \to 0 \text{ a.s. as } n \to \infty.$$

So it is enough to prove that $S_n^*/n \to E(X_1) < \infty$. Now X_n^* are independent but may not necessarily be identically distributed. We shall show that

$\displaystyle\sum_{n=1}^\infty \sigma^2(X_n^*)/n^2 < \infty$ and that will imply $(S_n^*/n - E(S_n^*/n))$ converges to zero almost surely.

$$E(X_n^*) = EX_n I_{[|X_n| \le n]} = EX_1 I_{[|X_1| \le n]} \to E(X_1) \text{ a.s.}$$

Therefore,

$$E(S_n^*/n) \to EX_1, \quad \sum_1^\infty \sigma^2\left(\frac{X_n^*}{n}\right) \le \sum_1^\infty E\, X_n^{*2}/n^2$$

$$= \sum_{n=1}^\infty \frac{1}{n^2} \int_{[|X_n|\le n]} X_n^2 dP = \sum_1^\infty \frac{1}{n^2} \sum_{k=1}^n \int_{[k-1<|X_n|\le k]} X_n^2 dP$$

$$= \sum_1^\infty \frac{1}{n^2} \sum_{k=1}^n \int_{[k-1<|X_1|\le k]} X_1^2 dP = \sum_{k=1}^\infty \sum_{n=k}^\infty \frac{1}{n^2} \int_{[k-1<|X_1|\le k]} X_1^2 dP$$

$$\le 2 \sum_{k=1}^\infty \frac{1}{k} k^2 \, P[k-1<|X_1|\le k]$$

$$= 2 \sum_{k=1}^\infty kP[k-1<|X_1|\le k] = 2 \sum_{k=1}^\infty (k-1)P[k-1<|X_1|\le k] + 2$$

$$\le 2 \sum_1^\infty \int_{[k-1<|X_1|<k]} X_1 dP + 2 = 2(E|X_1|+1) < \infty$$

8.4   UNIFORM INTEGRABILITY AND MEAN CONVERGENCE THEOREM

Let $\{X_n,\ n \ge 1\}$ and X are r.vs. on a probability space $(\Omega,\ \mathscr{F},\ P)$.

Definition 8.3   A sequence of r.vs. $\{X_n,\ n \ge 1\}$ is called *uniformly integrable* (u.i.) if for every $\varepsilon > 0$ there exists a $\delta > 0$ such that

$$\sup_n \int_A |X_n|\, dP < \varepsilon \tag{1}$$

whenever

$$P(A) < \delta \text{ and } \sup_{n\ge 1} E|X_n| \le C < \infty \tag{2}$$

are satisfied. In this connection it is worth mentioning that for every $\varepsilon > 0$, $\int_A |X|\, dP < \varepsilon$ holds whenever for $A \in \mathscr{F}$ there exists a $\delta > 0$ such that $P(A) < \delta$ and $E|X| < \infty$ (see Problem 5.4 at the end of Chapter 5).

From real analysis we know that if $f(x)$ is integrable on $(X,\ \mathscr{B},\ \mu)$, then $\int_A f(x)\, d_\mu(x) < \varepsilon$ if $\mu(A) < \delta$, which is known as *absolute continuity of integral*. So uniform integrability of r.vs. in a sense is the uniform absolute continuity.

Theorem 8.5 (*Equivalence relation of uniform integrability (u.i.)*).   Let $\{X_n\}$ be a sequence of r.vs. Then $\{X_n\}$ is uniformly integrable iff

$$\lim_{b\to\infty} \int_{[|X_n|\ge b]} |X_n|\, dP = 0 \text{ uniformly in } n \tag{3}$$

Proof. Let $\{X_n\}$ is u.i. Then $E\,|\,X_n\,| \le C < \infty$ for every n. Hence $P\,[|\,X_n\,| \ge b] \le \dfrac{1}{b}\,E(|\,X_n\,|)$ (by Markov's inequality)

$$\le \frac{C}{b} \to 0 \text{ as } b \to \infty \text{ uniformly in } n.$$

Therefore, from (1)

$$\sup_{n \ge 1} \int_{[|X_n| > b]} |\,X_n\,|\, dP \le \varepsilon \text{ if } b \text{ is large enough for a given } \varepsilon > 0.$$

So, (1) and (2) $\Rightarrow$ (3). Conversely, for $\varepsilon > 0$, let

$$\int_{[|X_n| > b]} X_n\, dP \le \varepsilon \text{ for some large } b > 0 \text{ and for every } n \ge 1$$

Then

$$E\,|\,X_n\,| = \int_{[|X_n| < b]} |\,X_n\,|\, dP + \int_{[|X_n| \ge b]} |\,X_n\,|\, dP < b + \varepsilon \qquad \text{(from (3))}$$

$$< \infty \text{ for all } n \ge 1.$$

Put $\delta = \varepsilon/b$. Then for $A \in \mathcal{F}$ with $P(A) < \delta$ (from (3) and $A[|\,X_n\,| \ge b] \subset A$)

$$\int_A |\,X_n\,|\, dP = \int_{A[|X_n| < b]} |\,X_n\,|\, dP + \int_{A[|X_n| \ge b]} |\,X_n\,|\, dP < bP(A)$$

$$+ \int_{[|X_n| \ge b]} |\,X_n\,|\, dP < b \cdot \frac{\varepsilon}{b} + \varepsilon = 2\varepsilon \text{ uniformly in } n.$$

So (1) holds.

Theorem 8.6 (*Mean convergence theorem*). Let $\{X_n\}$ be a sequence of r.vs. (integrable) and $X_n \overset{P}{\to} X$. Then $\{X_n\}$ converges in mean iff $\{X_n\}$ is uniformly integrable.

Corollary 8.1 (*Lebesgue dominated convergence theorem*). Let $\{X_n\}$ be a sequence of r.vs. $X_n \overset{P}{\to} X$ and $E(\sup_{n \ge 1} |\,X_n\,|) < \infty$. Then $E\,|\,X_n - X\,| \to 0$ as $n \to \infty$.

Proof. Note that $X_n \overset{P}{\to} X$, $|\,X_n\,| \le Y$ and Y is integrable, where $Y = \sup_{n} |\,X_n\,|$ $\Rightarrow EX_n \to EX$ as $n \to \infty$. Also note that $|\,EX_n - EX\,| \le E(|\,X_n - X\,|) \to 0$. Since $E\,(\sup |\,X_n\,|) < \infty$,

$$\int_{[|X_n| \ge b]} |\,X_n\,|\, dP \le \int_{[\sup|X_n| \ge b]} \sup_{n} |\,X_n\,|\, dP \to 0 \text{ as } b \to \infty$$

$$\text{(by absolute continuity of integral).}$$

Hence by mean convergence theorem the result follows.

Lemma 8.3 Let $\{X_n\}$ be a sequence of r.vs. which converges in mean (L_1) to ar.v. X. Then $E\,|\,X\,| < \infty$ (i.e. X is integrable).

Proof.　$E|X| \le E|X_n| + E|X_n - X|$ (Triangle inequality).

Proof of mean convergence theorem.　First note that the limiting r.v. in our convergence is unique. If possible let $X_n \to X'$ in L_1, then $X_n \overset{P}{\to} X'$ also. But $X_n \overset{P}{\to} X$ and hence $X = X'$ a.s. Now $E|X_n - X| \to 0$ as $n \to \infty$ and $E|X| < \infty$. We have to show that $X_n \xrightarrow{\;L_1\;} X \Rightarrow \{X_n\}$ is u.i. Now, $\sup E|X_n| \le \sup E|X_n - X| + E|X| < \infty$.

Let $\varepsilon > 0$. For $A \in \mathscr{F}$, $P(A) < \delta$ and $n \ge 0$,

$$\int_A |X_n|\, dP \le \int_A |X_n - X|\, dP + \int_A |X|\, dP \quad \text{(Triangle inequality)}$$

$$\le \varepsilon + \int_A |X|\, dP \le 2\varepsilon = \varepsilon^*$$

So we have, in other words $\int_A |X_n|\, dP < \varepsilon^*(> 0)$ if $n \ge n_0$ and $P(A) < \delta$. Since $E|X_n| < \infty$, $\int_A |X_n|\, dP < \varepsilon_n$ for a fixed $n \ge 1$ if $P(A) < \delta$. Let $\varepsilon_0 = \max \{\varepsilon_1, \ldots, \varepsilon_{n-1}, \varepsilon^*\}$. Then for all $n \ge 1 \int_A |X_n|\, dP < \varepsilon_0$ if $A \in \mathscr{F}$ and $P(A) < \delta$. Conversely, let $\{X_n\}$ is u.i. Then $E|X_n| \le C < \infty$. Since $X_n \overset{P}{\to} X$, there exists a subsequence $\{X_{nk}\}$ such that $|X_{nk}| \to |X|$ a.s. By Fatou's lemma $E|X| = E[\liminf_{k \to \infty} |X_{n_k}|] \le \liminf_{k \to \infty} E|X_{n_k}| \le C < \infty$.

For $\varepsilon > 0$, choose $\delta > 0$, $A \in f$, and $P(A) < \delta$ which will imply $\int_A |X|\, dP < \varepsilon$ and $\sup_n \int_A |X_n|\, dP < \varepsilon$. Let N be large enough such that $P[|X_n - X| > \varepsilon] < \delta$ for $n > N$ (since $X_n \overset{P}{\to} X$). Hence

$$E(|X_n - X|) = \int_{[|X_n - X| \le \varepsilon]} |X_n - X|\, dP + \int_{[|X_n - X| > \varepsilon]} |X_n - X|\, dP$$

$$\le \varepsilon + \int_{[|X_n - X| > \varepsilon]} |X_n|\, dP + \int_{[|X_n - X| > \varepsilon]} |X|\, dP < \varepsilon + \varepsilon + \varepsilon$$

$$= 3\varepsilon \text{ if } n > N.$$

Therefore,

$$X_n \overset{L_1}{\to} X \text{ (mean)}.$$

8.5　MEAN ERGODIC THEOREM

Theorem 8.7 (*Mean Ergodic theorem for i.i.d. r.vs.*).　Let $X_1, X_2, \ldots$ are i.i.d. r.vs. with $E|X_1| = C < \infty$. Then

$$E|S_n/n - C| \to 0 \text{ as } n \to \infty \text{ i.e. } S_n/n \overset{L_1}{\to} C.$$

Proof. By Kolmogorov's SLLN for i.i.d. r.vs. $S_n/n \to C$ a.s. Since $C = E(X_1) < \infty$. We only need to prove that S_n/n is u.i. (by mean convergence theorem)

$$E\,|\,S_n/n\,| \le \sum_1^n E(|\,X_i\,|)/n = E(|\,X_1\,|) < \infty \tag{1}$$

and hence the first condition of u.i. is satisfied. For $\varepsilon > 0$, choose N large enough such that

$$\int_{[|X_1|>N]} |X_1|\,dP < \varepsilon \text{ (since } E(|\,X_1\,|) < \infty \text{ and from absolute continuity) which}$$

implies $\displaystyle\int_{[|X_n|>N]} |X_n|\,dP < \varepsilon$ (because X_n's are i.i.d.)

Then for $A \in \mathcal{F}$ and $P(A) < \delta$ put $\delta = \varepsilon/N$.

$$\int_A |X_n|\,dP = \int_{A[|X_n|>N]} |X_n|\,dP + \int_{A \cap [|X_n| \le N]} |X_n|\,dP \le \varepsilon + NP(A) < \varepsilon + N.$$

$\varepsilon/N = 2\varepsilon$ for every $n \ge 1$.

Hence

$$\int_A |S_n/n|\,dP \le \int_A [|\,X_1\,| + \ldots + |\,X_n\,|]/n\,dP = \int_A |\,X_n\,|\,dP \le 2\varepsilon \tag{2}$$

for every $n \ge 1$, i.e. the second condition of u.i. is satisfied. Therefore $\{S_n/n\}$ is u.i. and the proof of the theorem is complete.

8.6 THE GLIVENKO–CANTELLI THEOREM

Often in the statistical literature one reads the statement "a random sample describes the population". What does this mean in crisp measure-theoretic terms? The problem is to reconstruct the so-called population or distribution function from the observable random variables $\{X_n\}$. The theorem which says that this can be done and in a uniform manner is the Glivenko–Cantelli theorem.

Definition 8.4 The nth empirical distribution function $F_n(x)$ is defined to be $1/n$ times the number of r.vs. among $X_1, \ldots, X_n$ which are $\le x$ or $F_n(x)$ is the proportion among the first n r.vs. which are $\le x$.

Theorem 8.8 (*The fundamental theorem of statistics (or Glivenko–Cantelli theorem*)).

$$P[\sup_{-\infty < x < \infty} |\,F_n(x) - F(x)\,| \to 0] = 1$$

Note that this implies weak convergence of $F_n(x)$ to $F(x)$.

Lemma 8.4 For every real x, $F_n(x) \to F(x)$ a.s. and $F_n(x-) \to F(x-)$ a.s. as $n \to \infty$.

Proof. Now $F(x-) = P[X < x]$, and $F(x) = P(X \le x)$. Also,

$$F_n(x) = \frac{1}{n} \sum_{k=1}^{n} I_{[X_k \le x]} \text{ and } F_n(x-) \frac{1}{n} \sum_{k=1}^{n} I_{[X < x]}.$$

Now

$$Y_k = I_{[X_k \le x]}$$

are i.i.d. r.vs. and $E(Y_k) = F(x)$. They $Y_k \sim B\,(1, p)$, $p = F(x)$. So by Kolmogorov's SLLN for i.i.d. r.vs. the result follows.

Let $D_n = \sup_x |\, F_n(x) - F(x)\,|$. Then the first question arises: is it a r.v.?

For each x, $F_n(x) = F_n(x, \omega)$, as a function of ω, is a r.v. By right continuity the supremum above is unchanged if x is restricted to the rationals (countable), and therefore, D_n is measurable. Obviously D_n is bounded by 2. Hence D_n is a r.v.

Proof of Glivenko–Cantelli theorem. Let r be any positive integer ≥ 2. For $k = 1, 2, \ldots, r - 1$ define $x_{r,k} = \min\,\{x : F(x) \ge k/r\}$—the k/rth population quantile (actually $F(x-) < k/r < F\,(x)$). The $-\infty = x_{r,0} < x_{r,1} < x_{r,2} < \ldots - < x_{r,r} = \infty$. We have to consider only those intervals $[x_{r,k}, x_{r,k+1})$ which are non-empty. If $x \in [x_{r,k}, x_{r,k+1})$ then

$$F_n(x) - F(x) \le F_n(x_{r,k+1}^-) - F(x_{r,k})$$

$$= F_n(x_{r,k+1}^-) - F(x_{r,k+1}^-) + (F(x_{r,k+1}^-) - F(x_{r,k}))$$

(note that $F(x_{r,k+1}^-) - F(x_{r,k}) \le \dfrac{k+1}{r} - \dfrac{k}{r} = \dfrac{1}{r}, F(x_{r,1}^-) \le \dfrac{1}{r}$, and $F(x_{r,r}) \ge 1 - \dfrac{1}{r}$

$$\le F_n(x_{r,k+1}^-) - F(x_{r,k+1}^-) + \frac{1}{r} \text{ for almost all } \omega,\ 1 \le k \le r - 1 \tag{1}$$

and

$$F_n(x) - F(x) \ge F_n(x_{r,k}) - F(x_{r,k+1}^-) = F_n(x_{r,k}) - F(x_{r,k}) - (F(x_{r,k+1}^-) - F(x_{r,k}))$$

$$\ge F_n(x_{r,k}) - F(x_{r,k}) - \frac{1}{r} \text{ for almost all } \omega,\ 1 \le k \le r - 1 \tag{2}$$

Lemma 8.4 says that for each x there is a set A_x with $P(A_x) = 0$ such that $\lim F_n(x, \omega) = F(x)$ for $x \in A_x$ and $\lim F_n(x-, \omega) = F(x-)$ except on a set B_x with $P(B_x) = 0$.

Similar arguments hold for $x < x_{r,1}$ and $x \ge x_{r,r-1}$. Now Glivenko–Cantelli theorem says that convergence holds for ω outside some set A of probability 0, where A does not depend on x, but as there are uncountably many of the sets A_x, it is conceivable *a priori* that their union might necessarily have positive measure. From (1) and (2), for almost all real x,

$$0 \le |F_n(x) - F(x)| \le \max_{\substack{1 \le k \le r \\ 1 \le j \le r}} \left\{ |\, F_n(x_{r,k}) - F(x_{r,k})\,|, |\, F_n(x_{r,j}^-) - F(x_{r,j}^-)\,| + \frac{1}{r} \right\}.$$

Let
$$D_{r,n}(\omega) = \max_{\substack{1 \le k \le r \\ 1 \le j \le r}} \{\, |F_n(x_{r,k}) - F(x_{r,k})|, |F_n(x_{r,j}^-) - F(x_{r,j}^-)| \,\}$$

and $\quad D_n = D_n(\omega) = \sup_x |F_n(x) - F(x)|$. Hence $D_n(\omega) \le D_{r,n}(\omega) + r^{-1}$.

If ω lies outside the union A of all the countably many $A_{x_r,k}$ and $B_{x_r,k}$ then by Lemma 8.4, $\lim_{n \to \infty} D_{r,n}(\omega) = 0$ for $\omega \in A$ and $P(A) = 0$. Therefore, $\limsup_{n \to \infty} D_n \le \dfrac{1}{r}$ a.s. Now by the arbitrariness and countability of the values of r the result follows by taking limit $r \to \infty$.

PROBLEMS AND COMPLEMENTS

8.1 Let $\{X_n\}$ be a sequence of independent r.vs. and $S_n = \sum\limits_{i=1}^{n} X_i$. Then

(a) convergence of S_n in probability and in law are equivalent.

(b) If $|X_n| \le C$ and $E(X_n) = 0$ for all $n \ge 1$, then convergences of S_n in q.m., in probability, in law and a.s. are all equivalent.

8.2 Let S_n be the number of heads in n tosses of a fair coin. Prove that for every $\delta > 0$,

$$\lim_{n \to \infty} n^{-(1+\delta)/2}\left(S_n - \frac{n}{2}\right) = 0 \quad \text{a.s.}$$

8.3 If the independent r.vs. $Y_n = n^{-1/4}X_n$, $n = 1, 2, \ldots$ are identically distributed, show that SLLN holds for $\{X_n\}$.

8.4 Let $\{X_n\}$ be independent r.vs. with $P(X_n = \pm n) = 1/2$. Show that SLLN does not hold for $\{X_n\}$.

8.5 Let $\{X_n\}$ be a decreasing sequence of positive r.vs. Then show that $X_n \overset{p}{\to} 0$ implies $X_n \to 0$ a.s.

8.6 (Converse of Kolmogorov's convergence theorem). If X_n's are a.s. uniformly bounded and $\sum\limits_{n} (X_n - EX_n)$ converges a.s., then $\sum\limits_{n=1}^{\infty} \sigma_n^2 < \infty$.

8.7 (Kolmogorov–Khintchine's theorem). If $\{X_n\}$ be a sequence of independent r.vs. then $\sum\limits_{n=1}^{\infty} \sigma_n^2 < \infty$ iff $\sum\limits_{n=1}^{\infty} (X_n - EX_n)$ converges in mean square.

NOTE. S_n converges in q.m. iff $\sum\limits_{n=1}^{\infty} \sigma_n^2 < \infty$ is also true even if X_n's are pairwise independent or uncorrelated only.

8.8 Let $\{X_n\}$ be i.i.d. r.vs. with $P(X_n = \pm 1) = 1/2$. Find the probability of the event $A\{\omega : \sum\limits_{n=1}^{\infty} X_n(\omega)/n$ is convergent$\}$.

8.9   (A generalization of Kolmogorov's inequality). If $X_1, \ldots, X_n$ are independent r.vs. with $EX_k = 0$, $E \mid X_k \mid^\alpha < \infty$ for all $k \geq 1$ and some $\alpha \geq 1$, then show

that $P(\max_{1 \leq k \leq n} \mid \sum_{j=1}^{k} x_j \mid \geq \varepsilon) \leq \sum_{k=1}^{n} E|X_k|^\alpha / \varepsilon^\alpha$ for all $\varepsilon > 0$.

NOTE 1.   Kolmogorov's inequality is a special case with $\alpha = 2$.

NOTE 2.   Obviously for any r.vs. $X_1, \ldots, X_n$ with $E \mid X_k \mid < \infty$ for all $k \geq 1$,

$$P\{\max_{1 \leq k \leq n} |S_k| \geq \varepsilon\} \leq \frac{\sum_{k=1}^{n} E \mid X_k \mid}{\varepsilon}, \varepsilon > 0$$

This follows from Markov's inequality and the fact that

$$\{\max_{1 \leq k \leq n} \mid S_k \mid \geq \varepsilon\} \subset \{\sum_{k=1}^{n} \mid X_k \mid \geq \varepsilon\}.$$

8.10   Let $\{X_k, k \geq 1\}$ be i.i.d. r.vs. such that $E|X_1|^r < \infty$ for some $r \geq 1$ and

set $S_n = \sum_{k=1}^{n} X_k$. Then show that

(a) $\left\{\left|\dfrac{S_n}{n}\right|^r, n \geq 1\right\}$ is uniformly integrable.

(b) $\dfrac{S_n}{n} \to EX_1$ in L_r as $n \to \infty$.

8.11   (Vitali's theorem): If $\{f_n\}$ is a sequence of non-negative integrable functions

tending to f a.e. and $\mu(E) < \infty$, then $\int_E f_n \, d\mu \to \int_E f \, d\mu$ iff $\{f_n\}$ is u.i.

NOTE.   A sequence $\{f_n\}$ of non-negative integrable functions is uniformly integrable (u.i.) if for each $\varepsilon > 0$ there exists a number k such that

$\int_{[f_n > k]} f_n \, d\mu < \varepsilon$ for all sufficiently large n (compare with the standard

definition for the probability measure P).

8.12   If $f_1, f_2, \ldots$ are non-negative and $g_n = \max \{f_1, \ldots, f_n\}$, show that

$$\int_{[g_n \geq \alpha]} g_n \, d\mu \leq \sum_{k=1}^{n} \int_{[f_k \geq \alpha]} f_k \, d\mu$$

If further $\{f_n\}$ is uniformly integrable and $\mu(\Omega) < \infty$, then show that $\int g_n \, d\mu$

$= o(n)$.

8.13 Let $\{X_n, n \geq 1\}$ be a sequence of i.i.d. r.vs. Let R_n be the relative rank of X_n among $X_1, \ldots, X_n$, i.e. $R_n = \sum_{i=1}^{n} I_{[X_i \geq X_n]}$, i.e. $R_n = r$ iff $X_i > X_n$ for exactly $r - 1$ values of i. Note $R_n = 1$ if $X_n = M_n = \max_{1 \leq i \leq n} X_i$. Let A_n be the event that a record occurs at time n, i.e. $\max_{k < n} X_k < n$. Show that

(a) $\{R_n, n \geq 1\}$ is an independent sequence of r.vs. with $P(R_n = r) = 1/n$, $1 \leq r \leq n$.

(b) $\{A_n, n \geq 1\}$ are independent events with $P(A_n) = 1/n$.

(c) No record stands forever.

(d) If N_n is the time of the first record after time n, then $P(N_n = n + k) = n(n + k - 1)^{-1}(n + k)^{-1}$.

8.14 Let $X_1, \ldots, X_n$ be the first n observations from a continuous d.f. $F(x)$. A_i is a record at time i if $\max_{k < i} X_k < X_i$. Let $\mu[1, n]$ be the number of records in the first n observations. Then show that $\mu[1, n]/\log n \to 1$ a.s.

8.15 (Polya's theorem). Let $\{F_n(x)\}$ be a sequence of d.fs. and $F_n(x) \to F(x)$ pointwise where $F(x)$ is a everywhere continuous d.f. Show that $F_n(x) \to F(x)$ uniformly in $\mathbb{R}$.

NOTE 1. The theorem is true for non-decreasing uniformly bounded functions F_n and F and for F in $\mathbb{R}^k$, $k > 1$ (T.K. Chandra, 1989).

NOTE 2. If $\{F_n(x)\}$ is a sequence of emperical distribution functions, then by SLLN for i.i.d. r.vs. $F_n(x) \to F(x)$ a.s. for all $x \in \mathbb{R}$.

Hence applying Polya's theorem we get a special case of Glivenko–Cantelli's theorem (Also known as Cantelli's theorem (1907)).

8.16 (Chung 1952). Let $\{X_n, n \geq 1\}$ be independent r.vs. with partial sums $S_n = \sum_{i=1}^{n} X_i, n \geq 1$. If $S_n \xrightarrow{p} 0$ and $S_{2^n}/2^n \xrightarrow{a.s.} 0$, show that $S_n/n \xrightarrow{a.s.} 0$.

9

Characteristic Functions

9.1 CHARACTERISTIC FUNCTIONS, MOMENTS AND APPLICATIONS

Let X be a random variable. The function ϕ(or ϕ_x) defined by $\phi(t) = E(e^{itx})$
$= \int_{-\infty}^{\infty} e^{itx} dF(x) = \int_{-\infty}^{\infty} \cos tx\, dF(x) + i \int_{-\infty}^{\infty} \sin\ tx\ dF(x)$ is called the characteristic
function (ch.f.) of X or of the corresponding d.fs. (since e^{itx} is a continuous function
of x, $\int_{-\infty}^{\infty} e^{itx}\ dF(x)$ can be taken in Riemann–Stieltjes sense as the ch.f. of d.f. F).

Theorem 9.1 (*Elementary Properties of* ch.f.).

    (a) $\phi(0) = 1$, (b) $|\phi(t)| \leq 1$, (c) $\phi(-t) = \overline{\phi(t)}$, the complex conjugate of $\phi(t)$.

Proof.

    (a) $\phi(0) = \int_{-\infty}^{\infty} dF(x) = 1 = F(\infty) - F(-\infty)$.

    (b) $\phi(t) = \int_{-\infty}^{\infty} e^{itx} dF(x) \leq \int |e^{itx}|\, dF(x) = \int_{-\infty}^{\infty} dF(x) = 1$

        (note that $|e^{itx}| = \sqrt{(\cos^2 tx + \sin^2 tx)} = 1$).

    (c) $\phi(-t) = \int_{-\infty}^{\infty} e^{-itx} dF(x) = \overline{\phi(t)}$.

Theorem 9.2 Every ch.f. is uniformly continuous on the whole real line.
    Proof.

$$|\phi(t+h) - \phi(t)| \leq \int_{-\infty}^{\infty} |e^{i(t+h)x} - e^{itx}|\, dF(x) = \int_{-\infty}^{\infty} |e^{itx}(e^{ixh} - 1)|\, dF(x)$$

$$\leq \int |(e^{ixh} - 1)|\, dF(x)$$

112

(since $|\,e^{itx}\,|$ is independent of t and $|\,2\sin x\,h/2\,| = |\,e^{ihx} - 1\,| \le 2$)

$$\le 2\int_{-\infty}^{\infty} dF(x) = 2.$$

Note that $|\,e^{ixh} - 1\,| \to 0$ as $h \to 0$ and by dominated convergence theorem for Riemann–Stieltjes integral we have

$$\int_{-\infty}^{\infty} |e^{ihx} - 1|\, dF(x) \to 0 \text{ as } h \to 0.$$

Theorem 9.3 Suppose a_1, a_2, ... are real numbers with $\sum_i a_i = 1$ and $\phi_1(t)$, $\phi_2(t)$, ... are ch.fs. Then $g(t) = \sum_j a_j \phi_j(t)$ is also a ch.f.

Proof. $a_j \ge 0$, $\sum_j a_j = 1 \Rightarrow F(x) = \sum_j a_j F_j(x)$ is also a d.f. if F_j is the d.f. corresponding to ϕ_j. Therefore,

$$\int_{-\infty}^{\infty} e^{itx} dF(x) = \sum_j a_j \int_{-\infty}^{\infty} e^{itx} dF_j(x) = \sum_j a_j \phi_j(t)$$

So $g(t) = \sum_j a_j \phi_j(t)$ is the ch.f. of the d.f. $F(x) = \sum_j a_j F_j(x)$.

NOTE. If F is a d.f. $a \ge 0$ and b is a real number, then

$$G(x) = F\left(\frac{x - a}{b}\right)$$

is the d.f. of the r.v. $a + bx = Y$ with ch.f. $g(t) = e^{ita}\,\phi(bt)$, where $\phi(t)$ is the ch.f. of X. If $b = -1$, $a = 0$ then $g(t) = \phi(-t) = \overline{\phi(t)}$ is the ch.f. of $-X$.

If X and $-X$ have same d.f. say F(or $F(x) = 1 - F(-x - 0)$) then X is called a symmetric r.v. or the d.f. F is called *symmetric* (about origin).

Theorem 9.4 A d.f. is symmetric iff its ch.f. is real and even.

Proof. If F is symmetric, i.e. X and $-X$ have same d.f. and hence have same ch.f. Therefore, $\phi(t) = \overline{\phi}(t)$. Hence $\phi(t)$ is real. Also,

$$\phi(t) = \int_{-\infty}^{\infty} \cos tx\, dF(x) + i \int \sin tx\, dF(x)$$

$$= \int_{-\infty}^{\infty} \cos tx\, dF(x) \text{ (since } \phi \text{ is real)}$$

which is an even function.

Conversely. If $\phi(t)$ is real then $\phi(t) = \overline{\phi}(t) = \phi(-t)$ i.e. if X and $-X$ have the same ch.f. and by the *uniqueness theorem* of ch.f. (to be proved subsequently) have the same d.f., and hence F is symmetric.

Corollary 9.1 If X and Y are two independent symmetric r.vs., then $(X + Y)$ is also symmetric (P. Lévy, 1937).

Proof.   Suppose $\phi_1(t)$ and $\phi_2(t)$ are ch.f. of X and Y respectively. Let $\phi(t)$ be the ch.f. of $Z = X + Y$. Then by convolution theorem $X + Y$ has ch.f. $\phi(t) = \phi_1(t)\phi_2(t)$. Now $\phi_1(t)$ and $\phi_2(t)$ are real and $\phi(t)$ is real. Therefore by previous theorem $Z = X + Y$ is symmetric.

Let $\phi(t)$ be a ch.f. then $\bar{\phi}(t)$ as well as Re $\phi(t)$ are ch.fs.

Since $\bar{\phi}(t) = \phi(-t)$ is the ch.f. of $-X$, Re $\phi(t) = 1/2 \, (\phi(t) + \bar{\phi}(t))$ is the convex combination of two ch.fs. and hence is a ch.f.

9.1.1   Convolution

Convolution of two d.fs. *F and G is defined by* $H(x) = (F * G)(x) = \displaystyle\int_{-\infty}^{\infty} F(x - y)\, dG(y)$. If we have r.vs. X and Y with d.fs. F and G, then convolution can be interpreted as the d.f. of $Z = X + Y$ if X and Y are independent. In discrete case

$$H(x) \sum_{0 < n \le x} \left(\sum_{k=0}^{n} p_k q_{n-k} \right), \text{ where } F(x) = \sum_{0 < k \le x} p_k, \; G(x) = \sum_{0 \le n \le x} g_n, \; p_k = P(X = k), \; g_n$$

$= P(Y = n)$ and $q_j = P(Y > j)$.

NOTES.

1. If F and G are d.fs., so is $H = F * G$, i.e. (a) H is non-decreasing, (b) Right continuous and (c) $H(\infty) = 1$ and $H(-\infty) = 0$. For $x' < x''$, $(F * G)(x'') - (F * G)(x')$ (From the properties of Stieltjes integral) $= \displaystyle\int_{-\infty}^{\infty} [F(x'' - y) - F(x' - y)]\, dG(y) \ge 0$.

2. $(G * F) = (F * G)(x)$ (i.e. commutative)

3. If F, G and H are d.fs., then
    $(F * G) * H = F * (G * H)$ (i.e. associative)

4. If F and G are d.fs. with f as the density of F (absolutely continuous), then $F * G$ is also absolutely continuous and its density is given by

    $$h(x) = \int_{-\infty}^{\infty} f(x - y)\, dG(y).$$

Convolution Theorem 9.5 $\phi_Z(t) = \phi_X(t) \cdot \phi_Y(t)$, where $Z = X + Y$ and X and Y are independent.

Proof.   Consider the integral $\phi_Z(t) = \displaystyle\int_{-\infty}^{\infty} e^{itz}\, dH(z)$ and the partition of the interval $[a, b]$ $(a = z_1^{(n)} < z_2^{(n)}, \ldots, z_{(n+1)}^{(n)} = b)$. By the definition of Stieltjes integral

$$\int_a^b e^{itz}\, dH(z) = \lim_{n \to \infty} \sum_{j=1}^{n} \exp\left(it \, (z_j^{(n)})\right) \{H(z_{j+1}^{(n)}) - H(z_1^{(n)})\}$$

$$= \lim_{n \to \infty} \int \{ \sum_1^n \exp\left(it \, z_j^{(n)} - x)\right) [F(z_{j+1}^{(n)} - x) - F(z_j^{(n)} - x)]\}(e^{itx}\, dG(x))$$

$$= \int_{-\infty}^{\infty} \left[\int_{a-x}^{b-x} e^{ity}\, dF(y) \right] e^{itx}\, dG(x) \text{ (taking limits inside the integral by}$$

dominated convergence theorem)

Now letting $a \to -\infty$ and $b \to \infty$, $\phi_Z(t) = \phi_X(t) \cdot \phi_Y(t)$.

9.1.2 Some Important Characteristic Functions and an Important Theorem

1. *Binomial distribution* denoted by $B(k; n, p)$ has the ch.f.

$$\phi(t) = \sum_{k=0}^{n} e^{itk}\, p^k q^{n-k} \binom{n}{k} = \sum_{k=0}^{n} \binom{n}{k}(pe^{it})^k q^{n-k} = (q + pe^{it})^n$$

2. *Negative binomial distribution* has

$$\phi(t) = \sum_{k=0}^{n} e^{itj}\, p^r \binom{-r}{j}(-q)^j, r > 0, p + q = 1$$

$$= \sum_{j} p^r \binom{-r}{j}(-e^{it}q)^j = \{p(1 - qe^{it})^{-1}\}^r.$$

3. *Normal* with $EX = \mu$, $V(X) = 1$, has

$$\phi(t) = \frac{1}{2\pi\sigma}\int_{-\infty}^{\infty} e^{itx} e^{-(x-\mu)^2/2\sigma^2}\, dx = e^{i\mu t - \sigma^2 t^2/2}.$$

4. *Degenerate distribution function* denoted by $\varepsilon(x - \xi)$. Here X is degenerate at $x = \xi$, i.e. $P(X = \xi) = 1$, ch.f. is $e^{it\xi}$.

5. *Laplace density:* $f(x) = 1/2\, e^{-|x|}$, $-\infty < x < \infty$, $\phi(t) = 1/(1 + t^2)$.

6. *Poisson:* $P(k; \lambda) = e^{-\lambda}\lambda^k/k!$, $k = 0, 1, 2, \ldots$, $\phi(t) = \exp(\lambda(e^{it} - 1))$.

7. *Cauchy density:* $f(x) = \dfrac{\theta}{\pi(\theta^2 + (x - \mu)^2)}$, $-\infty < x < \infty$ has the ch.f.

$\phi(t) = \exp[i\mu t - \theta|t|]$. If $\mu = 0$, $\theta = 1$ we get the standard form of Cauchy distribution and its ch.f. $e^{-|t|}$.

8. *Gamma distribution* (Γ): $f(x) = \dfrac{\theta^\lambda x^{\lambda-1} e^{-\theta x}}{\Gamma\lambda}$, $0 < x < \infty$ has ch.f.

$$\phi(t) = \frac{1}{(1 - it/\theta)^\lambda}.\; \lambda > 0,\, \theta > 0.$$

9. *Rectangular distribution* on $(a, -r, a + r)$:

$$F(x) = \frac{r - |x - a|}{r^2} \text{ if } |x - a| \le r,\; \phi(t) = e^{ita}(\sin tr/tr),\, r > 0$$

10. *Beta distributions:* $f(x) = \dfrac{\Gamma(m + n)}{\Gamma n\, \Gamma m} x^{m-1}(1 - x)^{n-1}$, $0 < x < 1$ has ch.f.

$$\phi(t) = \frac{\Gamma(m + n)}{\Gamma m} \sum_{j=0}^{\infty} \frac{\Gamma m + j}{\Gamma(m + n + j) + \Gamma(j + 1)}(it)^j$$

Theorem 9.6

(a) If the ch.f. of a d.f. ($F(x)$ has a derivative of order k at $t = 0$, then all the moments of $F(x)$ up to order k exists if k is even, respectively up to order $k - 1$ if k is odd.

Conversely, if the moment α_k of order k of a d.f. $F(x)$ exists the ch.f. $\phi(t)$ of $F(x)$ can be differentiated k times and

$$\phi^k(t) = i^k \int_{-\infty}^{\infty} e^{itx} \, dF(x) \text{ and } \alpha_k = i^{-k} \phi_{(0)}^{(k)}.$$

(b) Let $F(x)$ be a d.f. and assume that the nth moment of $F(x)$ exists. Then the characteristic function $\phi(t)$ of $F(x)$ admits the expansion

$$\phi(t) = \sum_{j=0}^{n} \alpha_j \, /(j!) \, (it)^j + o(t^n), \text{ as } t \to 0 \tag{1}$$

$$= 1 + \sum_{j-1}^{n-1} \alpha_j (it)^j /j! + t^n \int_0^1 \frac{(1-u)^{n-1}}{(n-1)!} \phi^{(n)}(tu)du$$

Proof of (a). $\Delta_1^h \phi(t) = \Delta^h \phi(t) = \phi(t + h) - \phi(t - h)$

$$\Delta_{k+1}^h \phi(t) = \Delta^h \Delta_k^h \phi(t) \text{ for } k = 1, 2, 3, \ldots$$

By induction,

$$\Delta_n^h \phi(t) = \sum_{k=0}^{n} (-1)^k \binom{n}{k} \phi(t + (n - 2k(h)))$$

$$\Delta_n^h e^{ixt} (e^{ixh} - e^{-ixh})^n = e^{ixt} (2i \sin xh)^n \tag{2}$$

Now we shall show that the $k = 2n$th central difference quotient of

$$\phi(t) \text{ at } t = 0 \text{ is } \left. \frac{\Delta_{2n}^h \phi(t)}{(2h)^{2n}} \right|_{t=0}$$

(Note that if a function g has ordinary mth derivative, then it has also the mth symmetric derivative and they are equal. But the converse is false). In fact, if $g(x)$ is $m(= 2n)$ times differentiable and $g^{(m)}(0) < \infty$, then

$$g_{(0)}^{(2n)} = \lim_{h \to 0} \left. \frac{\Delta_{2n}^h g(t)}{(2h)^{2n}} \right|_{t=0}$$

If $g(x)$ is differentiable m times and $g^{(m)}(0) < \infty$, then $g(x) = g(0) + g'(0)x + \ldots + x^m/m! \, [g^{(m)}(0) + o(1)]$ as $x \to 0$. Now.

$$\Delta_m^h g(x) = (g^{(m)}(0)/m! + o(1))\Delta_m^h (x^m)$$

$$= (g^{(m)}(0) + o(1))(2h)^m$$

Put $m = 2n$, then $(\Delta_{2n}^h g(0)) = (g^{(2n)}(0) + o(1))(2h)^{2n}$. Hence,

$$g^{(2n)}(0) = \lim_{h \to 0} (2h)^{-2n} \Delta_{2n}^h g(0) \tag{3}$$

Therefore,

$$\phi^{(2n)}(0) = \lim_{h \to 0} (2h)^{-2n} \Delta_{2n}^h \phi(t)\big|_{t=0}$$

Now

$$\Delta_{2n}^h \phi(t) = E(\Delta_{2n}^h e^{itx}) \text{ and by (2),}$$

$$\Delta_{2n}^h \phi(t) = \int_{-\infty}^{\infty} e^{ixt} (2i \sin xh)^{2n} \, dF(X).$$

Hence

$$\left| \frac{\Delta_{2n}^h \phi(t)}{(2h)^{2n}} \right|_{t=0} = \int_{-\infty}^{\infty} (\sin xh/h)^{2n} \, dF(X).$$

Now,

$$\infty > \phi^{(2n)}(0) \geq \int_{-\infty}^{\infty} \lim_{h \to 0} \inf (\sin xh/xh)^{2n} \, x^{2n} \, dF(X).$$

$$\leq E(X^{2n}) \text{ by Fatou's lemma.}$$

So if $\phi^{(2n)}(0) < \infty$ then $k = 2n$th absolute moment $v_{2n} = \alpha_{2n} < \infty$. To prove all the moments of $F(x)$ up to order k exists if k is even and $\phi^{(2n)}(0) < \infty$, it is enough to prove the following:

Lemma 9.1 $v_{r'} < \infty$ if $v_r < \infty$ for $r' \leq r$.

Proof.   Now $|x|^{r'} \leq 1 + |x|^r$ whenever $r' \leq r$ implies $E|X|^{r'} \leq 1 + E|X|^r$, i.e. $v_{r'} \leq 1 + v_r$. Hence $v_{r'} < \infty$ if $v_r < \infty$ for $r \geq r'$.

If k is odd, $k - 1$ is even and the same arguments works. Conversely, we have given $\int_{-\infty}^{\infty} x^k \, dF(x) < \infty$ we have to prove $\phi(t)$ can be differentiated k times under the integral sign and

$$\phi^{(j)}(t) = i^j \int_{-\infty}^{\infty} x^j e^{itx} \, dF(x) \qquad \text{for } j = 0, 1, 2, \ldots, k \tag{4}$$

We shall prove this by induction. Obviously (4) is true if $j = 0$. Suppose that it is true for $j = k$, i.e.

$$\alpha_k < \infty \Rightarrow \phi^{(k)}(t) = i^k \int_{-\infty}^{\infty} x^k e^{itx} \, dF(x)$$

Now

$$\lim_{h \to 0} \frac{\phi^{(k)}(t+h) - \phi^{(k)}(t)}{h} = \lim_{h \to 0} i^k \int_{-\infty}^{\infty} x^k \frac{\exp[i(t+h)x] - e^{itx}}{h} \, dF(x)$$

$$= \lim_{h \to 0} i^k \int_{-\infty}^{\infty} x^k e^{itx} \frac{(e^{ihx} - 1)}{h} \, dF(x)$$

[Note $\dfrac{e^{ixh}-1}{ih} = \displaystyle\int_0^x e^{iuh}\, du \le x$]

$$= i^{k+1} \int_{-\infty}^{\infty} e^{itx} x^{k+1} dF(x) \quad \text{(by dominated convergence theorem)}$$

$$< \infty \text{ if } \alpha_{k+1} < \infty.$$

i.e.
$$\phi^{(k+1)}(t) < \infty, \phi^{(k+1)}(t) = i^{k+1} \int_{-\infty}^{\infty} x^{k+1} e^{itx} dF(x)$$

and
$$\phi^{(k+1)}(0) = i^k \int_{-\infty}^{\infty} x^{k+1}\, dF(x) \text{ if } \alpha_{k+1} < \infty.$$

Hence by induction on k, $\alpha_{k+1} < \infty$ implies

$$\phi_{(t)}^{(k+1)} = i^{k+1} \int_{-\infty}^{\infty} e^{itx} x^{k+1}\, dF(x)$$

and all derivatives up to order $k + 1$ exist and

$$\phi^{(k)}(0) = i^k \int_{-\infty}^{\infty} x^k\, dF(x).$$

Proof of (b). $\phi^{(n)}(t + h) - \phi^{(n)}(t) = i^n E(X^n(e^{ihX} - 1))$ and since $|\, e^{ihx} - 1\,| \le 2$, once more by dominated convergence theorem $\phi^{(n)}(t + h) \to \phi^{(n)}(t)$ and $\phi(t)$ has a continuous nth derivative and hence $\phi(t)$ has Maclaurin's expansion (Taylor expansion)

$$\phi(t) = \sum_{j=0}^{n} \frac{\phi^{(i)}(0)}{j!}\ t^j + R_n(t), \text{ where the remainder term}$$

$R_n(t)/t^n = 1/n!\ [\phi^{(n)}(\theta t) - \phi^{(n)}(0)]$, $0 < \theta < 1 \to 0$ as $t \to 0$ (by continuity of $\phi^{(n)}(t)$).
Hence

$$\phi(t) = \sum_{j=0}^{n} \phi^{(j)}(0)/j! t^j + o(t^n) \qquad \text{as } t \to 0$$

$$= \sum_{j=0}^{n} \alpha_j (it)^{(j)}/j! + o(t^n) \qquad \text{as } t \to 0$$

The last expansion of $\phi(t)$ just follows from integral representation of the remainder term.

9.2   INVERSION THEOREM, CONTINUITY THEOREM AND THEIR APPLICATIONS

Inversion Theorem 9.6 (first form). Let ϕ and F be respectively the ch.f. and the d.f. of the r.v. X. Let $\alpha < \beta$ be two points of continuity of F. Then

$$F(\beta) - F(\alpha) = \lim_{\sigma \to 0} \frac{1}{2\pi} \int_{-\infty}^{\infty} \phi(t) \exp(-\sigma^2 t^2/2) \frac{e^{-it\beta} - e^{-it\alpha}}{-it} \, dt$$

P. Lévy's inversion theorem.

$$\lim_{T \to \infty} 1/2\pi \int_{-T}^{T} \frac{e^{-it\beta} - e^{it\alpha}}{-it} \phi(t) \, dt = \frac{F(\beta) + F(\beta-)}{2} - \frac{F(\alpha) + F(\alpha-)}{2}$$

$$= F(\beta) - F(\alpha) \quad \text{if } \alpha, \beta \in C(F)$$

Proof of the first form of inversion theorem.   Let

$$F_\sigma = F * \Phi_\sigma, \Phi_\sigma(x) = \int_{-\infty}^{x} 1/\sqrt{2\pi}\sigma \, e^{-t^2/2\sigma^2} \, dt$$

and $Z = X + Y_\sigma$, where Y_σ is $N(0, \sigma^2)$, X and Y_σ are independent.
Now, the ch.f. of Y_σ is $\exp(-t^2\sigma^2/2)$. Define

$$\phi_\sigma(t) = \exp(t^2\sigma^2/2) \cdot \phi(t)$$

which is integrable, since $|\phi(t)| \le 1$. Then,

$$1/2\pi \left(\int_{-\infty}^{\infty} e^{itx} \phi_\sigma(t) dt \right) = 1/2\pi \left(\int_{-\infty}^{\infty} e^{-itx} \exp(-t^2\sigma^2/2 \left[\int e^{itu} dF(u) \right] \right) dt$$

This double integral is absolutely integrable and hence by Fubini's theorem,

$$\int_{-\infty}^{\infty} 1/2\pi \int_{-\infty}^{\infty} \exp(it(u - x)) \exp(-t^2\sigma^2/2) \, dt \, dF(u) = \int_{-\infty}^{\infty} \frac{A(u, x)}{\sqrt{2\pi}\sigma} \, dF(u),$$

where $A(u, x)$ is the ch.f. of $N(0, 1/\sigma^2)$ at $t = u - x$. Therefore,

$$A(u - x) = \exp\{-(u - x)^2/2\sigma^2\}. \text{ Hence}$$

$$1/2\pi \int_{-\infty}^{\infty} e^{itx} \phi_\sigma(t) dt = \int_{-\infty}^{\infty} 1/(\sqrt{2\pi}\sigma) \exp\{-(u - x)^2/2\sigma^2\} \, dF(u)$$

which is the density of the convolution of ϕ_σ with F, i.e. the density of Z. Therefore,

$$1/2\pi \int_{-\infty}^{\infty} e^{-itx} \exp\{-t^2\sigma^2/2\} \phi(t) \, dt = \text{density of } F_\sigma = f_\sigma(x) \text{ (say)}.$$

Integrating both sides with respect to x from α to β and by Fubini's theorem

$$1/2\pi \int_{-\infty}^{\infty} \exp\{-t^2\sigma^2/2\}\phi(t) \left[\int_{\alpha}^{\beta} e^{-itx} dx \right] dt = \int_{\alpha}^{\beta} f_\sigma(x) \, dx = F_\sigma(\beta) - F_\sigma(\alpha).$$

Hence

$$F_\sigma(\beta) - F_\sigma(\alpha) = \frac{1}{2\pi} \int_{-\infty}^{\infty} \exp\{-t^2\sigma^2/2\}\phi(t) \frac{e^{-it\beta} - e^{-it\alpha}}{-it} \, dt$$

The proof will be complete by showing

$$F(\beta) - F(\alpha) = \lim_{\sigma \to 0} \frac{1}{2\pi} \int_{-\infty}^{\infty} \exp\{-t^2\sigma^2/2\}\, \phi(t)\, \frac{e^{-it\beta} - e^{-it\alpha}}{-it}\, dt$$

$$\text{and } \lim_{\sigma \to 0} F_\sigma(x) = F(x)$$

at each x for which F is continuous. Now,

$$F_\sigma(x) = P[X + Y_\sigma \le x] = P[X \le x - Y_\sigma]$$

$$= P[X \le x - Y_\sigma, |Y_\sigma| < \varepsilon] + P[X \le x - Y_\sigma, |Y_\sigma| \ge \varepsilon]$$

$$\le P[X \le x - Y_\sigma, |Y_\sigma| < \varepsilon] + P[|Y_\sigma| \ge \varepsilon]$$

$$\le P[X \le x - Y_\sigma, |Y_\sigma| < \varepsilon] + \varepsilon, \text{ is } \sigma \text{ is small enough}$$

$$\le F(x + \varepsilon) + \varepsilon \text{ (by Markov or Chebychev's inequality)}$$

Also,

$$P[X \le x - Y_\sigma] \ge P[X \le x - Y_\sigma, |Y_\sigma| < \varepsilon] \ge P[X \le x - \varepsilon, |Y_\sigma| < \varepsilon]$$

$$\ge P[X \le x - \varepsilon] - \varepsilon = F(x - \varepsilon) - \varepsilon.$$

Hence,

$$F(x - \varepsilon) - \varepsilon \le F_\sigma(x) \le F(x + \varepsilon) + \varepsilon \text{ if } \sigma \text{ is small}$$

Let $\varepsilon \to 0$. Then

$$F(x -) \le F_\sigma(x) \le F(x +) \text{ if } \sigma \text{ is small.}$$

Therefore,

$$\lim_{\sigma \to 0+} F_\sigma(X) = F(x) \quad \text{if } x \in C(F).$$

Corollary 9.2 If $\phi(t)$ is the ch.f. of the d.f. F and is absolutely integrable, i.e. $\int_{-\infty}^{\infty} |\phi(t)|\, dt < \infty$, then F is absolutely continuous and its density is given by

$$f(x) = 1/2\pi \int_{-\infty}^{\infty} e^{-itx} \phi(t)\, dt \tag{1}$$

Also, $f(x)$ is uniformly continuous.

NOTE. $\phi(t)$ is absolutely integrable is a sufficient condition for the corollary but not a necesary condition. If $\int_{-\infty}^{\infty} |\phi(t)|\, dt < \infty$, then we write $\phi(t) \in L_1$, e.g.

$$\left. \begin{array}{ll} f(x) = e^{-x} & \text{if } x \ge 0 \\ \quad\quad 0 & \text{if } x < 0 \end{array} \right\} \text{ has ch.f. } \phi(t) = 1/(1 - it), i = \sqrt{-1}, t > 0.$$

Note that $\phi(t)$ is not integrable (prove it). But by contour integration formula (1) gives the probability density $f(x)$.

$$\left(\text{note } \int_{-\infty}^{\infty} \frac{\sin x}{x}\, dx = \pi\right)$$

Proof of Corollary 9.2. Since $\phi \in L_1$, we can use the dominated convergence theorem to take the limit $\sigma \to 0+$, under the integral sign and the result is

$$F(\beta) - F(\alpha) = \frac{1}{2\pi} \int_{-\infty}^{\infty} \phi(t) \frac{e^{it\beta} - e^{-it\alpha}}{-it}\, dt \tag{2}$$

Since

$$\left|\phi(t)\frac{e^{it\beta} - e^{-it\alpha}}{-it}\right| = \left|\phi(t)\frac{e^{-it\alpha}(e^{-it\alpha(\beta-\alpha)} - 1)}{-it}\right| \le |\phi(t)|(\beta - \alpha)(\beta > \alpha)$$

Since $f(x)$ defined by (1) is continuous (to be proved later) and $e^{-itx}\phi(t)$ is absolutely integrable, by Fubini's theorem (integrating $f(x)$ with respect to x from α to β and changing the order of integration), we get

$$\int_{\alpha}^{\beta} f(x)\, dx = \frac{1}{2\pi} \int_{-\infty}^{\infty} \phi(t) \frac{e^{-it\beta} - e^{-it\alpha}}{-it}\, dt \tag{3}$$

Therefore,

$$F(\beta) - F(\alpha) = \int_{\alpha}^{\beta} f(x)\, dx \text{ (by (2) and (3))}$$

So, F is absolutely continuous with density $f(x)$ given by equation (1).

To prove f is uniformly continuous, it is necessary that

$$|f(x+h) - f(x)| \le \frac{2}{2\pi} \int_{-A}^{A} |\sin th/2||\phi(t)|\, dt + \frac{1}{\pi} \int_{|t|>A} |\sin th/2||\phi(t)|\, dt$$

$$\le \varepsilon + \varepsilon \quad \text{if } A \ge A_0 \text{ and } h \to 0.$$

The first integral is small due to the fact that sin function is uniformly continuous over a bounded interval and the second integral is small due to the fact that sin function is always bounded by 1 and by absolute integrability of

$$\phi, \int_{|t|>A} |\phi(t)|\, dt < \varepsilon \quad \text{if } A \ge A_0.$$

Proof of inversion theorem in P. Lévy's form (sketch of proof).   Let

$$I(T) = \frac{1}{2\pi} \int_{-T}^{T} \frac{e^{-it\alpha} - e^{-it\beta}}{-it} \phi(t)\, dt$$

$$= \frac{1}{2\pi} \int_{-T}^{T} \frac{e^{-it\alpha} - e^{-it\beta}}{-it} E(e^{itX})\, dt,$$

since

$$\frac{e^{-it\alpha} - e^{-it\beta}}{it} e^{itX}$$

is bounded for all ω and t and by Fubini's theorem,

$$I(T) = \frac{1}{2\pi} E\left(\int_{-T}^{T} \frac{e^{+it(X-\alpha)} - e^{+it(X-\beta)}}{it} \, dt \right)$$

$$= \frac{1}{\pi} E\left(\int_{0}^{T} \frac{\sin t(X - \alpha) - \sin t(X - \beta)}{t} \, dt \right)$$

(since cos $(t)/t$ is an odd function)

$$= \frac{1}{\pi} E\left\{ \int_{0}^{T(X-\alpha)} \frac{\sin \tau}{\tau} \, d\tau - \int_{0}^{T(X-\beta)} \frac{\sin \tau}{\tau} \, d\tau \right\}, \tau = t(X - \alpha), t(X - \beta)$$

$$= E(J_T(X)), \text{where } J_T(u) = \frac{1}{\pi} \int_{T(u-\beta)}^{T(u-\alpha)} \frac{\sin \tau}{\tau} \, d\tau = \begin{cases} 1 & \text{if } \alpha < u < \beta \\ 0 & \text{if } \alpha < u \text{ or } \beta < u \\ 1/2 & \text{if } u = \alpha \text{ or } \beta \end{cases}$$

and $J_T(u) \le 2$ for all $u, -\infty < u < \infty$. Therefore, by dominated convergence theorem,

$$\lim_{T \to \infty} I(T) = \lim_{T \to \infty} E(J_T(X)) = E(\lim_{T \to \infty} J_T(X))$$

$$= P[\alpha < X < \beta] + \frac{P(X = \alpha) + P(X = \beta)}{2}$$

$$= \frac{F(\beta+) + F(\beta-)}{2} - \frac{F(\alpha+) + F(\alpha-)}{2}$$

Corollary 9.3 (Uniqueness theorem for ch.f.).   Distribution function of a r.v. and its ch.f. determine each other.

Proof.   By definition $\phi(t) = \int_{-\infty}^{\infty} e^{itx} \, dF(x)$, if $F(x)$ is the d.f. of a r.v. and it determines the ch.f. uniquely. If F_1 and F_2 are two d.f.'s corresponding to a given $\phi(t)$, then from inversion theorem,

$$F_2(\beta) - F_2(\alpha) = F_1(\beta) - F_1(\alpha)$$

i.e. $F_2(\beta) - F_1(\beta) = F_2(\alpha) - F_1(\alpha)$ if $\alpha, \beta \in C(F_1) \cap C(F_2)$.

If β varies and α is fixed,

$$F_2(\beta) - F_1(\beta) = F_2(\alpha) - F_1(\alpha) = \text{constant, but } F_2(\infty) - F_1(\infty) = 0.$$

Hence

$$F_2(\alpha) - F_1(\alpha) = 0, \text{ i.e.}$$

$$F_2(\beta) - F_1(\beta) = 0, \beta \in C(F_1) \cap C(F_2).$$

Since the set of discontinuity points is at most countable and the two distribution functions agree on a dense set of the real line then they are identical and hence the result.

Theorem (continuity theorem for characteristic function).   Let $\{F_n\}$ be a sequence of d.f.'s with $\{\phi_n\}$ the corresponding sequence of characteristic functions. Suppose that $\lim\limits_{n\to\infty} \phi_n(t) = \phi(t)$ exists, $-\infty < t < \infty$ and ϕ is continuous at $t = 0$. Then there exists a d.f. F whose characteristic function is ϕ and $\lim\limits_{n\to\infty} F_n(x) = F(x)$ at all continuity points x of F. Conversely, if $F_n(x) \to F(x)$ at each $x \in C(F)$ then $\phi_n(t) \to \phi(t)$, the characteristic function of F, where $C(F)$ is the set of all continuity points of F. The proof will be given later at the end of Section 9.6.

Definition 9.1   A sequence $\{X_n\}$ of r.vs. converges to a r.v. X in distribution if $F_n(x) \to F(x)$ at each $x \in C(F)$, where F_n is the d.f. of X_n and F is the d.f. of X. We then write

$$F_n \xrightarrow{W} F \text{ or } F_n \xrightarrow{d} F \text{ or } X_n \xrightarrow{L} X.$$

Definition 9.2   A sequence $\{X_n\}$ of r.vs. converges in probability to a r.v. X if $P[\,|X_n - X| > \varepsilon\,] \to 0$ as $n \to \infty$, for every $\varepsilon > 0$.

We write $X_n \xrightarrow{P} X$ (note almost surely $\Rightarrow P \Rightarrow d$).

EXAMPLE 9.1 (*Weak law of large number due to Khintchine*).   Let $X_1, \dots X_n$, be i.i.d. r.vs. with common d.f. $F(x)$ and finite mean μ. Then $S_n/n \xrightarrow{P} \mu$.

Proof. If $\phi(t) = E(e^{itX})$ be the ch.f. of X_1, then $\phi(t) = 1 + \mu ti + o(t)$ as $t \to 0$ (Theorem 3.10).

$$E(e^{itS_n/n}) = \prod_{i=1}^{n} E(e^{it/nX_i}) = [\phi(t/n)]^n \text{ (since } X_i's \text{ are i.i.d. r.vs.)}$$

$$= (1 + \mu it/n + o(1/n)^n \to e^{i\mu t} \text{ as } n \to \infty$$

which is the ch.f. of the degenerate r.v. $X = \mu$ a.s. By the continuity of ch.f. $S_n/n \xrightarrow{d} \mu$.

Hence $S_n/n \xrightarrow{P} \mu$ (see Theorem 6.3).

EXAMPLE 9.2 (*CLT for i.i.d. r.vs*).   Let $\{X_n\}$ be a sequence of i.i.d. random variables with $E(X_n) = 0$, $0 < E(X_n^2) = \sigma^2 < \infty$ and $S_n = X_1 + \dots + X_n$. Then

$$S_n/\sigma\sqrt{n} \xrightarrow{d} N(0, 1).$$

Proof.   var $(S_n) = n\sigma^2(X_i's$ being i.i.d and $E(S_n) = 0)$

$$E(e^{itS_n/\sqrt{n}\sigma}) = [\phi(t/\sqrt{n}\sigma)]^n = (1 - \sigma^2/(2n\sigma^2)t^2 + o(t^2/n))^n \to e^{-t^2/2} \text{ as } n \to \infty,$$

which is the ch.f. of (by Theorem 9.6) $N(0, 1)$. So by continuity theorem of ch.f. $S_n/\sigma\sqrt{n} \xrightarrow{d} N(0, 1)$.

EXAMPLE 9.3 Let X_{n1}, X_{n2}, ..., X_{nn} be a triangular array of independent r.vs. with $P(X_{n1} = 1) = p_n$, $P(X_{n1} = 0) = 1 - p_n$ such that $0 < p_n < 1$ and $nP_n \to \lambda > 0$. Then

$$S_{nn} = X_{n1} + X_{n2} + \ldots + X_{nn} \xrightarrow{d} P(\lambda), \text{ the Poisson r.v.}$$

Proof. $E(e^{itS_{nn}}) = \prod_{k=1}^{n} E(e^{itx_{nk}}) = (p_n e^{it} + q_n)^n, \, q_n = 1 - p_n$

$$= (1 + p_n \, (e^{it} - 1))^n = \left[1 + \frac{\lambda}{n}(e^{it} - 1) + o(1/n) \right]^n$$

$$\to e^{\lambda(e^{it} - 1)} \text{ as } n \to \infty$$

which is the ch.f. of $P(\lambda)$. So again by continuity theorem for ch.f.

$$S_{nn} \xrightarrow{d} P(\lambda).$$

EXAMPLE 9.4 It X is an integer-valued r.v., show that $P(X = j) = 1/2\pi$ $\int_{-\infty}^{\infty} e^{-itj} \phi(t) \, dt$. This is the inversion theorem for discrete r.vs.

Proof.   Now

$$\phi(t) = \sum_{k=-\infty}^{\infty} p_k \exp(itk)$$

Consider

$$I_j = \int_{-\pi}^{\pi} \phi(t) \exp(-itj) \, dt = \int_{-\pi}^{\pi} \sum_{k=-\infty}^{\infty} p_k \exp[it(k-j)] \, dt$$

$$= \sum_{k=-\infty}^{\infty} p_k \int_{-\pi}^{\pi} \exp[it(k-j)] \, dt.$$

Since $|e^{it(k-j)}| \leq 1$ and $p_k \geq 0$, interchanging the order of integration by Fubini's theorem is valid here. But from trigonometry,

$$\int_{-\pi}^{\pi} \exp(itm) \, dt = \begin{cases} 0 & \text{if } m = \pm 1, \pm 2, \ldots \\ 2\pi & \text{if } m = 0 \end{cases}$$

Hence

$$\frac{1}{2\pi} \int_{-\pi}^{\pi} \phi(t) \exp(-itj) \, dt = \frac{1}{2\pi} I_j = p_j (k - j = m = 0 \Rightarrow I_j = 2\pi p_j)$$

9.3 THE MOMENT PROBLEM

We have seen that moments may not exist (see Theorem 9.6) but a ch.f. $\phi(t)$ always exists. If however moments of all orders are finite, then

$$\phi(t) = \sum_{j=0}^{\infty} \alpha_j (it)^j / j! \tag{1}$$

for all t for which the right-hand side converges, which is known from calculus to be equivalent to $|t| < 1/\rho$, where $\rho = \lim \sup \, (|\alpha_j|/j!)^{1/j}$, $j \to \infty$.

Theorem 9.7 If the series $\sum_j v_j t^j j!$ is convergent for some $t_0 > 0$, then the sequence of moments $\{\alpha_j\}$ determines the d.f. F uniquely. In particular, if $\rho = 0$ and all the moments are finite then the sequence of moments uniquely determines $F(x)$. Moreover if

$$\overline{\lim} \left(\frac{v_n^{1/n}}{n} \right) = \overline{\lim} \, (\alpha_n^{1/n}/n) = l < \infty$$

then also the set of moments determine the d.f. F uniquely.

Proof.   Since $\sum_j v_j t^j/j! < \infty, (v_n |h|^n /n!) \to 0$ as $n \to \infty$ for $|h| < t_0$. Thus, the absolute value of the remainder term in the Taylor series expansion in equation (1) of Theorem 9.6 tends to zero and $\phi(t)$ can be expanded in power series in (1) for $|t| < t_0$. Therefore,

$$\phi(t) = \sum_j (it)^j / j! \alpha_j, |t| < t_0.$$

Since $\phi(t)$ determines F uniquely, the first part of theorem follows. Now the radius of convergence of the power series $\sum_j \alpha_j t^j j!$ is $\rho^{-1} = (\overline{\lim} \, (\alpha_n/n!)^{1/n}]^{-1}$. If $\rho = 0$ and α_j's are finite, the power series $\phi(t)$ is absolutely convergent.

By Stirling's approximation $(n!)^{1/n} \sim n/e$. Then

$$1/\rho > 0 \text{ if } \overline{\lim} \, (\alpha_n^{1/n}/n!) < \infty.$$

Using Liapounov's inequality,

$$v_r^{1/r} \le v_r^{1/s}, r \le s. \text{ Since } v_{2n} = \alpha_{2n} \text{ and}$$

$$\frac{2n-1}{2n(2n-1)} v_{2n-1}^{1/(2n-1)} \le \frac{1}{2n} v_{2n}^{1/2n} \le \frac{2n+1}{2n(2n+1)} v_{2n+1}^{1/(2n+1)}$$

$$\Rightarrow \overline{\lim} \, (v_{2n-1}^{1/(2n-1)} /(2n-1)) \le \overline{\lim} \, (v_{2n}^{1/2n}/2n \le \overline{\lim} \, (v_{2n+1}^{1/(2n+1)} /(2n+1))$$

$$\Rightarrow \overline{\lim} \, (\alpha_{2n}^{1/(2n)} /n)) = \overline{\lim} \, (v_{2n}^{1/n}/n) = l$$

Then, the radius of convergence of two series $\sum_j v_j t^j/j!$ and $\sum_j \alpha_j t^j/j!$ are the same. If $l < \infty$, the later converges for certain $|t| \le t_0$, then the later also converges and determines $\phi(t)$ and hence the set of moments determine the d.f. F uniquely.

EXAMPLE 9.5 (*Normal distribution*).   If X is normal with $EX = 0$, var $(X) = 1$ then we know $\phi(t) = \exp(-t^2/2)$. In this example, $\alpha_j = 0$ if j is odd and $\alpha_{2j} = (2j)! /(j!2^j)$. Thus

$$\phi(t) = \sum_{j=0}^{\infty} (-1)^j (t^2/2)^j /j! = e^{-t^2/2} < \infty \text{ for all real } t.$$

So the moments uniquely determine the normal d.f.

***EXAMPLE* 9.6** (*Lognormal distribution*).   If Y is $N(0, 1)$, then $X = \exp(Y)$ is lognormal r.v. and

$E(X^r) = \exp(r^2/2)$, $r \geq 1$ integer. Heyde (1963) observed that these same moments are obtained for a r.v. whose density function $cg(x) = f(x)\{1 + (1/2) \sin [2\pi k \log x]\}$, where $f(x)$ is the density of lognormal distribution and k is any positive integer and $c > 0$ is then to make $g(x)$ a p.d.f.

Another important criterion for the set of moments to uniquely determine the d.f. F is the celebrated *Careleman Criterion C* (Shohat and Tamarkin, 1943) which asserts that a d.f. F is uniquely determined by its moments α_j if

$$\sum_{j=1}^{\infty} \frac{1}{(\alpha_{2j})^{1/2j}} = \infty.$$

9.4   TEST FOR CHARACTERISTIC FUNCTIONS AND POLYA'S THEOREM

***EXAMPLE* 9.7**   The only ch.f. which has the form $\phi(t) = 1 + o(t^2)$ as $t \to 0$ is the function $\phi(t) \equiv 1$.

Proof. $\phi(t) = 1 + o(t^2)$ as $t \to 0$ by Theorem 9.6 $\alpha_1 = \alpha_2 = 0$, where

$$\alpha_2 = \int_{-\infty}^{\infty} x^2 \, dF(x) = 0, \text{ i.e.}$$

$$F(x) = \varepsilon(x), \text{ Hence } \phi(t) \equiv 1.$$

As an example $\phi(t) = e^{-t^4}$ satisfies allo necessary conditions of a ch.f. but $\phi(t) = 1 + o(t^2)$ as $t \to 0$. So $\phi(t) \equiv 1$ but $1 \neq e^{-t^4}$. Hence e^{-t^4} is not a ch.f.

***EXAMPLE* 9.8**   Let $W(t) = o(t)$ as $t \to 0$ and suppose $W(-t) = -W(t)$. Then the only ch.f. of the form, $\phi(t) = 1 + W(t) + o(t^2)$ as $t \to 0$ is the function $\phi(t) \equiv 1$; e.g. $\exp(-t^r)$; $r \geq 2$, $(1 + t^4)^{-1}$ and $\exp(-P(t))$, where $P(t)$'s are polynomial of degree greater than 2, are not characteristic functions.

Proof.   Given that $\phi(t) = 1 + W(t) + o(t^2)$, as $t \to 0$ and $W(-t) = -W(t)$

$$|\phi(t)|^2 = \phi(t) \cdot \overline{\phi}(t) = \phi(t)\phi(-t) \quad (\because \ \phi(t) = \phi(-t))$$

$$= (1 + W(t) + o(t^2))(1 - W(t) + o(t^2))$$

$$= 1 + o(t^2) \text{ as } t \to 0 \text{ (since } W(t) = o(t) \text{ as } t \to 0)$$

So the new ch.f. $f = |\phi(t)|^2 = 1 + o(t^2)$ as $t \to 0$. Hence by the previous Example 9.7, $f(t) \equiv 1 \equiv |\phi(t)|^2$ and $|\phi(t) \equiv 1$. Therefore,

$\phi(t) = e^{iat}$ where a is some real number and $i = \sqrt{-1}$

$$= 1 + iat - 1/2a^2t^2 + o(t^2) \text{ as } t \to 0$$

$$= 1 + W(t) + o(t^2) \text{ as } t \to 0, \text{ where } W(t) = o(t) \text{ as } t \to 0 \text{ only if } a = 0.$$

Hence $\phi(t) \equiv 1$.

Let us state an important theorem due to George Polya (without proof) which helps us in many cases to know whether a function is a ch.f. or not.

Theorem 9.8 (*Polya's theorem*).   Let $\phi(t)$ be a real valued continuous function which is defined for all real t and which satisfies the following conditions: (i) $\phi(0) = 1$, (ii) $\phi(-t) = \phi(t)$, (iii) $\phi(t)$ is a convex function for $t > 0$, (iv) $\lim_{t \to \infty} \phi(t) = 0$. Then $\phi(t)$ is the ch.f. of an absolutely continuous d.f. $F(x)$ (i.e. it has density).

EXAMPLE 9.9   (a) $\phi(t) = e^{-|t|}$,   (b) $\dfrac{1}{1+t}$,   (c) $\begin{cases} 1 - t & \text{if } 0 \le t \le /2 \\ 1/(4\,|t|) & \text{if } t \ge 1/2 \end{cases}$

(d) $\begin{cases} 1 - |t| & \text{if } 0 \le |t| \le 1 \\ 0 & > 1 \end{cases}$

are all Polya type characteristic functions, i.e. they satisfy the above theorem due to G. Polya.

Note (a) is the ch.f. of Cauchy distribution

(a) and (d) are in L_1 (integrable ch.f.)

(b) and (c) are not in L_1 but still absolutely continuous

(d) has density $f(x) = \dfrac{1}{2\pi}\left[\dfrac{\sin x/2}{x/2}\right]^2$.

Criteria for continuous d.f.   Let

$$J_T = 1/2\pi \int_{-T}^{T} e^{-ixt} \phi(t)\, dt.$$

Then

$$\lim_{T \to \infty} \frac{J_T}{T} = \begin{cases} 0 \text{ if } F \text{ is also continuous} \\ \dfrac{\text{saltus at } x \text{ of } F(x)}{2\pi} = \dfrac{F(x+) - F(x-)}{2\pi} \text{ if } F \text{ has a jump at } x. \end{cases}$$

Note that if $\lim_{T \to \infty} J_T/T = 0$, then F is continuous and differentiable and

$$F'(x) = f(x) = 1/2\pi \int_{-\infty}^{\infty} e^{-ixt} \phi(t)\, dt.$$

9.5 NECESSARY AND SUFFICIENT CONDITION FOR A CHARACTERISTIC FUNCTION

So far we have given either necessary properties or sufficient properties for a ch.f., but they are not both necessary and sufficient. A sufficient and necessary condition is given by Bochner's theorem. It is very important in Time Series Analysis (stationary processes).

A complex valued function $f(t)$ of a real variable t is said to be non-negative definite (n.n.d.) for $-\infty < t < \infty$ if the following two conditions are satisfied: (i) $f(t)$ is continuous, and (ii) for any positive integer N, any real number $t_1, \ldots, t_N$ and complex number $\xi_1, \ldots, \xi_N$ the sum

$$S = \sum_{j=1}^{N} \sum_{k=1}^{N} f(t_j - t_k)\xi_j \bar{\xi}_k$$

is real and non-negative. Alternatively,

$$S = \sum_{j=1}^{N} \sum_{k=1}^{N} f(u_j - v_k)h(u_j)\bar{h}(v_k)$$

is real and non-negative, where h is a complex valued measurable function of real variables u_j, v_k's.

Theorem 9.9   Let $f(t)$ be n.n.d. Then

(a) $f(0)$ is real and $f(0) \geq 0$,

(b) $f(-t) = \overline{f(t)}$ (Hermitian property)

and (c) $|f(t)| \leq f(0)$
(Note that $|\phi(t)| \leq 1 = \phi(0)$.)

Proof.   Follows easily from definition and hence the proof is omitted.

Theorem 9.10 (*due to Bochner*).   A complex valued function of a real variable t is a characteristic function iff (i) $f(t)$ is n.n.d. (ii) $f(0) = 1$.

Proof.   Necessity

$$S = \sum_{j=1}^{N} \sum_{k=1}^{N} f(t_j - t_k)\xi_j \bar{\xi}_k$$

$$= \sum_{j} \sum_{k} \xi_j \bar{\xi}_k \int_{-\infty}^{\infty} \exp\left(i(t_j - t_k)x\right) dF(x)$$

$$= \int_{-\infty}^{\infty} \left(\sum_{1}^{N} \xi_j e^{it_j x}\right)\left(\sum_{1}^{N} \xi_k e^{-it_k x}\right) dF(x)$$

$$= \int_{-\infty}^{\infty} \left|\sum_{1}^{N} (\xi_j e^{it_j x})\right|^2 dF(x) \geq 0$$

and S is also a real number. $\phi(t)$ being a ch.f., it is continuous and $\phi(0) = 1$.
Now we shall prove the converse (*sufficiency part*). Let

$$h(u) = e^{-iux - u^2 \sigma^2} \tag{1}$$

Since f is n.n.d.

$$\sum_{j} \sum_{k} (u_j - v_k)h(u_j)\overline{h(v_k)} \geq 0$$

or

$$\sum_{j} \sum_{k} f(u_j - v_k) \exp\left[-i(u_j - v_k)x - (u_j^2 + v_k^2)\sigma^2\right] \geq 0 \tag{2}$$

Since f is continuous, the summation tends to

$$\int_{-\infty}^{\infty} \int_{-\infty}^{\infty} f(u - v) \exp\left[-i(u - v)x - (u^2 + v^2)\sigma^2\right] du\, dv \geq 0 \tag{3}$$

for each x and $\sigma^2 > 0$ because the above double integral is the limit of Riemann sums, each of which is non-negative. Putting $t = u - v$ and $s = u + v$ so that $u^2 + v^2 = (t^2 + s^2)/2$ and integrating with respect to s in (3), we get

$$g(x, \sigma) = \int_{-\infty}^{\infty} \frac{f(t)}{2\pi} \exp\left[-itx - \frac{t^2\sigma^2}{2}\right] dt \geq 0 \tag{4}$$

By monotone convergence theorem,

$$\int_{-\infty}^{\infty} g(x, \sigma)\, dx = \lim_{\beta \to 0} \int_{-\infty}^{\infty} g(x, \sigma) \exp\left(\frac{x^2\beta^2}{2}\right) dx$$

$$= \lim_{\beta \to 0} \int_{-\infty}^{\infty} \int_{-\infty}^{\infty} \frac{f(t)}{2\pi} \exp\left(-itx - t^2\sigma^2/2 - x^2\beta^2/2\right) dx\, dt$$

noting the fact $0 \leq g(x, \sigma) \exp(-x^2\beta^2/2) = h(x, \beta) \downarrow$ in β.
Here change of order of integration is justified by Fubini's theorem. Hence integrating over x and using continuity of f, we obtain

$$\int_{-\infty}^{\infty} g(x, \sigma)\, dx = \lim_{\beta \to 0} \int_{-\infty}^{\infty} f(t) \exp(t^2\sigma^2/2 - t^2/2\beta^2)\frac{dt}{\beta\sqrt{2\pi}}$$

$$= \lim_{\beta \to 0} \int_{-\infty}^{\infty} f(u\beta) \exp\left[-u^2/2 - \frac{u^2\beta^2\sigma^2}{2}\right] \frac{du}{\sqrt{2\pi}} \quad (\text{Put } t = \beta u)$$

$$= f(0) \int_{-\infty}^{\infty} e^{-u^2/2} \frac{du}{\sqrt{2\pi}} = f(0) = 1$$

(since f is continuous and $\lim_{\beta \to 0} f(u\,\beta) = f(0)$). So for fixed σ, $g(x, \sigma)$ as a function of x is the density function of a d.f. Now we shall prove that $f(t) \exp(-t^2\sigma^2/2)$ is the ch.f. of $g(x, \sigma)$ and then by continuity theorem of ch.f.

$$\lim_{\sigma \to 0} f(t)e^{-t^2\sigma^2/2} = f(t) \text{ is also a ch.f.}$$

Integrating with respect to x from $-\infty$ to ∞ and applying Fubini theorem again we get

$$\int_{-\infty}^{\infty} \exp(i\tau x)\, g(x, \sigma)\, dx = \lim_{\beta \to 0} \int_{-\infty}^{\infty} g(x, \sigma) \exp\left[i\tau x - x^2\beta^2/2\right] dx$$

$$= \lim_{\beta \to 0} \int_{-\infty}^{\infty} \frac{f(\tau + u\beta)}{\sqrt{2\pi}} \exp\left[-\frac{(\tau + u\beta)^2 \sigma^2}{2} - u^2/2\right] du$$

$$= f(\tau) \exp(-\tau^2\sigma^2/2)$$

$$\text{(due to continuity of } f\text{)}.$$

Here we substitute $(t = \tau + u\beta)$ and use the value of normal ch.f. The proof is now complete.

9.6   WEAK CONVERGENCE AND HELLY'S THEOREMS

Let μ_n be a family of finite measure on $(R, \mathcal{B})$. We say that μ_n converges weakly to another finite measure $\mu (\mu_n \Rightarrow \mu)$ if $\lim\limits_{n \to \infty} \int f \, d\mu_n \int f \, d\mu$ for bounded continuous function f. Specifically, $\mu_n = P_n$, a probability measure on R.

EXAMPLE 9.10   (a) If $f = 1$, then $P_n(E) \to P(E)$ and $\mu_n(E) \to \mu(E)$, $E \in \mathcal{B}$, (b) Let μ_n be a unit mass at the points x_n, $n \geq 1$ i.e.

$$\mu_n(E) = \begin{cases} 1 & \text{if } x_n \in E \\ 0 & \text{if } x_n \notin E \end{cases}, \text{ then } \mu_n \Rightarrow \mu \text{ iff } x_n \to x$$

exists and

$$\mu(E) = \begin{cases} 1 & \text{if } x \in E \\ 0 & \text{if } x \notin E \end{cases},$$

since $x_n \to x$ implies $f(x_n) \to f(x)$, f being continuous and hence

$$\int f \, d\mu_n = f(x_n) = x_n \text{ if } f \equiv 1 \text{ and}$$

$$f(x_n) \to \int f \, d\mu_n = f(x) = x \text{ if } f \equiv 1.$$

(c) Let $X = [0, 1]$ and $\{\mu_n\}$ is a sequence of discrete measure which put mass $\dfrac{1}{n+1}$ at 0, $1/n$, ..., 1. Then $\mu_n \Rightarrow \mu$, Lebesgue measure on $[0, 1]$, since

$$\int_{[0,1]} f \, d\mu_n = \int_0^1 f \, d\mu_n = \frac{1}{n+1} \sum_{i=0}^{n} f(i/n) \to \int_0^1 f(x) \, dx$$

for every bounded continuous function on $[0, 1]$. Let us state a relevant theorem without proof.

Theorem 9.11   Let μ_n and μ be finite measures on the Borel subsets of real line and F_n and F their corresponding d.fs. Then $\mu_n \Rightarrow \mu$ iff $F_n \xrightarrow{d} F$. (The proof may be attempted as solution to Problem 6.3.)

NOTE. $F_n(x) = \mu_n(-\infty, x]$ and $F(x) = \mu(-\infty, x]$. Sets $(-\infty, x]$ are Borel sets of real line and they form a class of generating sets for Borel σ-field. Taking $f \equiv 1$. it is easy to prove $\mu_n \Rightarrow \mu \Rightarrow F_n \xrightarrow{d} F$. Let us state a lemma which will be very useful later.

Lemma 9.2   Let $\{F_n(x)\}$ be a sequence of non-decreasing functions of real variable x and let D be a dense subset of the real line. Suppose that the sequence $\{F_n(x)\}$ converges to some function $F(x)$ at all points of D, then $\{F_n\}$ converges in distribution to F.

Proof.   Let x be an arbitrary point on the real line. Choose two points x', $x'' \in D$ such that $x' \le x \le x''$. Then $F_n(x') \le F_n(x) \le F_n(x'')$. Hence $\lim_{n \to \infty} F_n(x')$ $\le \lim_{n \to \infty} \inf F_n(x) \le \lim_{n \to \infty} \sup F_n(x) \le \lim_{n \to \infty} F_n(x'')$. Therefore, $F(x') \le \lim \inf F_n(x)$ $\le \lim \sup F_n(x) \le F(x'')$. Since D is dense in $R(x'_m \uparrow x, \; x''_m \downarrow x$ and $x'_m \le x \le x''_m)$, $F(x-) \le \lim \inf F_n(x) \le \lim \sup F_n(x) \le F(x+)$.

If $x \in c(F)$, then $F(x-) = F(x) = F(x+)$ and hence $F(x) \le \lim \inf F_n(x) \le \lim \sup F_n(x) = F(x)$. Therefore $\lim F_n(x) = F(x)$ if $x \in c(F)$.

Let us consider following examples concerning convergence of d.fs.

EXAMPLE 9.11   Let

$$F_n(x) = \begin{cases} 0 & \text{if } x < -n \\[2mm] \dfrac{n+x}{2n} & -n \le x \le n \\[2mm] 1 & x \ge n \end{cases}$$

be a sequence of rectangular distributions. Obviously, $F_n(x) \to 1/2 \equiv F(x)$ for all $x \in R$ as $n \to \infty$. But obviously F is not a d.f. of a r.v., as $F(-\infty) \ne 0$ and $F(\infty) \ne 1$. Alternatively if ϕ_n be the ch.f. of F_n, then

$$\phi_n(t) = \frac{\sin nt}{nt} \to \phi(t) = \left\{ \begin{array}{ll} 0 & \text{if } t \ne 0. \\ 1 & \text{if } t = 0. \end{array} \right\}.$$

Since ϕ is not continuous at $t = 0$, ϕ cannot be a ch.f. and hence by continuity theorem for ch.fs. F is not a d.f.

EXAMPLE 9.12   A sequence of distributions $\{F_n\}$ may not converge at all, e.g. $F_n(x) = \varepsilon(x - n)$ with unit mass at $x = n$. When $n \to \infty$ the mass of $F_n(x)$ escapes to $+\infty$. Hence no limit exists.

EXAMPLE 9.13 $F_n(x) = \dfrac{n}{\sqrt{2\pi}} \displaystyle\int_{-\infty}^{x} \exp(-n^2 y^2/2)$, i.e. $X_n \sim N(0, 1/n)$

$$= \frac{1}{\sqrt{2\pi}} \int_{-\infty}^{nx} e^{-z^2/2} \, dz$$

$$= \Phi(nx) \to \begin{cases} 0 & \text{if } x < 0 \\ 1/2 & \text{if } x = 0 \\ 1 & \text{if } x > 0 \end{cases}$$

Hence $F_n(x)$ converges to a degenerate d.f. $\varepsilon(x)$, where

$$\varepsilon(x) = \begin{cases} 0 & \text{if } x < 0 \\ 1 & \text{if } x \ge 0 \end{cases}.$$

Note that $\lim_{n \to \infty} F_n(0) = 1/2$ but $\varepsilon(0) = 1$, i.e. at 0, $F_n(x)$ does not converge to $\varepsilon(x)$ and 0 is a discontinuity point of $\varepsilon(x)$. In the definition of convergence in distribution only convergence at continuity points of the limiting distribution is sought.

The above examples show that a whole sequence of distribution functions either does not converge at all, or it may converge but converges to a function which is not a distribution function and last of all convergence at all points is not necessary.

The next theorem which is one of the most frequently used results in analysis says that there always exists a sequence of d.fs. which always converges to a d.f.

Theorem 9.12 (*Montell-Helly's theorem or Helly's first theorem or Helly selection principle*—discovered in 1912). Every sequence $\{F_n(x)\}$ of uniformly bounded non-decreasing functions, contains a subsequence $\{F_{n_k}(x)\}$ which converges weakly to a non-decreasing bounded function $F(x)$. If, moreover, F_n's are d.fs. so is F whenever $F_n(\infty) \to F(\infty)$ and $F_n(-\infty) \to F(-\infty)$ as $n \to \infty$.

Proof. The proof is carried out by standard diagonal method and uses the fact that the set of rationals is denumerable and can be arranged in a sequence $\{r_j\}$. The sequence of numbers $\{F_n(r_1)\}$ is bounded, hence by the Bolzano-Weierstrass theorem there exists a subsequence $\{F_{1n}\}$ such that $\lim\limits_{n \to \infty} F_{1n}(r_1) = G(r_1)$, say. Next, the sequence of numbers $\{F_{1n}\}(r_2)\}$ is bounded and hence there exists a subsequence $\{F_{2n}\}$ such that $\lim\limits_{n \to \infty} F_{2n}(r_2) = G(r_2)$. Note that $F_{2n}(x)$ converges at both r_1 and r_2 and $\lim\limits_{n \to \infty} F_{2n}(r_1) = G(r_1)$. Continue this process a countable number of times. Finally, let $\{F_{n_k}\}$ consist of the diagonal elements, i.e. use the first member of $\{F_{1n}\}$, second of $\{F_{2n}\}$ and so on. So we have proved the existence of an infinite subsequence $\{n_k\}$ and a function G defined and increasing on $D = \{r_1, r_2, \ldots\}$ such that for all $r \in D$, $\lim\limits_{k \to \infty} F_{n_k}(r) = G(r)$. G is also bounded since F_n's are uniformly bounded. Define F on R as

$$F(x) = \inf_{x < r \in D} G(r) = \lim_{r_n \downarrow x} G(r_n), \, r_n \in D.$$

By the last lemma $\lim\limits_{k \to \infty} F_{n_k}(x_2) = F(x)$ at $x \in C(F)$, since the set of rationals D is dense in R. Clearly F is non-decreasing and bounded. Now from the definition of greatest lower bound $G(r_i) \geq F(x)$ for all $r_i \geq x$ and $F(x) + \varepsilon > G(r_i)$ for some $r_i \in D$ and $\varepsilon > 0$. Hence we can choose $r_i \geq x$ such that $0 < h < r_i - x < \delta$ for which $|F(r_i) - F(x)| < \varepsilon$ (since $G(r_i) = F(x)$) and $|F(x + h) - F(x)| \leq |F(r_i) - F(x)| < \varepsilon$. Hence F is right continuous. (An alternative proof of convergence without the lemma is as follows. Now choose rational r and s so that $r < x < s$ and $G(s) < F(x) + \varepsilon$. From $F(x) - \varepsilon < G(r) \leq G(s) < F(x) + \varepsilon$ and $F_{n_k}(r) \leq F_{n_k}(x) \leq F_{n_k}(s)$, it follows that as $k \to \infty$, $F_{n_k}(x)$ has lim sup and lim inf within ε of $F(x)$. Hence

$$\lim_{k \to \infty} F_{n_k}(x) = F(x) \quad \text{if } x \in C(F))$$

The F in this theorem necessarily satisfies $0 \leq F(x) \leq 1$, since if $F_n(-\infty) = 0$, $F_n(\infty) = 1$, then $F_{n_k}(-\infty) = 0$, $F_{n_k}(\infty) = 1$ and hence $\lim\limits_{k \to \infty} F_{n_k}(-\infty) = 0$ and $\lim\limits_{k \to \infty} F_{n_k}(\infty) = 1$. Hence $0 \leq F(-\infty) \leq F(\infty) \leq 1$. So F need not be a proper d.f. In Example 9.12, F_n has a unit jump at n, for example $F(x) \equiv 0$ is the only possibility. Now when $F_n(\infty) \to F(\infty)$ and $F_n(-\infty) \to F(-\infty)$ as $n \to \infty$, then $F(-\infty) = 0$ and $F(\infty) = 1$. Hence F is a proper d.f.

Theorem 9.13 (*Helly's second theorem* (*1919*)).   Let $f(x)$ be a continuous function and assume that $\{F_k(x)\}$ is a sequence of uniformly bounded non-decreasing functions which converges to some function $F(x)$ at all points $x \in C(F) \cap [a, b]$ and $a, b \in C(F)$, then

$$\lim_{k \to \infty} \int_a^b f(x) \, dF_k(x) = \int_a^b f(x) \, dF(x).$$

Proof. Since f is uniformly continuous on the bounded interval $[a, b]$ it is possible to construct a partition $a = x_0 < x_1 < \ldots < x_n = b$ of $[a, b]$ which is so fine that

$$| f(x) - f(x_j) | \leq \varepsilon \text{ for all } \varepsilon > 0 \text{ and } x_j \leq x \leq x_{j+1}, \, 0 \leq j \leq n - 1 \qquad (1)$$

Since the set of continuity points of $F(x)$ is dense in $[a, b]$ we can select partition points to be continuity points of F. Let

$$M = \max_{a \leq x \leq b} f(x). \text{ Then}$$

$$| F_k(x_j) - F(x_j) | \leq \varepsilon/(Mn) \text{ if } k \geq k_0 \text{ and for all } n \geq j \geq 0 \qquad (2)$$

Define $f_\varepsilon(x)$ the step function on $[a, b]$ as

$$f_\varepsilon(x) = f(x_j) \text{ for } x_j \leq x < x_{j+1}$$

By (1), $| f(x) - f_\varepsilon(x) | \leq \varepsilon$. Now

$$\left| \int_a^b f(x) \, dF(x) - \int_a^b f(x) \, dF_k(x) \right|$$

$$\leq \left| \int_a^b f(x) \, dF(x) - \int_a^b f_\varepsilon(x) \, dF(x) \right|$$

$$+ \left| \int_a^b f_\varepsilon(x) \, dF(x) - \int_a^b f_\varepsilon(x) \, dF_k(x) \right|$$

$$+ \left| \int_a^b f_\varepsilon(x) \, dF_k(x) - \int_a^b f(x) \, dF_k(x) \right| \qquad (3)$$

$$= I_1 + I_2 + I_3$$

By Helly's first theorem the limit function F is bounded and hence

$$\left| \int_a^b f(x) \, dF(x) - \int_a^b f_\varepsilon(x) \, dF(x) \right| \leq c_1 \varepsilon \qquad (4)$$

Also, by definition of $f_\varepsilon(x)$ and rearranging the terms, we get from (2),

$$\left| \int_a^b f_\varepsilon(x) \, dF(x) - \int_a^b f_\varepsilon(x) \, dF_k(x) \right| = \left| \sum_{j=0}^{n-1} f(x_j)[F(x_{j+1}) - F(x_j)] \right.$$

$$-\sum_{j=0}^{n-1} f(x_j)\,[F_k(x_{j+1}) - F(x_j)]\Bigg| \le n\left[M\cdot\frac{\varepsilon}{Mn} + M\cdot\frac{\varepsilon}{Mn}\right] = 2\varepsilon \ \text{ if } k \ge k_0 \tag{5}$$

Since $\{F_k\}$ is uniformly bounded,

$$\left|\int_a^b f_\varepsilon(x)\,dF_k(x) - \int_a^b f(x)\,dF_k(x)\right| \le c_2\varepsilon \tag{6}$$

By (3–6), we get

$$\left|\int_a^b f(x)\,dF(x) - \int_a^b f(x)\,dF_k(x)\right| \le (2 + c_1 + c_2)\,\varepsilon$$

for all $\varepsilon > 0$ and if $k \ge k_0$.

Theorem 9.14 (*Helly-Bray theorem* [Extended Helly's theorem, 1919]).   Let $f(x)$ be continuous and bounded in the infinite interval $-\infty < x < \infty$ and $\{F_k(x)\}$ be a sequence of non-decreasing, uniformly bounded functions, which converges to some function $F(x)$ at each $x \in C(F)$. Suppose that $\lim\limits_{k\to\infty} F_k(-\infty) = F(-\infty)$ and $\lim\limits_{k\to\infty} F_k(\infty) = F(\infty)$ (or $f(x) \to 0$ as $x \to \infty$ or $-\infty$; of course then f is bounded is superfluous). Then

$$\lim_{k\to\infty} \int_{-\infty}^{\infty} f(x)\,dF_k(x) = \int_{-\infty}^{\infty} f(x)\,dF(x)$$

Proof.
$$\left|\int_{-\infty}^{\infty} f(x)\,dF_k(x) - \int_{-\infty}^{\infty} f(x)\,dF(x)\right|$$

$$\le \left|\int_{-\infty}^{a} f(x)\,dF_k(x) - \int_{-\infty}^{a} f(x)\,dF(x)\right| + \left|\int_a^b f(x)\,dF_k(x) - \int_a^b f(x)\,dF(x)\right|$$

$$+ \left|\int_b^{\infty} f(x)\,dF_k(x) - \int_b^{\infty} f(x)\,dF(x)\right|, \text{ where } a < 0 < b,\ a, b \in C(F)$$

$$= I_1 + I_2 + I_3.$$

Now, $|f(x)| \le M$ for all $-\infty < x < \infty$. By Helly's second theorem the second term $I_2 \le \varepsilon$ if $k \ge k_0$. Now

$$|F_k(a) - F(a)| \le \varepsilon/2M \text{ if } a \in C(F) \text{ and } k \ge k_0. \text{ Hence}$$

$$I_1 = \left|\int_{-\infty}^{a} f(x)\,dF_k(x) - \int_{-\infty}^{a} f(x)\,dF(x)\right| \le M\left|\int_{-\infty}^{a} dF_k(x) - \int_{-\infty}^{a} dF(x)\right|$$

$$\le M[F_k(a) - F_k(-\infty) - F(a) + F(-\infty)] = M\cdot\frac{2\varepsilon}{2M} = \varepsilon \text{ if } k \ge k_0.$$

Similarly $I_3 \le \varepsilon$ if $k \ge k_0$. Therefore,

$$\left| \int_{-\infty}^{\infty} f(x)\,dF_k(x) - \int_{-\infty}^{\infty} f(x)\,dF(x) \right| \le 3\varepsilon \text{ if } k \ge k_0 \text{ and } \varepsilon > 0.$$

NOTE. This proof is classical and based on classical real analysis but a modern proof can be given with the help of dominated convergence theorem and powerful Skorohod representation theorem (see Billingsley, 1968).

Motivated by Helly-Bray theorem we have a slightly stronger form of convergence of the sequence of d.fs.

Definition 9.3 A sequence of distribution functions converges to a d.f. F *completely* if $F_n \overset{d}{\to} F$ (i.e. $F_n(x) \to F(x)$, $x \in C(F)$) and $F_n(\infty) \to F(\infty)$ and $F_n(-\infty) \to F(-\infty)$.

9.6.1 Application of Helly's Theorem

Theorem 9.15 (*Continuity theorem of* ch.fs.). Let $\{F_n\}$ be a sequence of d.fs. and $\{\phi_n\}$ be the corresponding sequence of ch.fs. Suppose that $\lim_{n \to \infty} \phi_n(t) = \phi(t)$ exists for $-\infty < t < \infty$ and ϕ is continuous at $t = 0$. Then there exists a d.f. F which has ϕ as its ch.f. and F_n converges to F completely. Conversely if $\{F_n\}$ converges to F completely, then $\phi_n(t) \to \phi(t)$, the ch.f. of F.

Proof. (a) *Necessity part* follows as for extended Helly's theorem if we choose $f(x) = e^{itx}$ and if we note that here $\lim_{k \to \infty} F_k(\infty) = F(\infty)$ and $\lim_{k \to \infty} F_n(-\infty) = F(-\infty)$ are satisfied.

(b) *Sufficiency part.* Let $\phi_n(t)$ be the ch.f. of F_n and $\phi_n(t) \to \phi(t)$ and $\phi(t)$ is continuous at $t = 0$. Now, $|\phi_n(t)| \le 1$ and $\phi_n(t) \to \phi(t)$ imply

$$\int_0^{\upsilon} \phi_n(t)\,dt \to \int_0^{\upsilon} \phi(t)\,dt, \ \upsilon \text{ finite}, \ n \to \infty. \tag{1}$$

Now for any subsequence $\{n_k\}$,

$$\int_0^{\upsilon} \phi_{n_k}(t)\,dt = \int_0^{\upsilon} \left\{ \int_{-\infty}^{\infty} e^{itx}\,dF_{n_k}(x) \right\} dt = \int_{-\infty}^{\infty} dF_{n_k} \int_0^{\upsilon} e^{itx}\,dt$$

(change of order of integration is valid by Fubini Theorem)

$$= \int_{-\infty}^{\infty} \frac{e^{i\upsilon x} - 1}{ix}\,dF_{n_k}(x), \tag{2}$$

where the last integrand is defined by continuity at $x = 0$. By Helly's first theorem, F_n has a subsequence $F_{n_k} \to G$ weakly. By Helly's second theorem (2) yields, as $k \to \infty$,

$$\int_0^{\upsilon} \phi_{n_k}(t)\,dt \to \int_{-\infty}^{\infty} \frac{e^{i\upsilon x} - 1}{ix}\,dG(x) \tag{3}$$

Let $\phi_G(t)$ be the ch.f. of G. Since G is the weak limit of a subsequence of F_n, it might be an improper d.f. Similar to the derivation of (2) we get for ϕ_G and G

$$\int_0^v \phi_G(t)\,dt = \int_{-\infty}^{\infty} \frac{e^{ivx}-1}{ix}\,dG(x) \tag{4}$$

By (1), (3) and (4), we get

$$\int_0^v \phi(t)\,dt = \int_{-\infty}^{\infty} \frac{e^{ivx}-1}{ix}\,dG(x) = \int_0^v \phi_G(t)\,dt \tag{5}$$

Both extreme sides are differentiable at $v = 0$ because the integrands are continuous at $t = 0$, since $\phi_n(0) = 1$ and $\phi_n(t) \to \phi(t)$, we have by differentiation of extreme sides at $v = 0$, $1 = \phi(0) = \phi_G(0) = G(\infty) - G(-\infty)$ and since $0 \le G(-\infty) \le G(\infty) \le 1$, we must have $G(\infty) = 1$ and $G(-\infty) = 0$. Therefore G is a proper d.f. Now right hand side of (5) is differentiable for all v, with derivative $\phi_G(t)$. But then the left hand side of (5) which does not depend on G, is differentiable for all v. Hence $\phi(t) = \phi_G(t)$. By uniqueness theorem, G is the unique proper d.f. for every subsequence of F_n.

PROBLEMS AND COMPLEMENTS

9.1   Find the ch.f. of the following p.d.fs.:

(a) $f(x) = \dfrac{1}{2a}\, e^{-|x-b|/a}, a > 0, -\infty < x < \infty$

(b) $f(x) = \dfrac{C^r}{\Gamma r}\, e^{-cx} x^{r-1}, x > 0, c > 0, r < 0$

(c) $f(x) = \dfrac{1}{\pi}\, \dfrac{a}{a^2 + (x-b)^2}, a > 0, -\infty < x < \infty$

9.2   Show that $\dfrac{1}{n+1} \sum_{k=0}^{n} e^{ikt/n}$ is a ch.f. of a r.v. To which r.v. the sequence of r.vs. corresponding to the above ch.fs. converge in distribution?

9.3   Show that $\phi(t) = \dfrac{1}{8}(1 + 7e^{it})$, $t \in \mathbb{R}$ is a ch.f. but $|\phi(t)|$ is not.

9.4   Show that if $\phi(t)$ is a ch.f. then so is $|\phi(t)|^2$ and, in general, $|\phi(t)|^{2n}$, $n \ge 1$ integer. The statement, however, is not true for odd exponents. In particular, show that

$$\phi(t) = \frac{1}{4} + \frac{3}{4}\, e^{it}$$

is a ch.f. but $|\phi(t)|$ is not.

9.5 Give an example to show that the ratio of two ch.fs. may not be a ch.f.

9.6 Give examples of distributions without first moments but with a differentiable ch.f.

9.7 Let
$$f(x) = \begin{cases} c\,|x|^{-4}\,(\log|x|)^{-1} & \text{if } |x| \geq 2 \\ 0 & \text{if } |x| < 2 \end{cases}$$

Show that although $\phi^{(3)}(0)$ is finite, $E\,|\,X\,|^3$ does not exist.

9.8 Let $\phi(t)$ be the ch.f. of a d.f. $F(x)$. Then for every x,
$$F(x) - F(x-) = \lim_{U \to \infty} \frac{1}{2U} \int_{-U}^{U} e^{itx}\phi(t)\,dt.$$

9.9 Let $F(x)$ be a d.f. with discontinuities at the points x_j, $j \geq 1$. Let $p_j = F(x_j) - F(x_j -)$. Then show that
$$\lim_{U \to \infty} \frac{1}{2U} \int_{-U}^{U} |\phi(t)|^2\,dt = \sum_j p_j^2,$$
where $\phi(t)$ is the ch.f. of F.

9.10 Prove that if $\phi(t)$ is a ch.f. then
$$\lim_{T \to \infty} \frac{1}{2T} \int_{-T}^{T} |\phi(t)|^2\,dt \leq 1$$

with equality iff $\phi(.)$ corresponds to a degenerate law.

9.11 Give an example of a sequence of ch.fs. $\{\phi_n(t)\}$ which may converge for some t but not all $t \in \mathbb{R}$.

9.12 Let $X_1, X_2, \dots$ be a sequence of i.i.d. r.vs. with $P(X_1 = 1) = P(X_1 = -1) = 1/2$. If $Y_n = 2^{-n} \sum_{k=1}^{n} X_k$, find its ch.fs. and show that $Y_n \to Y$ a.s., where Y is uniformly distributed on the interval $[-1, 1]$.

9.13 Investigate whether the following are ch.fs:

(a) $\phi_1(t) = \dfrac{1}{t} \displaystyle\int_0^t \phi(u)\,du$ (b) $\phi_2(t) = e^{\phi(t)-1}$

(c) $\phi_3(t) = e^{-|\,t\,|3}$ (d) $\phi_4(t) = (1 + |\,t\,|)^{-1}$

where $\phi(t)$ is a ch.f.

10

Central Limit Theorem (CLT)

10.1 INTRODUCTION AND I.I.D. CASE

10.1.1 Introduction

Central limit theorem popularly known as CLT has played a prominent role in statistics and probability theory starting—in the case of independent random variables—with DeMoivre-Laplace version and culminating with that of Lindeberg-Feller. The central limit problem of probability theory is the problem of convergence of laws of sequences of sums of r.vs. It says roughly that the sum of many independent r.vs. will be approximately normally distributed if each summand has high probability of being small. For more than two centuries this classical limit problem has been the limit problem of probability theory. The precise formulation of this and its solution were obtained in the second quarter of this century. Central limit theorems also govern various classes of dependent r.vs. including Martingales and Markov processes. But we shall not deal with them in this book.

10.1.2 Central Limit Theorem for Independent Identically Distributed r.vs.

(i) *Lindeberg-Lévy theorem.* Let $\{X_n\}_1^\infty$ be a sequence of i.i.d. r.vs. such that $EX_k = 0$, var $(X_k) = \sigma^2 < \infty$. Then as $n \to \infty$

$$\frac{S_n}{\sigma\sqrt{n}} \xrightarrow{d} N(0, 1)$$

Proof. Done before in Section 9.3.

NOTE. Lindeberg-Lévy theorem implies that the normalized sum

$$S_n^* = \frac{S_n - n\mu}{\sqrt{n}\sigma} = (\bar{X}_n - \mu)(\sqrt{n}/\sigma) \xrightarrow{d} N(0,1)$$

and that implies

$$P(|S_n^*| \le x) \to \Phi(x) - \Phi(-x)$$

or
$$P(|X_n - \mu| \le x\,\sigma/\sqrt{n}) \to \Phi(x) - \Phi(-x)$$

where $\phi(x)$ is the c.d.f. of $N(0, 1)$. Taking $x\sigma/\sqrt{n} = \varepsilon$, i.e. $x = \varepsilon\sqrt{n}/\sigma$, we get $P(|\overline{X}_n - \mu| \le \varepsilon) \to 1$, i.e. $\{X_n\}$ satisfies WLLN. Thus, CLT gives the probability bound for $\overline{X}_n - \mu$ while WLLN gives only the limiting value.

10.2   CLT FOR INDEPENDENT NON-IDENTICALLY DISTRIBUTED RANDOM VARIABLES

10.2.1   Liapounov's and Lindeberg's Theorems

(ii) *Bounded Liapounov's CLT.*   Let $\{X_n\}_1^\infty$ be a sequence of independent r.vs. with $EX_k = 0$, $|X_k| \le C < \infty$ a.s., $\sigma_k^2 = EX_k^2 < \infty$. If moreover $s_n^2 = \sum_{k=1}^{n} \sigma_k^2 \to \infty$ as $n \to \infty$, then $S_n/s_n \xrightarrow{d} N(0, 1)$.

Proof.   Proof will be given as corollary to Liapounov's CLT.

(iii) *Liapounov's CLT.*   Let $\{X_n\}_1^\infty$ be a sequence of independent r.vs. with $EX_k = 0$.

$$\sigma_k^2 = EX_k^2 < \infty,\ s_n^2 = \sum_{k=1}^{n} s_k^2 \text{ and } \sum_{k=1}^{n} E|X_k|^{2+\delta}/s_n^{2+\delta} \to 0 \text{ for } 0 < \delta \le 1 \tag{1}$$

Then $S_n/s_n \xrightarrow{d} N(0, 1)$.

Proof.   Condition (1) is called Liapounov's condition and this implies $E|X_k|^{2+\delta} < \infty$ and (since $\sigma_k^2 > 0$) X_k is not degenerate for at least one k and also $s_n^2 \to \infty$ as $n \to \infty$. Since $(E|X_k|^r)^{1/r}$ is increasing in r,

$$\max_{k \le n} \left(\frac{\sigma_k}{s_n}\right)^{2+\delta} \le \max_{k \le n} \left(\frac{E|X_k|^{2+\delta}}{s_n^{2+\delta}}\right) \le s_n^{-2-\delta} \sum_1^n E|X_k|^{2+\delta} \tag{2}$$

The ch.f. of S_n/s_n is

$$\phi(t) = \prod_1^n \phi_k(t/s_n),$$

where $\phi_k(t)$ is the characteristic function of X_k. As in the proof of Theorem 9.6 in Section 9.1 expanding $\phi_k(t)$ in Taylor series for non-integral term for $|\theta_{nk}| < 1$ (see for example Loéve)

$$\phi_k(t/s_n) = 1 - t^2/2\,(\sigma_k^2/s_n^2) + 2^{1-\delta}(1+\delta)^{-1}(2+\delta)^{-1}\theta_{nk}|t|^{2+\delta} E|X_k|^{2+\delta} \cdot s_n^{-2-\delta}$$

$\to 1$ uniformly in $k \le n$ (by (1) and (2)), as $n \to \infty$. Since $\log(1+x) = x + o(x) = x(1 + o(1))$, for $|x| < 1$,

$$\log \phi_k(t/s_n) = -t^2/2\,(\sigma_k^2/s_n^2)(1 + o(1)) + 2\theta_{nk}|t|^{2+\delta} E|X_k|^{2+\delta} \cdot s_n^{-2-\delta}$$

for large n and where $o(1)$ and θ_{nk} do not depend on k. Therefore,

$$\sum_1^n \log \phi_k(t/s_n) = -(t^2/2)(1 + o(1)) + 2\theta_{nk} |t|^{2+\delta} \sum_{k=1}^n E|X_k|^{2+\delta} s_n^{-2-\delta}$$

$$\to -(t^2/2) \text{ as } n \to \infty.$$

Hence $\phi(t) \to e^{-t^2/2}$ as $n \to \infty$. Therefore, by continuity theorem for ch.f. the proof is complete.

NOTES:

(a) $L_{n,\delta} = \sum_1^n E|X_k|^{2+\delta} / s_n^{2+\delta}$, $\delta > 0$ is called Liapounov's function and $L_{n,\delta} \to 0$ as $n \to \infty$ is called Liapounov's condition.

(b) The theorem will continue to hold if condition (a) is satisfied for a certain $\delta > 0$.

(c) Bounded Liapounov's theorem is a corollary of Liapounov's theorem.

Proof. It is enough to show that conditions of bounded Liapounov's theorem imply Liapounov's condition.

$$\text{If } |X_k| \leq C < \infty \text{ a.s., then}$$

$$E|X_k|^{2+\delta} \leq C^\delta E|X_k|^2 = C^\delta \sigma_k^2$$

and hence

$$s_n^{-2-\delta} \sum_{k=1}^n E|X_k|^{2+\delta} \leq C^\delta \cdot s_n^{-\delta} \to 0 \text{ as } n \to \infty$$

Corollary 10.1 (*Particular case of Liapounov's CLT for $\delta = 1$*). Let $\{X_n\}$ be a sequence of independent r.vs. with

$$E|X_k|^3 = \upsilon_{3k} \text{ and } \sum_{k=1}^n \upsilon_{3k} = \upsilon_n. \text{ Then } \upsilon_n^{1/3} s_n^{-1} \to 0 \text{ as } n \to \infty$$

implies $S_n \cdot s_n^{-1} \xrightarrow{d} N(0, 1)$. This corollary may be proved directly (see Cramer 1946, p. 215).

(iv) *Lindeberg-Feller CLT.* Liapounov's theorem is not satisfactory, since moments of higher order than those figure in the formulation of the problem is used. Lindeberg in 1928 came up with a CLT for independent r.vs. for which Liapounov's condition for finiteness of more than second moment is not necessary. Feller in 1936 proved that Lindeberg's condition is also necessary for CLT and for negligibility of individual contribution of variance of the r.vs. compared to the total variance.

Lindeberg-Feller theorem. Let $\{X_n\}_1^\infty$ be a sequence of independent r.vs. with $EX_k = 0$, $\sigma_k^2 = \text{var}(X_k)$ and $s_n^2 = \text{var}(S_n)$. Then

$$S_n \cdot s_n^{-1} \xrightarrow{d} N(0,1) \text{ and } \max_{1 \leq k \leq n} \left(\frac{\sigma_k}{s_n} \right) \to 0 \text{ as } n \to \infty \tag{3}$$

iff for every $\varepsilon > 0$,

$$g_n(\varepsilon) = s_n^{-2} \sum_{k=1}^{n} \int_{[|x| \geq \varepsilon s_n]} x^2 \, dF_k \to 0 \text{ as } n \to \infty \tag{4}$$

where F_k is the d.f. of X_k.

NOTE.   (1) The condition (3) is called Feller's condition. $g_n(\varepsilon)$ is called Lindeberg's function and the condition (4) is called Lindeberg's condition.

$$g_n^*(\varepsilon) = s_n^{-2} \sum_{k=1}^{n} \int_{[|x| < \varepsilon s_n]} x^2 \, dF_k \to 1 \text{ as } n \to \infty \text{ is equivalent to (4).}$$

Proof of Lindeberg-Feller's theorem.   Let us first prove that Lindeberg's condition implies Feller's condition.

$$\sigma_k^2 = \int_{[|x| \leq \varepsilon s_n]} x^2 \, dF_k + \int_{[|x| > \varepsilon s_n]} x^2 \, dF_k \leq \varepsilon^2 s_n^2 + \int_{[|x| > \varepsilon s_n]} x^2 \, dF_k$$

Hence

$$\max_{k \leq n} \left(\frac{\sigma_k}{s_n} \right)^2 \leq \varepsilon^2 + s_n^{-2} \sum_{1}^{n} \int_{[|x| \geq \varepsilon s_n]} x^2 dF_k \to 0 \text{ as } n \to \infty.$$

Lindeberg's proof follows by the method of truncation and then by applying Liapounov's theorem. Since for $\varepsilon > 0$, $g_n(\varepsilon) \to 0$ as $n \to \infty$, it is possible to choose $\varepsilon_n \to 0$ as $n \to \infty$ such that $\varepsilon_n^{-2} g_n(\varepsilon_n) \to 0$ and $\varepsilon_n^{-1} g_n(\varepsilon_n) \to 0$ as $n \to \infty$. Define

$$X_{n_k} = X_k I_{[|x_k| < \varepsilon_n s_n]} \text{ and } S_{nn} = \sum_{1}^{n} X_{n_k}. \text{ Then}$$

$$P\left[\frac{S_{nn}}{s_n} \neq S_n / s_n \right] \leq \sum_{1}^{n} P[X_{n_k} \neq X_k]$$

$$= \Sigma \int_{[|x| \geq \varepsilon_n s_n]} dF_k \leq (\varepsilon_n s_n)^{-2} \sum_{1}^{n} \int_{[|x| \geq \varepsilon_n s_n]} x^2 \, dF_k = g_n(\varepsilon_n) \varepsilon_n^{-2} \to 0 \text{ as } n \to \infty.$$

Since S_{nn}/s_n and S_n/s_n are tail equivalent, (S_{nn}/s_n) and (S_n/s_n) have the same limit in law as $n \to \infty$. So it is enough to prove $S_{nn}/s_n) \xrightarrow{d} N(0, 1)$.

Put $s_{nn} = \sigma(S_{nn})$, then

$$0 \leq 1 - s_{nn}^2 / s_n^2 = [\text{var}(S_n) - \sum_{k=1}^{n} (EX_{n_k}^2 - E^2 X_{n_k})] / s_n^2$$

$$= s_n^{-2} \sum_{1}^{n} (EX_k^2 - EX_{n_k}^2) + s_n^{-2} \sum_{1}^{n} E^2 X_{n_k}$$

$$= s_n^{-2} \sum_{1}^{n} \int_{[|x| > \varepsilon_n s_n]} x^2 \, dF_k + s_n^{-2} \sum_{k=1}^{n} E^2 (X_{n_k})$$

$$= g_n(\varepsilon_n) + s_n^{-2} \sum_1^n \left| \int_{[|x| > \varepsilon_n s_n]} x \, dF_k \right|^2 \text{ (since } EX_k = 0)$$

$$\leq g_n(\varepsilon_n) + s_n^{-2} \sum_1^n \int_{[|x| > \varepsilon_n s_n]} x^2 \, dF_k \to 0 \text{ as } n \to \infty$$

Hence

$$(s_{nn}/s_n) \to 1 \text{ as } n \to \infty.$$

Therefore, it is enough to prove $(S_{nn}/s_n) \overset{d}{\to} N(0, 1)$ and it will be proved by showing that Liapounov's condition is satisfied in the case. Now

$$|X_{n_k} - EX_{n_k}| \leq 2(\varepsilon_n s_n).$$

As $\varepsilon_n \to 0$,

$$s_{nn}^{-3} \sum_1^n E|X_{n_k} - EX_{n_k}|^3 \leq 2\varepsilon_n s_n \cdot s_{nn}^{-3} \sum_1^n E(X_{n_k} - EX_{n_k})^2$$

$$\leq 2\varepsilon_n s_n \cdot s_{nn}^{-1} \to 0.$$

Since

$$EX_k = 0,$$

$$\int_{[x < \varepsilon_n s_n]} x \, dF_k = - \int_{[x \geq \varepsilon_n s_n]} x \, dF_k \leq (\varepsilon_n s_n)^{-1} \int_{[|x| \geq \varepsilon_n s_n]} x^2 \, dF_k$$

Therefore,

$$s_n^{-1} \sum_1^n |EX_{n_k}| \leq (\varepsilon_n s_n^2)^{-1} \sum_1^n \int_{[|x| \geq \varepsilon_n s_n]} x^2 \, dF_k$$

$$= \varepsilon_n^{-1} g_n(\varepsilon_n) \to 0 \text{ as } n \to \infty.$$

Hence

$$s_n^{-1} ES_{nn} \to 0 \text{ as } n \to \infty \text{ and } E(S_{nn}/s_{nn}) \to 0 \text{ as } n \to \infty.$$

Proof of the converse (i.e. Feller's condition and CLT implies Lindeberg's condition). Let $\phi_k(t)$ be the ch.f. of X_k. Then by Theorem 9.6 of Section 9.1

$$|\phi_k(t/s_n) - 1| \leq t^2 \, \sigma_k^2 \cdot \frac{1}{2} s_n^{-2}$$

and hence by Feller's condition

$$\max_{k \leq n} |\phi_k(t/s_n) - 1| \to 0 \text{ as } n \to \infty \tag{5}$$

Also,

$$\sum_{k=1}^n |\phi_k(t/s_n) - 1|^2 \leq \max_{k \leq n} |\phi_k(t/s_n) - 1| \sum_1^n |\phi_k(t/s_n) - 1|$$

$$\leq t^4/4 \max_{k \leq n} (\sigma_k/s_n)^2 \to 0 \text{ as } n \to \infty.$$

By (5), $|\phi_k(t/s_n) - 1| < \delta$ for $n \geq n_o$ and $k \leq n$. Hence

$$\sum_1^n |\log \phi_k(t/s_n) - (\phi_k(t/s_n) - 1| \leq \sum_1^n |\phi_k(t/s_n) - 1|^2 = o(1) \text{ as } n \to \infty \quad (6)$$

Since

$$\text{CLT holds } \prod_1^n \phi_k(t/s_n) \to e^{-t^2/2}, \text{ whence } \sum_1^n \log \phi_k(t/s_n) \to -t^2/2. \text{ Hence by (6),}$$

$$-t^2/2 = \sum_1^n [\phi_k(t/s_n) - 1] + o(1) \quad (7)$$

Taking the real part on both sides of (7)

$$\sum_1^n \int_{-\infty}^{\infty} (1 - \cos(tx/s_n)) \, dF_k + o(1) = t^2/2 = \sum_1^n \int_{[|x|<\varepsilon s_n]} (1 - \cos(tx/s_n)) \, dF_k$$

$$+ \sum_1^n \int_{[|x|\geq\varepsilon s_n]} (1 - \cos(tx/s_n)) \, dF_k + o(1)$$

Now

$$\sum_1^n \int_{[|x|<\varepsilon s_n]} (1 - \cos(tx/s_n)) \, dF_k \leq \frac{1}{2} t^2 s_n^{-2} \sum_1^n \int_{[|x|<\varepsilon s_n]} x^2 dF_k$$

$$= \frac{1}{2} t^2 s_n^{-2} (s_n^2 - \sum_1^n \int_{[|x|\geq\varepsilon s_n]} x^2 dF_k).$$

Hence

$$\frac{1}{2} t^2 g_n(\varepsilon) + o(1) \leq \sum_1^n \int_{[|x|\geq\varepsilon s_n]} (1 - \cos(tx/s_n)) \, dF_k$$

$$\leq 2 \sum_1^n \int_{[|x|\geq\varepsilon s_n]} dF_k \leq 2(\varepsilon s_n)^{-2} \sum_{k=1}^n \int_{[|x|\geq\varepsilon s_n]} x^2 \, dF_k$$

$$= 2 \varepsilon^{-2} g_n(\varepsilon).$$

Choose $t \leq 4/\varepsilon$. Then $\varepsilon^{-2} g_n(\varepsilon) = o(1)$ for $\varepsilon > 0$.
Therefore $g_n(\varepsilon) \to 0$ as $n \to \infty$ for all $\varepsilon > 0$.

10.2.2 A Counter Example to CLT (Non-universality of the Normal Law)

Theory asserts and observation confirms the assertion that the normal law is to be expected in a very great number of frequency distributions but not in all. Lippmann remarked to Poincare "Everybody believes in the normal law of errors, the experimenters because they think it can be proved by mathematics and the mathematicians because they believe it has been established by observation". Following example shows that even in case of continuous r.vs. a normalized partial sum can converge in distribution to some non-normal distribution.

Let X_1, X_2, ... be a sequence of i.i.d. r.vs. with p.d.f.

$$f(x) = \frac{1}{2} e^{-|x|}$$

$$= \begin{cases} 1/2 e^{-x} & \text{if } x \geq 0 \\ 1/2 e^{x} & \text{if } x \leq 0 \end{cases}$$

Then $EX_k = 0$ for all $k \geq 1$ and $EX_k^2 = 2$. Ch.f. of X_k is

$$\phi_k(\theta) = \int_{-\infty}^{\infty} e^{i\theta x} f(x)\, dx$$

$$= \int_{-\infty}^{0} e^{i\theta x} f(x)\, dx + \int_{0}^{\infty} e^{i\theta x} f(x)\, dx$$

$$= \frac{1}{2} \int_{0}^{\infty} e^{-x(1-i\theta)}\, dx + \frac{1}{2} \int_{-\infty}^{0} e^{x(1+i\theta)}\, dx$$

$$= \frac{1}{2} \frac{1}{1-i\theta} + \frac{1}{2} \frac{1}{1+i\theta} = \frac{1}{1+\theta^2}$$

Consider the new r.v.

$$Y \overset{d}{=} \frac{2}{\pi} \left(X_1 + \frac{1}{3} X_2 + \frac{1}{5} X_3 + \dots \right)$$

where

$$Z_k = \frac{2}{\pi} \frac{X_k}{2k-1}, \ k \geq 1 \text{ and } S_n = Z_1 + Z_2 + \dots + Z_n \overset{d}{\to} Y \text{ as } n \to \infty.$$

Since $1 + (1/3)^2 + (1/5)^2 + \dots = \pi^2/8$, $ES_n = 0$, var $(Y) = 1$ and var $(S_n) \sim 1$ as $n \to \infty$. Now $\{Z_k\}$ is an independent non-identically distributed sequence of r.vs. So the ch.f. of the sum

$$S_n = Z_1 + Z_2 + \dots + Z_n \text{ is } \phi_{[n]}(\theta) = \phi_1(2/\pi\theta) \dots \phi_n \left(\frac{2\theta}{(2n-1)\pi} \right)$$

$$= \frac{1}{1+\dfrac{2^2 \theta^2}{\pi^2}} \frac{1}{1+\dfrac{2^2 \theta^2}{3^2 \pi^2}} \frac{1}{+\dfrac{2^2 \theta^2}{5^2 \pi^2}} \dots \frac{1}{1+\dfrac{2^2 \theta^2}{(2n-1)^2 \pi^2}}$$

$$\text{(By Gregory series)} \to 2(e^{\theta} + e^{-\theta})^{-1}$$

So the ch.f. of Y is $\phi(\theta) = \lim_{n \to \infty} \phi_{[n]}(\theta) = \dfrac{1}{\cosh \theta}$

By inversion theorem p.d.f. of Y is $f_Y(y) = \dfrac{1}{2\pi} \int_{-\infty}^{\infty} \phi(\theta)\, e^{-i\theta y}\, d\theta$

Making the transformation (in contour integration) $\theta = x + iy$

$$f_Y(y) = \frac{1}{\pi} \int_{-\infty}^{\infty} \frac{e^{-i\theta y}}{e^{\theta} + e^{-\theta}} \, d\theta$$

$$= \frac{1}{e^{1/2\pi y} + e^{-1/2\pi y}}$$

which is the logistic p.d.f.

10.2.3   Exercises on Univariate CLT

Exercise 10.1   Let $\{X_k\}$ are independent r.vs. with $P(X_k = \pm k^{\lambda}) = 1/2$. Then weak law of large numbers holds iff $\lambda < 1/2$ and CLT holds iff $\lambda \geq -1/2$.

Proof. $\mu_k = 0$ for $k \geq 1$, $\sigma_k^2 = k^{2\lambda}$,

$$s_n^2 = 1 + 2^{2\lambda} + \ldots + n^{2\lambda} \sim \int_0^n x^{2\lambda} \, dx = (n^{2\lambda+1})/(2\lambda + 1)$$

Then

$$s_n^2/n^2 = \frac{1}{2\lambda + 1} n^{2(\lambda - 1/2)} \to 0 \text{ if } \lambda < 1/2$$

Therefore, WLLN holds if $\lambda < 1/2$. Now for $k = 1, 2, \ldots, n$, $|X_k| = k^{\lambda} \leq n^{\lambda}$ for all $k \leq n$.

Also

$$s_n^2 = \frac{n^{2\lambda+1}}{2\lambda + 1} \to \infty \text{ if } 2\lambda + 1 > 0 \text{ or } \lambda > -1/2.$$

To complete the proof of Exercise 10.1 let us state and prove the following result as a lemma.

Lemma 10.1   For a sequence of uniformly bounded independent r.vs. a necessary and sufficient condition for the central limit theorem to hold is $s_n^2 \to \infty$ as $n \to \infty$.

Proof.   Suppose that if possible $\lim_{n \to \infty} s_n^2 < \infty$ and the Lindeberg condition holds, the $s_n^2 \to B^2$. For any fixed j, we can find an $\varepsilon > 0$ such that $P\,[|X_j - \mu_j| > \varepsilon B] > 0$. Then for $n \geq j$,

$$s_n^{-2} \sum_{k=1}^{n} \int_{[|x - \mu_k| \geq \varepsilon s_n]} (x - \mu_k)^2 \, dF_k(s) \geq \varepsilon^2 \sum_{k=1}^{n} P[|X_k - \mu_k| > \varepsilon s_n]$$

$$\geq \varepsilon^2 P\,[|X_j - \mu_j| > \varepsilon B] > 0$$

and the Lindeberg condition does not hold. Hence $s_n^2 \to \infty$ is a necessary condition. Alternately

$$\sum_1^n (X_k - \mu_k)/s_n = \frac{X_1 - \mu_1}{s_n} + \frac{1}{s_n}\sum_2^n (X_k - \mu_k)$$

If CLT holds the second term in the right must converge to $N(0, 1)$ but the first term on the right is a bounded normal r.v. Hence $s_n^2 \to \infty$ as $n \to \infty$. Take $s_n > 2c\varepsilon^{-1}$ if n is large enough. Then $\varepsilon s_n > 2c$ if n is large enough. Also $|X_k - \mu_k| \le 2c$ and there exists an N_ε such that for $n \ge N_\varepsilon$ and $|X_k - \mu_k| \le \varepsilon s_n$ for $k = 1, 2, 3, \ldots, n$ with probability one and hence

$$\sum_1^n \int_{|X_k - \mu_k| \le \varepsilon s_n} (X_k - \mu_k)^2 \, dP = \mathrm{var}\,(S_n) = s_n^2$$

So Lindeberg condition (4) is trivially satisfied, which completes the proof of the lemma.

Now

$$s_n^2 = \frac{n^{2\lambda+1}}{2\lambda+1},\, |X_k| \le n^\lambda \text{ for all } 1 \le k \le n$$

so that Lindeberg condition holds if

$$\varepsilon s_n \sim \frac{n^{\lambda+1/2}}{\sqrt{(2\lambda+1)}} \varepsilon > n^\lambda \ \text{ or } \ n^{1/2} > (\sqrt{(2\lambda+1)}\varepsilon^{-1}) \ \text{ or } \ n > (2\lambda+1)\,\varepsilon^{-2} \ge 0$$

holds. This requires $2\lambda + 1 \ge 0$, i.e. $\lambda \ge -1/2$, so CLT holds. The case $\lambda = -1/2$, should be treated separately.

If $\lambda = -1/2$, $s_n^2 = 1 + 1/2 + \ldots + 1/n$, $s_n \varepsilon \sim \sqrt{(\log n)}\,\varepsilon > $, $|X_k| \le 1$ for $1 \le k \le n$. So the lemma applies here. To show that WLLN holds in the same example if $\lambda < 1/2$, we have proved that WLLN holds if $\lambda < 1/2$. Now we shall show that WLLN does not hold if $\lambda \ge 1/2$.

We claim that $S_n = 0\,(n^{\lambda+1/2})$ with probability tending to one, i.e. $S_n = 0_p(n^{\lambda+1/2})$. Since CLT holds for $\lambda \ge 1/2$,

$$P\left(C_1 < \frac{S_n}{n^{\lambda+1/2}}\sqrt{(2\lambda+1)} < C_2\right) \to \Phi(C_2) - \Phi(C_1),$$

where Φ is the c.d.f. of $N(0, 1)$. This implies $S_n = 0(n^{\lambda+1/2})$ with probability close to 1 if we choose C_2 large and C_1 small. Therefore $S_n/n = 0\,(n^{\lambda-1/2})$ with probability close to 1. So S_n/n tend to a non-zero constant or ∞ with probability close to 1 if $\lambda \ge 1/2$. So WLLN does not hold if $\lambda \ge 1/2$.

Exercise 10.2   Let $\{X_k\}$ be a sequence of independent r.vs. with

$$P[X_k = k] = 1/2\, k^{-\lambda} = P(X_k = -k), P(X_k = 0) = 1 - k^{-\lambda}, \lambda \ge 0$$

show that CLT holds if $0 \le \lambda < 1$.

Solution.   Here $EX_k = 0$, $\sigma_k^2 = k^{2-\lambda}$, $s_n^2 = \sum_1^n k^{2-\lambda} \sim C n^{3-\lambda}$.

Now $s_n \to \infty$ only if $\lambda < 3$. Using Liapounov's condition, CLT holds for $0 < \lambda < 1$. In fact, since $s_n = 0(n^{(3-\lambda)/2})$, $\varepsilon s_n > n$ for n sufficiently large, $0 < \lambda < 1$ and $\varepsilon > 0$. Therefore,

$$s_n^2 g_n(\varepsilon) = \sum_{k \geq [\varepsilon s_n]} (k^{2-\lambda})$$

will not contain any term for large n and is finite. Hence $g_n(\varepsilon) \to 0$. So for $0 < \lambda < 1$ CLT holds. For $\lambda = 1$,

$$\sigma_k^2 = k, \ s_n^2 = n(n+1)/2, s_n \sim cn, c > 0,$$

$$s_n^2 g_n(\varepsilon) \sim \sum_{k \geq [\varepsilon s_n]}^{n} k \sim n^2 (1-\varepsilon^2), \varepsilon > 0$$

$g_n(\varepsilon) \sim 1 - \varepsilon^2 \not\to 0$. Obviously

$$\max_k (\sigma_k/s_n)^2 = \frac{2}{n+1} \to 0$$

and CLT does not hold. For $\lambda = 0$, CLT holds by previous Exercise 10.1.

Exercise 10.3 Let X_1, X_2, ... be independent r.vs. with $P(X_1 = \pm 1) = 1/2$ and for $k > 1$, $P(X_k = \pm 1) = 1/(2c)$, $P(X_k = \pm k) = 1/(2k^2)(1 - 1/c)$, $P(X_k = 0) = 1 - 1/c - 1/k^2(1 - 1/c)$, $1 < c < \infty$.

Then show that $\{X_k\}$ does not obey Lindeberg's condition and CLT does not hold.

Solution. $EX_k = 0$, var $X_k = 1$, $s_n = \sqrt{n}$, and for $\varepsilon > 0$ take n so large that $\varepsilon\sqrt{n} > 1$. Then

$$\frac{1}{s_n^2} \sum_{k=1}^{n} \int_{[|x| \geq \varepsilon s_n]} x^2 \, dF_k(x) = 1/n \sum_{k=1}^{n} E[X_k^2 I_{[x_k| \geq \varepsilon\sqrt{n}]}] >$$

$$1/n \sum_{k=[\varepsilon\sqrt{n}]}^{n} k^2 P[|X_k| = k] \sim 1/n (n - \varepsilon\sqrt{n})(1 - 1/c) \to (1 - 1/c) > 0$$

So Lindeberg's condition does not hold, but Feller's condition $\max_{1 \leq k \leq n} (\sigma_k/s_n) = 1/\sqrt{n} \to 0$ holds. So CLT does not hold.

Exercise 10.4 With same $\{X_k\}$ as in Exercise 10.3 define:

$$Y_k = X_k I_{[|x_k| \leq \sqrt{n}]}$$

satisfying Lindeberg's condition and hence show that CLT holds.

Solution. $\sigma^2(Y_k) = EX_k^2 = 1$ if $k \leq \sqrt{n}$

$$= P[|X_k| = 1] = 1/c \text{ if } k > \sqrt{n}$$

Let $S_n^* = \sum_1^n Y_k$. Then $\sigma^2(S_n^*) = \sum_1^n \sigma^2(Y_k) = [\sqrt{n}] + 1/c(n - [\sqrt{n}]) = 0(n/c)$

$$g_n(\varepsilon) = \frac{1}{\sigma^2(S_n^*)} \sum_{k=1}^n E(Y_k)^2 \, I_{[|Y_k| \geq \varepsilon\sigma(S_n^*)]}$$

$$\sim c/n \sum_{k=[\sigma(S_n^*)\varepsilon]}^n k^2 P[|X_k| = k]$$

$$\sim c/n \, (\sqrt{n} - \varepsilon\sqrt{n}/c) \, (1 - 1/c) \to 0$$

Hence Lindeberg's condition holds.

Exercise 10.5   Let $X_1, \ldots, X_n$ be independent r.vs., $P(X_k = \pm a_k) = 1/2$, $a_k > 0$, k 1, 2, 3, ... Find the condition when CLT holds.

Solution. $EX_k = 0, \sigma_k = a_k, s_n = \sqrt{(\sum_1^n a_k^2)} \to \infty, m_n = \max_{1 \leq k \leq n} a_k$

$$a_k^{2+\delta} = E|X_k|^{2+\delta} \leq m_n^\delta a_k^2, \sum_k a_k^{2+\delta} \geq \max a_k^{2+\delta} = m_n^{2+\delta}$$

$$(m_n/s_n)^{2+\delta} \leq \frac{\sum_{k=1}^n a_k^{2+\delta}}{(s_n^{2+\delta})} = \frac{1}{(s_n^{2+\delta})} \sum_1^n E|X_k|^{2+\delta} \leq m_n^\delta/s_n^\delta$$

Therefore, $\max_{1 \leq k \leq n} (\sigma_k/s_n) \to 0$ (Feller's condition holds) iff Liapounov's condition holds. Since the latter implies Lindeberg's condition to hold, the condition $m_n/s_n \to 0$ is necessary and sufficient for CLT to hold.

Exercise 10.6   Let $\{X_n\}$ be an independent sequence of r.vs. such that each X_n only attains the values 0 and 1 with positive probability. Prove that $\{X_n\}$ satisfies the CLT if the series $\sum_{n=1}^\infty P(X_n = 0) P(X_n = 1)$ diverges.

Proof.   Since r.vs. are independent,

$$s_n^2 = \sum_{k=1}^n \text{var}(X_n) \text{ and hence } s_n \to \infty \text{ iff } \sum_{n=1}^\infty \text{var}(X_n) \text{ diverges.}$$

Since r.vs. are uniformly bounded, by bounded Liapounov's theorem it is enough

to show that $\sum_{n=1}^\infty \text{var}(X_n)$ diverges. Now

$$\text{var}(X_n) = P(X_n = 1) - P^2(X_n = 1) = P(X_n = 1)(1 - P(X_n = 1))$$

$$= P(X_n = 1) P(X_n = 0).$$

Hence CLT holds if $\sum_{n=1}^\infty \text{var}(X_n) = \sum_{n=1}^\infty P(X_n = 0)P(X_n = 1)$ diverges.

Exercise 10.7   Let $\{X_n\}$ be a sequence of independent r.vs. with

$$P(X_n = \pm \, n^\lambda) = 1/2. \text{ Show that}$$

(a) For $\lambda < - 1/2$ the sequence $\{X_n\}$ does not satisfy the Feller's condition.

(b) $\{X_n\}$ satisfies the CLT for all $\lambda \geq - 1/2$.

Solution.   (a) $s_n^2 = 1^{2\lambda} + 2^{2\lambda} + \ldots + n^{2\lambda} \sim n^{2\lambda+1}/(2\lambda + 1)$

Then $\max\limits_{1 \leq k \leq n} (\sigma_k/s_n) = \max\limits_{1 \leq k \leq n} \dfrac{k^\lambda}{(n^{\lambda+1/2})} \sqrt{(2\lambda + 1)}$

$$= \max\limits_{1 \leq k \leq n} c \cdot k^{-\eta} n^\varepsilon \text{ for some } 0 < \eta < 1/2 \text{ and } \varepsilon > 0 \text{ if } \lambda < -1/2$$

$$= c_1 \cdot n^\varepsilon \to \infty \text{ as } n \to \infty \text{ if } \lambda < -1/2.$$

So the Feller's condition does not hold.

(b) See Exercise 10.1.

10.3   CRÁMER-WOLD DEVICE AND MULTIVARIATE CLT

By means of a simple device due to Crámer and Wold problems involving random vectors in R^k can often reduced to problems involving only ordinary r.vs. in R^1. The following result, due to Crámer and Wold (1936) allows the question of convergence of multivariate d.f.s. to be reduced to that of convergence of univariate d.f.

Theorem 10.1   In R^k the random vectors $\mathbf{X}_n$ converge in distribution to the random vector $\mathbf{X}$ iff each linear combination of the components of $\mathbf{X}_n$ converges in distribution to the same linear combination of the components of $\mathbf{X}$.

Proof.   Put $\mathbf{X}_n = (X_{n1}, \ldots, X_{nk})$ and $\mathbf{X} = (X_1, \ldots, X_k)$ and denote the corresponding ch. f. by ϕ_n and ϕ. Assume now that for any real $l_1, \ldots, l_k$, $l_1 X_{n1} + \ldots + l_k X_{nk} \xrightarrow{d} l_1 X_1 + \ldots + l_k X_k.$

Then by continuity theorem for ch.f.

$$\lim_{n \to \infty} \phi_n(tl_1, \ldots, tl_k) = \phi(tl_1, \ldots, tl_k) \text{ for all } t.$$

With $t = 1$ and since $l_1, \ldots, l_k$ are arbitrary, it follows by the multivariate version of continuity theorem $\mathbf{X}_n \xrightarrow{d} \mathbf{X}$.

The converse is proved by a similar argument.

Exercise 10.8   Show that $\mathbf{X} = (X_1, \ldots, X_k)$ is multivariate normal iff $l_1 X_1 + \ldots + l_k X_k$ is univariate normal for every real $l_1, \ldots, l_k$.

Proof.   The characteristic function of $\mathbf{X}$ is

$$\phi(u_1, \ldots, u_k) = E\,(\exp\, i(u_1 X_1 + \ldots + u_k X_k))$$

Now

$$Y = \sum_{j=1}^{n} u_j X_j \text{ is univariate normal and hence}$$

$$E(e^{itY}) = \exp\,(it\mu - t^2\sigma^2/2), \text{ where } \mu = E(Y) = \sum_{j=1}^{k} u_j\, EX_j$$

and

$$\sigma^2 = \mathrm{var}\,(Y) = \sum_{i,j=1}^{k} u_i\, \mathrm{cov}\,(X_i, X_j)\, u_j.$$

Set $t = 1$, $b_j = EX_j$ and $\sigma_{ij} = \mathrm{cov}\,(X_i, X_j)$ to obtain

$$\phi(u_1, \ldots, u_k) = \exp\Big[i \sum_{j=1}^{k} u_j b_j - 1/2 \sum_{i,j=1}^{k} u_i \sigma_{ij} u_j\Big]$$

which is the characteristic function of multivariate normal. By uniqueness theorem of multivariate characteristic function $\mathbf{X}$ is multivariate normal. Converse can be done similarly and the proof is well known. Hence we omit the proof.

10.3.1   Multivariate CLT for Independent Random Variables (Multivariate Version of Lindeberg's CLT)

Let $\mathbf{X}_1, \ldots, \mathbf{X}_n$ be independent (k-dimensional) random vectors with mean vector

$$EX_i = 0 \text{ and } D(\mathbf{X}_i)^{k\times k} = \sum_{i}^{k\times k} \text{ and } \mathbf{S}_n = \mathbf{X}_1 + \ldots + \mathbf{X}_n.$$

If

$$\frac{1}{n}\sum_{i=1}^{n}\sum_{i}^{k\times k} \to \Sigma \neq \mathbf{O} \text{ and for every } \varepsilon > 0, \text{ and}$$

$$\frac{1}{n}\sum_{1}^{n} \int_{[\|\mathbf{x}\| \geq \varepsilon\sqrt{n}]} \|\mathbf{x}\|^2\, dF_i \to 0 \text{ as } n \to \infty,$$

where F_i is the k-dimensional d.f. of $\mathbf{X}_i$ and $\|\,\mathbf{x}\,\| = (\sum_{1}^{k} x_i^2)^{1/2}$, then $\mathbf{S}_n/n^{1/2} \to N_k$

$(\mathbf{O}, \Sigma^{k\times k})$.

Proof.   The proof follows from Univariate Lindeberg's CLT and Crámer-Wold device.

10.4   APPLICATION TO LARGE SAMPLE THEORY OF STATISTICS

10.4.1   Order in Probability (Stochastic O(·) and o(·))

Let $\{a_n\}$ be an infinite sequence of real numbers and $\{g_n\}$ be a sequence of positive real numbers.

Definition 10.1 $\{a_n\}$ is of smaller order than $\{g_n\}$ and write

$$\{a_n\} = o(g_n) \text{ if } \lim_{n\to\infty} a_n/g_n = 0$$

Definition 10.2 $\{a_n\}$ is atmost of order $\{b_n\}$ and write $a_n = O(g_n)$ if there exists a real $M > O$ such that $|a_n|/g_n \leq M$ for all $n \geq n_o$.

Lemma 10.2   Let $\{a_n\}$ and $\{b_n\}$ be real sequences.
Let $\{f_n\}$ and $\{g_n\}$ be positive real sequences.

(a) If $a_n = o(f_n)$ and $b_n = o(g_n)$, then

$$a_n b_n = o(f_n g_n),$$

$$|a_n|^r = o(f_n^r) \text{ for } r > 0,$$

$$a_n + b_n = o(\max(f_n, g_n)).$$

(b) If $a_n = O(f_n)$ and $b_n = O(g_n)$, then

$$a_n b_n = O(f_n g_n),$$

$$|a_n|^r = O(f_n^r) \text{ for } r \geq 0,$$

and $$a_n + b_n = O(\max(f_n, g_n)).$$

(c) If $a_n = o(f_n)$ and $b_n = O(g_n)$, then

$$a_n b_n = o(f_n g_n).$$

Order in probability was introduced by Mann and Wald (1943). Let $\{X_n\}$ be a sequence of r.vs. and $\{g_n\}$ a positive real sequence.

Definition 10.3 $\{X_n\}$ is of smaller order in probability than $\{g_n\}$ and write $X_n = O_p(g_n)$ if $X_n/g_n \xrightarrow{P} 0$.

Definition 10.4   We say $\{X_n\}$ is atmost of order $\{g_n\}$ in probability and write $X_n = O_p(g_n)$, if for all $\varepsilon > 0$, there exists $M > 0$ such that $P(|X_n| \geq Mg_n) \leq \varepsilon$ for all $n \geq n_0$.

Note that $X_n = o_p(g_n) \Rightarrow X_n = O_p(g_n)$. More generally, for two sequences $\{X_n\}$ and $\{Y_n\}$, the notation $X_n = O_p(Y_n)$ denotes that the $\{X_n/Y_n\}$ is $O_p(1)$. Further, the notation $X_n = o_p(Y_n)$ denotes that $X_n/Y_n \xrightarrow{P} 0$. Also note that $X_n = o_p(Y_n) \Rightarrow X_n = O_p(Y_n)$.

Definition 10.5   If $\{\mathbf{X}_n\}$ is a sequence of k-dimensional r.vs. $\{\mathbf{X}_n\}$ is atmost of order $\{g_n\}$ in probability and write $\mathbf{X}_n = O_p(g_n)$ if, for every $\varepsilon > 0$, there exists $M_\varepsilon > 0$ such that

$$P[|X_{jn}| \geq M_\varepsilon g_n] \leq \varepsilon, j = 1, 2, ..., k,$$

for all n. $\{\mathbf{X}_n\}$ is of smaller order in probability than $\{g_n\}$ and we write $\mathbf{X}_n = o_p(g_n)$ if for all $\varepsilon > 0$ and $\delta > 0$ there exists an N such that for all $n \geq N$

$$P[|X_{jn}| \geq \varepsilon g_n] \leq \delta, j = 1, 2, ..., k$$

(note k might be a function of n)

NOTES.

(a) If $\mathbf{X}_n = O_p(g_n)$, we also say that $\{\mathbf{X}_n\}$ is bounded in probability by $\{g_n\}$. It is readily seen that $\mathbf{X}_n \xrightarrow{d} \mathbf{X} \Rightarrow \mathbf{X}_n = O_p(1)$.

(b) A matrix r.v. may be viewed as a vector r.v. with the elements displayed in a particular manner or as a collection of vector random variables.

(c) $\mathbf{X}_n = o_p(1). \Rightarrow \lim \mathbf{X}_n \overset{P}{=} \mathbf{X}$ if dimension k is not fixed; e.g. if $k = n$ and $X_{jn} = n^{-1/2}$ a.s. $j = 1, 2, \ldots, n$.

Definition 10.6   A $k \times r$ matrix B_n of r.vs. is atmost of order $\{g_n\}$ in probability and we write

$$B_n = O_p(g_n) \text{ if, for every } \varepsilon > 0,$$

there exists a positive number M_ε such that $P[|b_{ijn}| \geq M_\varepsilon g_n] \leq \varepsilon$ for $i = 1, 2, \ldots,$ $k, j = 1, 2, \ldots, r$ for all large n, where the b_{ijn} are elements of B_n. Similarly, smaller order in probability for matrix B_n is defined. An analogue of Lemma 10.2 for r.vs. is the following.

Lemma 10.3   Let $\{f_n\}$ and $\{g_n\}$ be sequences of positive real numbers and let $\{X_n\}$ and $\{Y_n\}$ be sequences of r.vs.

(a) If $X_n = o_p(f_n)$ and $Y_n = o_p(g_n)$, then

$$X_n Y_n = o_p(f_n g_n),$$

$$|X_n|^s = o_p(f_n^s) \text{ for } s > 0 \text{ and}$$

$$X_n + Y_n = o_p(\max (f_n, g_n))$$

(b) If $X_n = O_p(f_n)$ and $Y_n = O_p(g_n)$, then

$$X_n Y_n = O_p(f_n g_n),$$

$$|X_n|^s = O_p(f_n^s) \text{ for } s \geq 0 \text{ and}$$

$$X_n + Y_n = O_p(\max (f_n, g_n))$$

(c) If $X_n = o_p(f_n)$ and $Y_n = O_p(g_n)$, then

$$X_n Y_n = o_p(f_n g_n).$$

Proof.   (a) Now for $\varepsilon > o$ there exists $\delta > 0$ such that $P[|X_n| > \varepsilon f_n] < \delta/2$ and

$$P[|Y_n| > \varepsilon g_n] < \delta/2 \text{ for } n \geq N_o \qquad (1)$$

Therefore

$$P[|X_n Y_n| > \varepsilon^2 f_n g_n]$$

$$\leq P[|X_n/f_n| > \varepsilon] + P[|Y_n/g_n| > \varepsilon] \leq \delta \text{ for } n \geq N_o.$$

(b) $P[|X_n| > \varepsilon f_n] = P[|X_n|^s > \varepsilon^s f_n^s]$, for all $\varepsilon > 0$. So $|X_n|^s = O_p(f_n^s)$ for $s > 0$.

From writing $q_n = \max (f_n, g_n)$, we get, given $\varepsilon > 0$ and $\delta > 0$ there exists integer n such that

$$P[|X_n| > 1/2 \, \varepsilon \, q_n] < \delta/2 \text{ and } P[|Y_n| > 1/2 \, \varepsilon \, q_n] < \delta/2$$

for all $n \geq N_o$. So $P\,[\,|\,X_n + Y_n\,| > \varepsilon\,q_n] \leq P\,[\,|\,X_n\,| > \varepsilon/2q_n] + P\,[\,|\,Y_n\,| > \varepsilon/2q_n]$ $< \delta/2 + \delta/2 = \delta$ for all $n \geq N_o$.

Result 1.   Let $\{X_n\}$ be a sequence of r.vs. and $\{a_n\}$ be a sequence of positive real numbers such that $EX_n^2 = O(a_n^2)$. Then $X_n = O_p(a_n)$.

Proof.   By assumption there exists $M_2 > 0$ such that $EX_n^2 \leq M_1^2 a_n^2$ for all $n \geq N_o$.

By Chebychev's inequality

$$P\,[\,|\,X_n\,| \geq M_2 a_n] \leq EX_n^2 / (M_2^2 a_n^2) \text{ (by choosing } M_2 \geq M_1 \varepsilon^{-1/2})$$

$$\leq EX_n^2 / (M_1^2 \varepsilon^{-1} a_n^2) < \varepsilon \text{ for all } n \geq N_o.$$

So $$X_n = O_p(a_n).$$

Result 2   Let $\{X_n\}$ be a sequence of r.vs. satisfying

$$\sigma_n^2 = E(X_n - EX_n)^2 = O(a_n^2) \text{ and } EX_n = O(a_n)$$

Then, $$X_n = O_p(a_n).$$

Proof. $EX_n^2 = \sigma_n^2 + [E(X_n)]^2 = O(a_n^2)$ by Lemma 9.2(b).

So by result 1, $X_n = O_p(a_n)$.

Mann and Wald (1943) demonstrated that the algebra of the common order relationship holds for order in probability.

Theorem 10.2 (*Pratt, 1959; Mann and Wald, 1943*).   Let $\{X_n\}$ be a sequence of k-dimensional r.vs. with elements $\{X_{jn}, j = 1, 2, \ldots, k\}$ and let $\{r_n\}$ be a sequence of k-dimensional vectors with positive real elements $\{r_{jn}, j = 1, 2, \ldots, k\}$ such that

$$X_{jn} = O_p(r_{jn}) \; j = 1, 2, \ldots, l$$

$$X_{jn} = o_p(r_{jn}) \; j = l + 1, l + 2, \ldots, k$$

Let $g_n(X)$ be a sequence of real-valued (Borel measurable) functions defined on k-dimensional space and let $\{s_n\}$ be a positive real sequence. Let $\{a_n)$ be a non-random sequence of k-dimensional vectors.

If $g_n(a_n) = O(s_n)$ for all sequences $\{a_n\}$ such that

$$a_{jn} = O(r_{jn}), j = 1, 2, \ldots, l$$

$$a_{jn} = o(r_{jn}), j = l + 1, \ldots, k, \text{ then } g_n(\mathbf{X}_n) = O_p(s_n).$$

If $g_n(a_n) = o(s_n)$ instead of $g_n(a_n) = O(s_n)$ in the hypothesis, then $g_n(X_n) = o_p(s_n)$.

10.4.2 Taylor Expansion in Probability

We know that if g is continuous at a and $X_n = a + o_p(1)$ then $g(X_n) = g(a) + o_p(1)$. If we strengthen the assumption on g to include the existence of derivatives, it is possible to derive probabilistic analogies of the Taylor expansions of non-random functions about a given point a.

Proposition 10.1   Let $\{X_n\}$ be a sequence of r.vs. such that $X_n = a + O_p(r_n)$ and $0 < r_n \to 0$ as $n \to \infty$. If g is a function with s derivatives at a, then

$$g(X_n) = \sum_{j=0}^{s} g^{(j)}(a)/j!\,(X_n - a)^j + o_p(r_n^s),\ \text{where}\ g^{(j)}$$

is the jth derivative of g and $g^{(o)} = g$.

Proof.   Let $h(X) = [g(x) - \sum_{j=0}^{s} g^{(j)}(a)/j!\,(X_n - a)^j]/\left[\dfrac{(x-a)^s}{s!}\right],\ X \neq a$

$$= 0\ \text{if}\ x = a.$$

Then h is continuous at a, so that $h(X_n) = h(a) + o_p(1)$. This implies $h(X_n) = o_p(1)$. By Lemma 10.3

$$(X_n - a)^s\, h(X_n) = o_p(r_n^s)$$

which proves the result.

EXAMPLE 10.1   Let $\{X_n\}$ be a sequence of i.i.d. r.vs. with mean $\mu > 0$ and variance σ^2. By Chebychev's inequality $P(n^{1/2}\,|\,\bar{X}_n - \mu\,| > \varepsilon) \leq \sigma^2 \varepsilon^{-2}$ and hence $\bar{X}_n - \mu = O_p(n^{1/2})$. Since in (X) has a derivative at $\mu > 0$, by Proposition 10.1, we obtain the expansion

$$\ln(\bar{X}_n) = \ln(\mu) + \mu^{-1}(\bar{X}_n - \mu) + o_p(n^{-1/2}).$$

We give a multivariate analogue of Proposition 10.1.

Proposition 10.2   Let $\{X_n\}$ be a sequence of random k-vectors such that $\mathbf{X}_n - \mathbf{a} = O_p(r_n)$ and $r_n \to 0$ as $n \to \infty$. If g is a function from R^k to R such that the derivatives $\partial g/\partial x_i$ are continuous in a neighbourhood of $\mathbf{a}$, then

$$g(\mathbf{X}_n) = g(\mathbf{a}) + \sum_{i=1}^{k} \frac{\partial g}{\partial x_i}(\mathbf{a})(X_{ni} - a_i) + o_p(r_n).$$

Proof.   By Taylor expansion of several variables we have, as $\mathbf{x} \to \mathbf{a}$

$$g(\mathbf{X}) = g(\mathbf{a}) + \sum_{i=1}^{k} \frac{\partial g}{\partial x}(\mathbf{a})(X_i - a_i) + o(|\mathbf{X} - \mathbf{a}|)$$

Define $h(\mathbf{X}) = [g(\mathbf{X}) - g(\mathbf{a}) - \sum_{i=1}^{k} \dfrac{\partial g}{\partial x_i}(\mathbf{a})(X_i - a_i)]/(|\,\mathbf{X} - \mathbf{a}\,|)\ \text{if}\ \mathbf{X} \neq \mathbf{a}$

$$= 0\ \text{if}\ \mathbf{X} = \mathbf{a}.$$

Then h is continuous at $\mathbf{a}$ and hence $h(\mathbf{X}_n) = o_p(1)$. By Lemma 10.3 this implies $h(\mathbf{X}_n)\,|\,\mathbf{X} - \mathbf{a}\,| = o_p(r_n)$, which proves the result.

Definition 10.7   A sequence of r.vs. $\{X_n\}$ is said to be asymptotically normal with "mean" μ_n and "standard deviation" σ_n if $\sigma_n > 0$ for all sufficiently large n and

$$\sigma_n^{-1}(X_n - \mu) \Rightarrow Z,\ \text{where}\ Z \sim N(0, 1).$$

In the notation of Serfling (1980) we shall write this as

$$X_n \text{ is } AN(\mu_n, \sigma_n^2).$$

NOTE.   If X_n is AN $(\mu_n,\ \sigma_n^2)$ it is not necessarily the case that $\mu_n = E(X_n)$ or that $\sigma_n^2 = \text{var } (X_n)$.

Theorem 10.3   If X_n is AN$(\mu_n,\ \sigma_n^2)$ where $\sigma_n \to 0$ as $n \to \infty$, and if g is a function which is differentiable at μ, then $g(X_n)$ is AN$(g(\mu),\ [g'(\mu)]^2 \sigma_n^2)$.

Proof.   Since $Z_n = \sigma_n^{-1} (X_n - \mu) \overset{d}{\Rightarrow} Z$ which is $N(0, 1)$ and hence $Z_n = O_p(1)$ as $n \to \infty$. Hence $X_n = \mu + O_p(\sigma_n)$. By Proposition 10.1 we have $\sigma_n^{-1} [g(X_n) - g(\mu)]$ $= \sigma_n^{-1} g'(\mu)(X_n - \mu) + o_p(1)$, which implies the result.

NOTE.   This is known as the δ-method.

Definition 10.8   The sequence $\{X_n\}$ of random k-vectors is asymptotically normal with "mean vector" μ_n and "covariance matrix" Σ_n if (1) Σ_n has no zero diagonal elements for all sufficiently large n, and

(2) $\lambda' X_n$ is AN$(\lambda' \mu_n, \lambda' \Sigma_n \lambda)$ for every $\lambda \in R^k$ such that $\lambda' \Sigma_n \lambda > 0$ for all sufficiently large n.

The following theorem is the multivariate analogue of previous theorem.

Theorem 10.4   Suppose $\mathbf{X}_n$ is AN$(\mu,\ c_n^2 \Sigma)$ where Σ is a symmetric non-negative definite matrix and $c_n \to 0$ as $n \to \infty$. If $g(\mathbf{X}) = (g_1(\mathbf{X}), \ldots, g_m(\mathbf{X}))'$ is a mapping from R^k into R^m such that each g_i is continuously differentiable in a neighbourhood of μ and if $\mathbf{D\Sigma D}'$ has all of its diagonal elements non-zero, where $\mathbf{D}$ is the $m \times k$ matrix $[(\partial g_i/\partial x_j)\ (\mu)]$, then

$$g(\mathbf{X}_n) \text{ is } AN(g(\mu),\ c_n^2 \mathbf{D\Sigma D}')$$

Proof.   Since $\mathbf{X}_n$ is AN$(\mu,\ c_n^2 \Sigma)$, each X_{nj} is AN$(\mu_j,\ c_n^2 \sigma_{jj})$ were σ_{jj} is the jth diagonal element of Σ and $\sigma_{jj} > 0$ by Definition 10.8.

Since $c_n^{-1} (X_{nj} - \mu_j)$ converges in distribution it is bounded in probability and hence $X_{nj} = \mu_j + O_p(c_n)$.

By multivariate Taylor expansion in probability (Proposition 10.2), we can write, for $i = 1, 2, \ldots, m$,

$$g_i(\mathbf{X}_n) = g_i(\mu) + \sum_{j=1}^{k} \partial g_i/\partial x_j (\mu)(X_{nj} - \mu_j) + o_p(c_n)$$

which is the same as

$$g(\mathbf{X}_n) - g(\mu) = D(\mathbf{X}_n - \mu) + o_p(c_n).$$

Dividing both sides by c_n we obtain

$$c_n^{-1}[g(\mathbf{X}_n) - g(\mu)] = c_n^{-1} D(\mathbf{X}_n - \mu) + o_p(1)$$

and since $c_n^{-1} D(\mathbf{X}_n - \mu)$ is AN$(\mathbf{O},\ \mathbf{D\Sigma D}')$, by Slutsky's theorem the same is true for $c_n^{-1} [g(\mathbf{X}_n) - g(\mu)]$.

EXAMPLE 10.2 (The sample coefficient of variation).   Suppose that $\{X_n\}$ is an i.i.d. sequence of r.vs. with common mean μ and s.d. $\sigma > 0$ and $EX_n^4 = \mu_4 < \infty$, $EX_n^3 = \mu_3$, $EX_n^2 = \mu_2 = \mu^2 + \sigma^2$ and $EX_n = \mu_1 = \mu \neq 0$. The sample coefficient of variation is defined as $v_n = s_n/\overline{X}_n$ where $\overline{X}_n = n^{-1}(X_1 + \ldots + X_n)$ and $s_n^2 = n^{-1}\sum_1^n (X_i - \overline{X}_n)^2$. It is well known from CLT of sample moments (see Problem 10.15),

$$\begin{bmatrix} \overline{X}_n \\ n^{-1}\sum_1^n X_i^2 \end{bmatrix} \text{ is AN} \left\{ \begin{bmatrix} \mu_1 \\ \mu_2 \end{bmatrix}, n^{-1}\Sigma \right\}$$

where Σ is the matrix with components $\underset{ij}{\Sigma} = \mu_{i+j} - \mu_i\mu_j$, $i, j = 1, 2$. Now $v_n = g(\overline{X}_n, n^{-1}\sum_1^n X_1^2)$ where $g(x, y) = x^{-1}(y - x^2)^{1/2}$. By previous theorem with

$$\mathbf{D} = \left[\frac{\partial g}{\partial x}(\mu), \frac{\partial g}{\partial y}(\mu) \right] = [-\mu_2/(\sigma\mu_1^2), 1/(2\sigma/\mu_1)],$$

we have v_n AN$(\sigma/\mu_1, n^{-1}\mathbf{D}\Sigma\mathbf{D}')$.

10.4.3   The Asymptotic Joint Distribution of Quantiles

Let $\{X_i\}_1^n$ be a random sample from an absolute continuous distribution with p.d.f. $f(x)$. We consider the asymptotic joint distribution of the k sample quantiles $X(n_j)$, $(j = 1, 2, \ldots, k)$, where $n_j = [n_{p_j}] + 1$ and $0 < p_1 < p_2 < \ldots < p_k < 1$. The population quantile corresponding to $X_{(nj)}$, is denoted by ξ_{p_j}.

Theorem 10.5 (*Ghosh, 1971*).   If $0 < f(\xi_{p_j}) < \infty$, $j = 1, 2, \ldots, k$, then the asymptotic joint distribution of $n^{1/2}(X_{(n_1)} - \xi_{p_1}), \ldots, n^{1/2}(X_{(n_k)} - \xi_{p_k})$ is k-dimensional normal with zero mean vector and covariance matrix

$$\frac{p_i(1 - p_j)}{f(\xi_{p_i})f(\xi_{p_j})}, i \leq j$$

Before proving this theorem, we establish a lemma which allows us to obtain, under the assumption of the theorem, the following elegant representation of $X_{(n_j)}$, first introduced by Bahadur (1966).

$$X_{(n_j)} = \xi_{p_j} - \frac{F_n(\xi_{p_j}) - p_j}{f(\xi_{p_j})} + R_n(j) \tag{1}$$

where $F_n(\xi_{p_j})$ is the empirical c.d.f. of $X_1, X_2, \ldots, X_n$ evaluated at ξ_{p_j} (namely the proportion of X's, $\leq \xi_{p_j}$) and where $n^{1/2} R_n(j) \overset{P}{\rightarrow} 0$ as $n \rightarrow \infty$.

Lemma 10.4   Let V_n and W_n be two sequences of r.vs. such that (a) $W_n = O_p(1)$ (b) for every y and every $\varepsilon > 0$

(i) $\lim_{n \to \infty} P(V_n \le y, W_n \ge y + \varepsilon) = 0$

(ii) $\lim_{n \to \infty} P(V_n \le y + \varepsilon, W_n \le y) = 0$

Then

$$V_n - W_n = \xrightarrow{P} 0 \text{ as } n \to \infty.$$

Proof.   Fix $\varepsilon > 0$, $\delta > 0$. Since (a) holds, it is possible to choose integers m and n_o (both depending on ε and δ) such that

$$P\left(|W_n| > m\varepsilon\right) < \delta \text{ for } n \ge n_o.$$

Hence

$$P\left(|V_n - W_n| > 2\varepsilon\right) < \delta + P\left(|W_n| \le m\varepsilon, |V_n - W_n| > 2\varepsilon\right)$$

$$\le \delta + \sum_{j=-m+1}^{m} [P((j - 1)\varepsilon \le W_n \le j\varepsilon, V_n < (j - 2)\varepsilon)]$$

$$\to 0 + P((j - 1)\varepsilon \le W_n \le j\varepsilon, V_n > (j + 1)\varepsilon)]$$

as $n \to \infty$ (by condition (b)).

To obtain Bahadur's representation of quantile, let

$$V_n = n^{1/2} (X_{(n_j)} - \xi_{p_j}).$$

Then

$$[V_n \le y] \text{ iff } [X_{(n_j)} \le \xi_{p_j} + yn^{-1/2}]$$

$$\text{iff } [nF_n(\xi_{p_j} + yn^{-1/2}) \ge n_j] \tag{2}$$

$$\text{iff } [Z_n \le y_n]$$

where

$$Z_n = n^{1/2}[F(\xi_{p_j} + yn^{-1/2}) - F_n(\xi_{p_j} + yn^{-1/2})]/f(\xi_{p_j}),$$

$$y_n = n^{1/2}[F(\xi_{p_j} + yn^{-1/2} - n_j n^{-1})]/f(\xi_{p_j}).$$

We note that $y_n \to y$ as $n \to \infty$ since by mean value theorem

$$y_n = n^{1/2}[F(\xi_{p_j}) + yn^{-1/2}\{f(\xi_{p_j}) + 0(1)\} - n_j n^{-1}]/f(\xi_{p_j})$$

Now writing $W_n = n^{1/2}[p_j - F_n(\xi_{p_j})]/f(\xi_{p_j})$

we have $E(Z_n - W_n)^2 = \dfrac{n}{f^2(\xi_{p_j})} \dfrac{1}{n^2} E(V_n - EV_n)^2$

where V_n has a binomial distribution $B(\pi_n, n)$ with

$$\pi_n = |F(\xi_{p_j} + yn^{-1/2}) - F(\xi_{p_j})|.$$

It follows that $E(Z_n - W_n)^2 = \pi_n(1 - \pi_n)/f^2(\xi_{p_j}) \to 0$ as $n \to \infty$. Thus,

$$Z_n - W_n \xrightarrow{P} 0$$

Hence from (2), for every $\varepsilon > 0$,

$$P(V_n \le y, \ W_n \ge y + \varepsilon) = P(Z_n \le y, \ W_n \ge y + \varepsilon)$$

$$\le P[\ |W_n - Z_n| > \varepsilon] \to 0 \text{ as } n \to \infty$$

This shows that part b (i) of the lemma is satisfied. Similarly, condition b (ii) of the lemma is satisfied. So the lemma establishes $V_n - W_n = n^{1/2} R_n(j) \xrightarrow{P} 0$.

Proof of the Theorem 10.4 By (1) the asymptotic joint distribution of $n^{1/2}$ $(X_{(n_1)} - \xi_{p_1})$, ..., $n^{1/2} (X_{(n_k)} - \xi_{p_k})$ is the same as that of

$$n^{1/2}[p_1 - F_n(\xi_{p_1})]/f(\xi_{p_1}), \ldots, \frac{n^{1/2}[p_k - F_n(\xi_{p_k})]}{f(\xi_{p_k})}$$

Since $F_n(\xi_{p_j}) = \dfrac{1}{n} \sum_{i=1}^{n} I[X_i \le \xi_{p_j}]$, we have for $i \le j$

$$n \text{ cov } [F_n(\xi_{p_i}), F_n(\xi_{p_j})] = p_i - p_i p_j = p_i(1 - p_j).$$

The result now follows from the multidimensional central limit theorem for i.i.d. r.vs. and Slutsky's theorem.

PROBLEMS AND COMPLEMENTS

10.1 Show that $\displaystyle \lim_{n \to \infty} e^{-n} \sum_{k=0}^{n} \frac{n^k}{k!} = \frac{1}{2}$.

10.2 Let $\{X_k\}$ be a sequence of r.vs. with $P(X_k = \pm k) = 1/2 \ k^{-\lambda}$ and $P(X_k = 0) = 1 - k^{-\lambda}$, $\lambda \ge 0$. Show that CLT holds for $0 \le \lambda \le 1$.

10.3 Let $\{X_n\}$ be a sequence of independent r.vs. such that $P(X_n = n^\alpha) = P(X_n = -n^\alpha) = n^{-2\alpha}/2$ and $P(X_n = 0) = 1 - n^{-2\alpha}$. Show that CLT holds for $\alpha < 1/2$.

10.4 If $\{X_n\}$ is a sequence of i.i.d. r.vs. with mean 0 and variance 1, then show that

$$\xi_n = \frac{1}{\sqrt{n}} \sum_{i=1}^{n} X_i + \frac{1}{n} \sum_{i=1}^{n} X_i$$

converges in law to a standard normal variable.

10.5 If $\{X_k\}$ be a sequence of independently distributed r.vs. such that $X_n = \pm \sqrt{2n-1}$ with probability 1/2 in each case. Show that the CLT holds but not the WLLN.

10.6 $\{X_n\}$ is a sequence of independent normal r.vs. with $E(X_n) = 0$ and $V(X_n) = 1/n^2$, $n \ge 1$. Show that the Lindeberg-Feller condition is not satisfied by this sequence but the sequence obeys CLT notwithstanding.

10.7 $\{X_n\}$ is a sequence of independent normal r.vs. with $E(X_n) = 0$ and $\sigma_n^2 = 2^{n-2}$ for $n \geq 2$ and $\sigma_1^2 = 1$. Show that Lindeberg's condition is not satisfied by this sequence but the sequence obeys CLT notwithstanding.

10.8   Let $(X_{(1)} < X_{(2)} < \ldots < X_{(n)}$ be the order statistics from a random sample X_1, X_2, ..., X_n with d.f. F and

$$F_n(x) = \begin{cases} 0 & \text{if } x < X_{(1)} \\ \dfrac{k}{n} & \text{if } X_{(k)} \leq x < X_{(k+1)} \ k = 1, \ldots, n-1 \\ 1 & \text{if } x > X_{(n)} \end{cases}$$

be the empirical d.f. Prove that

$$\frac{\sqrt{n}(F_n(x) - F(x))}{\sqrt{F(x)(1 - F(x))}} \xrightarrow{d} N(0,1) \ \text{ as } n \to \infty.$$

10.9   Let $\{X_n\}$ be i.i.d. r.vs. with a continuous common d.f. defined on $(\Omega, \mathscr{F}, P)$. Let Y_i be the relative rank of X_i among X_1, ..., X_i, i.e. $Y_j = j$ if $X_k < X_i$ for exactly $j - 1$ values of $k = 1$, ..., $j - 1$. Show that $\{Y_n\}$ satisfies CLT.

10.10   Show that the sequence of r.vs. $\{X_n\}$ satisfies the classical Lindeberg condition iff $s_n^2 > 0$ for some n and

$$\frac{1}{s_n^2} \sum_{i=1}^{n} \int_{|X - EX_j| > \varepsilon s_j} (x - E(X_j))^2 \, dF_j(x) \to 0 \text{ for all } \varepsilon > 0$$

10.11   Let $\phi_{nj}(t)$ be the ch.f. of X_{nj}. Then show that $\lim\limits_{n \to \infty} \max\limits_{i \leq j \leq k_n} P(|X_{nj}| > \varepsilon) = 0$, $\varepsilon > 0$ iff for all $t \in R$, $\lim\limits_{n \to \infty} \max\limits_{i \leq j \leq k_n} |\phi_{nj}(t) - 1| = 0$.

10.12   Show that Feller's condition implies

$$\lim\limits_{n \to \infty} \max\limits_{1 \leq k \leq n} P[|X_{nk}| \geq \varepsilon] = 0, \varepsilon > 0 \ldots (*)$$

where $X_{nk} = X_k/s_n$. Also show that $\max\limits_{k \leq n} |\phi_{nk}(t) - 1| \to 0$ if (*) holds, where $\phi_{nk}(t)$ is the ch.f. of X_{nk}.

NOTE.   Condition (*) is known as *uniform asymptotic negligibility* (UAN) condition. It can be shown that Lindeberg's condition is equivalent to CLT and UAN condition.

10.13   Let $\{X_n\}$ be a sequence of independent r.vs. with

$$P(X_1 = \pm 1) = 1/2 \text{ and for } k \geq 2, P(X_k = \pm 1) = 1/2c$$

$$P(X_k = \pm k) = \frac{1}{k^2}\left(1 - \frac{1}{c}\right), \ P(X_k = 0) = 1 - \frac{1}{c} - \frac{1}{k^2}\left(1 - \frac{1}{c}\right)$$

for some constant c. Show that Lindeberg's condition does not hold but UAN condition holds.
NOTE.   CLT does not hold.

10.14 Let (X_n, Y_n) be an i.i.d. sequence of two-dimensional random vectors with $EX_1 = EY_1 = 0$, var $(X_1) = \text{var}(Y_1) = 1$ and cov $(X_1, Y_1) = \text{corr}(X_1, Y_1) = \rho$. Show that $\sqrt{n}\ (r - \rho) \xrightarrow{d} N(0, 1)$, where

$$r = \frac{\sum\limits_{i=1}^{n} X_i Y_i - n\overline{X}_n \overline{Y}_n}{\sqrt{(\sum\limits_{i=1}^{n} X_i^2 - n\overline{X}_n^2)(\sum\limits_{i=1}^{n} Y_i^2 - n\overline{Y}_n^2)}}, \text{ the sample correlation.}$$

You can assume $EX_1^4 < \infty$ and $EY_1^4 < \infty$.

10.15 Let $a_k = \dfrac{1}{n}\sum\limits_{i=1}^{n} X_i^k, k \geq 1$ and $m_k = \dfrac{1}{n}\sum\limits_{i=1}^{n}(X_i - \overline{X})^2$, $k \geq 2$, $m_1 = 0$ be the sample moments statistics corresponding to population moments $\alpha_k = E(X_1^k)$ and $\mu_k = E(X_1 - \mu)^k$, $k \geq 2$, $\mu_1 = 0$, respectively ($\mu = \alpha_1 = EX_1$) for a random sample (i.i.d. r.vs.) $X_1, \ldots, X_n$. Then

(a) if $\alpha_{2k} < \infty$, the random vector $n^{1/2}(a_1 - \alpha_1, \ldots, a_k - \alpha_k)$ converges in distribution to k-variate $N_k(\mathbf{0}, (\sigma_{ij}))$, where $\sigma_{ij} = \alpha_{i+j} - \alpha_i \alpha_j$.

(b) if $\mu_{2k} < \infty$, the random vector $n^{1/2}\ (m_2 - \mu_2, \ldots, m_k - \mu_k) \xrightarrow{d} N_{k-1}$

$(\mathbf{0}, (\sigma_{ij}^*))$, where

$$\sigma_{ij}^* = \mu_{i+j+2} - \mu_{i+1}\mu_{j+1} - (i+1)\mu_i\mu_{j+2} - (j+1)\mu_{i+2}\mu_j + (i+1)(j+1)\mu_i\mu_j\mu_2.$$

(c) $(a_1, m_2) = (\overline{X}_n, s_n^2)$ is AN $\left((0,0), \begin{pmatrix} \sigma^2 & \mu_3 \\ \mu_3 & \mu_4 - \sigma^4 \end{pmatrix}\right)$.

11

Product Space

11.1 PRODUCT MEASURE AND FUBINI THEOREMS

The set of all pairs (ω_1, ω_2), where $\omega_1 \in \Omega_1$ and $\omega_2 \in \Omega_2$, is called the Cartesian product of Ω_1 and Ω_2 and is denoted by $\Omega_1 \times \Omega_2$. Suppose that $(\Omega_1, \mathcal{A}_1)$ and $(\Omega_2, \mathcal{A}_2)$ are two measurable spaces. The set of all pairs $\{(\omega_1, \omega_2): \omega_1 \in A_1, \omega_2 \in A_2\}$ is the rectangle $A_1 \times A_2$. If $A_1 \in \mathcal{A}_1$ and $A_2 \in \mathcal{A}_2$, then $A_1 \times A_2$ is a *measurable rectangle*. In general, the class $\{A_1 \times A_2: A_1 \in \mathcal{A}_1, A_2 \in \mathcal{A}_2\}$ is not a σ-field (but they form a semi-algebra). Now the minimal σ-field containing the above class is called the *product σ-field* and is denoted by $\mathcal{A}_1 \times \mathcal{A}_2$. Then $(\Omega_1 \times \Omega_2, \mathcal{A}_1 \times \mathcal{A}_2)$ is a measurable space. Instead of pairs we can have n-tuples $(\omega_1, \omega_2, \ldots, \omega_n)$; $\omega_i \in \Omega_i$, $i = 1, 2, \ldots, n$ and $(\Omega_1 \times \ldots \times \Omega_n, \mathcal{A}_1 \times \ldots \times \mathcal{A}_n)$ is a measurable space where $(A_1 \times \ldots \times A_n)$, $A_i \in \mathcal{A}_i$ is a measurable rectangle.

We shall consider the bivariate case in more details. If $A \subset \Omega_1 \times \Omega_2$ then $A_{\omega_1} = [\omega_2: (\omega_1, \omega_2) \in A]$ is called ω_1-section of A. Similarly, $A_{\omega_2} = [\omega_1: (\omega_1, \omega_2) \in A]$ is the ω_2-section of A; e.g. if A is a rectangle $A_1 \times A_2$, then $A_{\omega_1} = A_2$ and $A_{\omega_2} = A_1$. Let f be a function on $\Omega_1 \times \Omega_2$ or on its subset A. Then for $\omega_1 \in \Omega_1$, ω_1-section of f is a function $f(\cdot)$ defined on A_{ω_1} such that $f_{\omega_1}(\omega_2) = f(\omega_1, \omega_2)$. Similarly, $f_{\omega_2}(\cdot)$ is defined on A_{ω_2}.

Lemma 11.1 Every section $\omega_2(\omega_1)$ of a simple function on $\Omega_1 \times \Omega_2$ is a simple function on $\Omega_1(\Omega_2)$.

Proof. For $A \subset \Omega_1 \times \Omega_2$ denote by $I_A(\cdot)$ the indicator function of A, similarly for $A_{\omega 1}$ and $A_{\omega 2}$ sections, $I_{A_{\omega_1}}(\cdot)$ and $I_{A_{\omega_2}}(\cdot)$ be their indicator functions respectively. Since

$$A_{\omega_1} = \{\omega_2: (\omega_1, \omega_2) \in A\} = \{\omega_2: I_A(\omega_1, \omega_2) = 1\}$$

$$= \{\omega_2: I_{A_{\omega_1}}(\omega_2) = 1\}, \, I_{A_{\omega_1}}(\omega_2) = 1 \text{ iff } I_A(\omega_1, \omega_2) = 1.$$

Thus, sections of indicator functions are indicator functions. Hence, a simple function being a linear combination of indicator functions, it has the same property like indicator functions.

Lemma 11.2   Section of measurable function is measurable.

Proof.   If $A = A_1 \times A_2$ is a measurable rectangle in $\Omega_1 \times \Omega_2$, $I_{A_{\omega_1}} = I_{A_2}$ and $I_{A_{\omega_2}} = I_{A_1}$, so that $I_A = I_{A_{\omega_1}} \times I_{A_{\omega_2}}$. In general, sections of a measurable set is measurable and hence by Lemma 11.1 section of a measurable function (being pointwise limit of simple functions) is measurable. So far we have considered with integrals (expectations) of one dimensional measurable functions (r.vs.). Now we are in a position to discuss about bivariate (and multivariate) integrals on product measurable spaces.

11.1.1   Product Measure Space

Let $(\Omega_1, \mathscr{A}_1, \mu_1)$ and $(\Omega_2, \mathscr{A}_2, \mu_2)$ be two σ-finite measure spaces. We would like to define a σ-finite measure λ on $\mathscr{A}_1 \times \mathscr{A}_2$ so that for measurable ractangles $A_1 \times A_2$, $\lambda(A_1 \times A_2) = \mu_1(A_1)\mu_2(A_2)$. Such a measure exists on account of the following basic theorem for double integral.

Theorem 11.1   Let $(\Omega_1, \mathscr{A}_1, \mu_1)$ and $(\Omega_2, \mathscr{A}_2, \mu_2)$ be two σ-finite measure spaces and A be any measurable subset of $\Omega_1 \times \Omega_2$. Then the function $f(\omega_1) = \mu_2(A_{\omega_1})$ and $g(\omega_2) = \mu_1(A_{\omega_2})$ defined on Ω_1 and Ω_2, respectively are non-negative measurable functions such that

$$\int f \, d\mu_1 = \int g \, d\mu_2 = \lambda(A), \text{ say,} \tag{1}$$

where λ is a unique σ-finite measure on $\mathscr{A}_1 \times \mathscr{A}_2$. If $A = A_1 \times A_2$, $A_i \in \mathscr{A}_i$, $i = 1, 2$ then $\lambda(A_1 \times A_2) = \mu_1(A_1)\mu_2(A_2)$.

NOTE. λ is called *product measure* and is denoted by $\mu_1 \times \mu_2$. If h is a measurable function on $(\Omega_1 \times \Omega_2, \mathscr{A}_1 \times \mathscr{A}_2)$ then $\int h\,(\omega_1, \omega_2)\, d\lambda$ is called the *double integral* of h with respect to product measure λ. If $f\,(\omega_1) = \int h\,(\omega_1, \omega_2)\, d\mu_2(\omega_2)$, the integrals $\int f \, d\mu_1 = \int \left(\int h(\omega_1, \omega_2)\, d\mu_2(\omega_2) \right) d\mu_1(\omega_1)$ and $\int g \, d\mu_2 = \int \left(\int h(\omega_1, \omega_2)\, d\mu_1(\omega_1) \right) d\mu_2(\omega_2)$ are called iterated (or repeated) integrals of h.

Under certain conditions double integrals equal either of the two repeated integrals. This forms the content of the following theorems, the so-called "Fubini theorems".

11.1.2   Fubini Theorems

Theorem 11.2   A necessary and sufficient condition that a measurable subset A of $\Omega_1 \times \Omega_2$ have measure zero is that almost every ω_1-section (or every ω_2-section) have μ_1 measure (or μ_2-measure) zero.

Theorem 11.3   If h is a non-negative measurable function on $\Omega_1 \times \Omega_2$ then

$$\int h \, d\lambda = \iint h \, d\mu_1 \, d\mu_2 = \iint h \, d\mu_2 \, d\mu_1 \tag{1}$$

Theorem 11.4   If h is an integrable function on $\Omega_1 \times \Omega_2$, then almost every section of h is integrable. If $f(\omega_1) = \int h(\omega_1, \omega_2)\, d\mu_2(\omega_2)$ and $g(\omega_2) = \int h(\omega_1, \omega_2)\, d\mu_1(\omega_1)$, then f and g are integrable and

$$\int h\, d\lambda = \int f\, d\mu_1 = \int g\, d\mu_2 \qquad (2)$$

Remarks.   Taking one of the measures, say μ_1, as a counting measure, the first iterated integral in (1) becomes an integral of the sum, while the second becomes the sum of the integrals. These two expressions would be equal under the conditions of Fubini theorem. In case μ_2-measure is a probability measure equation (1) implies the well known result that the expectation of sums is the sum of the expectations.

Proof of Theorem 11.0   We suppose first that the result holds if μ_1 and μ_2 are finite measures. We may write $\Omega_1 = \overset{\infty}{\underset{n=1}{\cup}} \Omega_{1n}$, $\Omega_2 = \overset{\infty}{\underset{m=1}{\cup}} \Omega_{2m}$ decomposing Ω_1 and Ω_2 into disjoint sequences of sets of finite measure, and the result is therefore assumed true for each rectangle $\Omega_{1n} \times \Omega_{2m}$ where we are considering μ_1 and μ_2 restricted to measurable subsets of Ω_{1n} and Ω_{2m}, respectively. Let $A \in \mathcal{A}_1 \times \mathcal{A}_2$ and write $A_{nm} = A \cap (\Omega_{1n} \times \Omega_{2m})$, then for each ω_1, $A_{\omega_1} = \underset{n,m}{\cup} (A_{nm})_{\omega_1}$. By assumption $\mu_2((A_{nm})_{\omega_1})$ is a measurable function of ω_1 on Ω_{1n} for each m, so $\overset{\infty}{\underset{m=1}{\Sigma}} \mu_2((A_{nm})_{\omega_1})$ is measurable on Ω_{1n}. Hence $f_{\omega_1}) = \mu_2(A_{\omega_1}) = \overset{\infty}{\underset{m,n=1}{\Sigma}} \mu_2((A_{nm})_{\omega_1})$ measurable with respect to $\mathcal{A}_1$.

Similarly,

$$g(\omega_2) = \overset{\infty}{\underset{m,n=1}{\Sigma}} \mu_1((A_{nm})_{\omega_2})$$

is measurable with respect to $\mathcal{A}_2$. Since f is non-negative,

$$\int f\, d\mu_1 = \overset{\infty}{\underset{n=1}{\Sigma}} \int_{\Omega_{1n}} f\, d\mu_1 = \overset{\infty}{\underset{n=1}{\Sigma}} \int_{\Omega_{1n}} \overset{\infty}{\underset{m=1}{\Sigma}} \mu_2((A_{nm})_{\omega_1})\, d\mu_1$$

$$= \overset{\infty}{\underset{n=1}{\Sigma}}\, \overset{\infty}{\underset{m=1}{\Sigma}} \int_{\Omega_{1n}} \mu_2((A_{nm})_{\omega_1})\, d\mu_1 = \overset{\infty}{\underset{n=1}{\Sigma}}\, \overset{\infty}{\underset{m=1}{\Sigma}} \int_{\Omega_{2n}} \mu_1((A_{nm})_{\omega_2})\, d\mu_2$$

by assumption and this last term is similarly equal to $\int g\, d\mu_2$. So, we suppose that μ_1 and μ_2 are finite measures on $\mathcal{A}_1$ and $\mathcal{A}_2$, respectively, and write $\mathcal{G} = \{A \mid A \in \mathcal{A}_1 \times \mathcal{A}_2$ for which (1) holds$\}$. Then $\mathcal{G}$ contains every measurable rectangle $A_1 \times A_2$, since $\mu_2 ((A_1 \times A_2)_{\omega_1}) = I_A(\omega\, \mu\, A\, \mu\, A \times A)_{\omega_2}) I_{A_2}\, \omega\, \mu\, A$.

Then it is easy to see that (1) holds for any finite union of disjoint measurable rectangles which form a field and call it α. Therefore $\mathcal{G} \supset \alpha$. We shall show that $\mathcal{G}$ is a monotone class. Let $B_n^{\uparrow}$ sequence and $B_n \in \mathcal{G}$, $n \geq 1$ and $\underset{n\to\infty}{\lim} B_n = \overset{\infty}{\underset{1}{\cup}} B_i$. Then write $f_i(\omega_1) = \mu_2((B_i)_{\omega_1})$, $g_i(\omega_2) = \mu_1((B_i)\omega_2)$ and then they are measurable functions. $f_n(\omega_1) \uparrow f(\omega_1) = \mu_2(B_{\omega_1})$, $g_n(\omega_2) \uparrow g(\omega_2) = \mu_1(B_{\omega_2})$. (Prove it!. Hence f and g are measurable and $\int f\, d\mu_1 = \lim_{n\to\infty} \int f_n\, d\mu_1 \lim_{n\to\infty} \int g_n\, d\mu_2 = \int g\, d\mu_2$. Let B_n

$\downarrow$ and $B_n \in \mathcal{G}$, $n \geq 1$ and $\lim B_n = \overset{\infty}{\underset{1}{\cap}} B_i$. Then we obtain similarly sequences $f_n \downarrow f$ and $g_n \downarrow g$. Since $f_n \leq \mu_2(\Omega_2)$, a function integrable over Ω_1, $g_n \leq \mu_1(\Omega_1)$ which is integrable over Ω_2 and μ_1 and μ_2 are finite, we can apply dominated convergence theorem to get (1) holds for f and g. So $\mathcal{G}$ is a monotone class contained in $\mathcal{A}_1 \times \mathcal{A}_2$ and containing the field α and hence $\mathcal{G} = \sigma(\alpha)$. Hence (1) holds for all $A \in \mathcal{A}_1 \times \mathcal{A}_2$.

Now
$$f(\omega_1) = \mu_2(A_{\omega_1}) = \int_{\Omega_2} I_A(\omega_1, \omega_2)\, d\mu_2(\omega_2) \text{ and}$$

$$g(\omega_2) = \mu_1(A_{\omega_2}) = \int_{\Omega_1} I_A(\omega_1, \omega_2)\, d\mu_1(\omega_1)$$

give
$$\int_{\Omega_1} d\mu_2(\omega_1) \int_{\Omega_2} I_A(\omega_1, \omega_2)\, d\mu_2(\omega_2)$$

$$= \int_{\Omega_2} d\mu_2(\omega_2) \int_{\Omega_1} I_A(\omega_1, \omega_2)\, d\mu_1(\omega_1).$$

By monotone convergence theorem, now it is clear that λ is a measure on $\mathcal{A}_1 \times \mathcal{A}_2$. To show that λ is σ-finite and to prove the last part of the theorem note that for any set A in $\alpha = \{$the field of finite disjoint union of measurable rectangles$\}$ $\lambda(A)$

$$= \overset{n}{\underset{i=1}{\Sigma}} \mu_1(A_i)\mu_2(B_i), \text{ where } A = \overset{n}{\underset{i=1}{\cup}} (A_i \times B_i).$$ Now λ clearly takes the correct value

on measurable rectangles and we have proved that it is a measure on α so it takes the correct value on the sets of α and indeed is clearly a σ-finite measure on the field α. But the extension from α to $\sigma(a) = \mathcal{A}_1 \times \mathcal{A}_2$ is unique by the uniqueness of Carathéodory extension.

11.1.3 Proof of Fubini's Theorems

Proof of Theorem 11.1   By the definition of product measure we have

$$\lambda(A) = \begin{cases} \int \int \mu_1(A_{\omega_2})\, d\mu_2(\omega_2) \\ \int \int \mu_2(A_{\omega_1})\, d\mu_1(\omega_1) \end{cases}$$

If $\lambda(A) = 0$, then the integrals on the right are in particular finite and their non-negative integrands must be 0 a.e. Conversely, if either of the integrands is 0 a.e., then $\lambda(A) = 0$.

Proof of Theorem 11.2.   If $h = I_A$, and A is measurable $\mathcal{A}_1 \times \mathcal{A}_2$, then

$$\int h(\omega_1, \omega_2)\, d\mu_2(\omega_2) = \mu_2(A_{\omega_1}) \text{ and}$$

$$\int h(\omega_1, \omega_2)\, d\mu_1(\omega_1) = \mu_1(A_{\omega_2}),$$

and the desired result follows from Lemma 11.2. In general by Theorem 4.4 we can find an increasing sequence $\{h_n\}$ of non-negative simple functions converging to h everywhere. Since a simple function is a finite linear combination of indicator functions, the theorem holds for every h_n in place of h. By monotone convergence theorem $\lim_{n\to\infty} \int h_n\, d\lambda = \int h\, d\lambda$. If $f_n(\omega_1) = \int h_n(\omega_1, \omega_2)\, d\mu_2(\omega_2)$, then it follows from the properties of the non-negative simple functions $\{h_n\}_1^\infty$ that $\{f_n\}$ is a non-decreasing sequence of non-negative measurable functions and by monotone convergence theorem, $f_n(\omega_1) \to f(\omega_1) = \int h(\omega_1, \omega_2)\, d\mu_2(\omega_2)$ for every $\omega_1 \in \Omega_1$. Hence $f \geq 0$ and is measurable. Again applying monotone convergence theorem, $\lim_{n\to\infty} \int f_n\, d\mu_1 = \int f\, d\mu_1$. This proves first equality equation (1). The second equality follows similarly.

Proof of Theorem 11.3 Since $h = h^+ - h^-$ is integrable iff h^+ and h^- are both integrable, it is enough to consider only the case $h \geq 0$. Then equation (2) follows in this case from Theorem 11.2. From the proof of Theorem 11.2 it follows that $f \geq 0$ and $g \geq 0$ and they are measurable having finite integrals, and hence they are integrable. Since, finally, this implies that $0 \leq f < \infty$, $0 \leq g < \infty$ a.e., the section of h have the desired integrability properties, and the proof of Theorem 11.3 (the so-called *Fubini theorem*) is complete.

PROBLEMS AND COMPLEMENTS

11.1 Suppose the first moments of X and Y exist. Then show that $E(Y) - E(X)$

$$= \int_{-\infty}^{\infty} \{P(X < t \leq Y) - P(Y < t \leq X)\}\, dt.$$

11.2 Show that if

$$f(x, y) = \frac{x^2 - y^2}{(x^2 + y^2)^2}, \quad (x, y) \neq (0, 0) \text{ and } 0 \text{ otherwise}$$

then $\int_0^1 dx \int_0^1 f(x, y)\, dy = \pi/4$

and $$\int_0^1 dy \int_0^1 f(x, y)\, dx = -\pi/4$$

Also show that this does not contradict Fubini's theorem.

11.3 Let $$f(x, y) = \frac{xy}{(x^2 + y^2)^2}, |x| \leq 1, |y| \leq 1.$$

Defining $f(0, 0) = 0$, show that the iterated integrals of f over the square are equal but f is not integrable.

12

Conditional Expectation

12.1 LEBESGUE DECOMPOSITION AND RADON-NIKODYM THEOREM

Let $(\Omega, \mathcal{F}, \mu)$ be a measure space all throughout this section. Integrating a non-negative function over the sets of $\mathcal{F}$ produces a new measure from the original one (μ) and in Radon-Nikodym theorem we show that any new measure continuous in a certain way (absolutely continuous) can be formed in this manner. This gives rise to derivatives of one measure with respect to another (the so-called Radon-Nikodym derivative) and later we describe the calculus of derivatives which gives rise to and give further decomposition results.

Definition 12.1   A set function v defined on $\mathcal{F}$ to $\overline{R}$ (extended real line) is said to be *absolutely continuous* with respect to μ if

$$\mu(E) = 0 \Rightarrow v(E) + 0 \text{ for every } E \in \mathcal{F} \tag{1}$$

We write $v \ll \mu$ and sometimes v is simply called μ-*continuous*.

EXAMPLE 12.1   Let $f: \Omega \to \overline{R}$ and μ-integrable (i.e. $\int |f|\, d\mu < \infty$) then define

$$\lambda(E) = \int_E f\, d\mu \text{ for } E \in \mathcal{F} \tag{2}$$

Then λ is a finite valued absolutely continuous set function defined on $\mathcal{F}$. We have seen that λ is countably additive (see Problem. 5.3). Then given $\varepsilon > 0$, there exists $\delta > 0$ such that for $E \in \mathcal{F}$, $\mu(E) < \delta \Rightarrow$ that total variation of λ,

$$|\lambda(E)| = \lambda^+(E) + \lambda^-(E) < \varepsilon \tag{3}$$

where $\lambda^+(E) = \sup \{\lambda(B) : B \in \mathcal{F}, B \subset E\}$, and $\lambda^-(E) = \inf \{\lambda(B) : B \in \mathcal{F}, B \subset E\}$ and $\lambda(E) = \lambda^+(E) - \lambda^-(E)$. (This follows from absolute continuity of indefinite integral $\lambda(E)$ defined by (2) see Problem 12.7.) Any set function λ satisfying the condition is equivalent to absolute continuity when the measure μ is finite but not in general (see Problem 12.8). But there is a partial converse, namely,

166

if v is finite valued countably additive and absolutely continuous with respect to a finite signed measure μ then v satisfies (3) (see Problem 12.7).

Definition 12.2   An extended real valued set function v defined on a σ-field $\mathcal{F}$ is said to be *singular* with respect to μ if there exists $E_0 \in \mathcal{F}$ for which $\mu(E_0) = 0$ and $v(E) = v(E \cap E_0)$ for all $E \in \mathcal{F}$. This means that points of Ω outside the μ-null set E_0 make no contribution to the mass of v. In fact if v is also a measure we see that Ω can be dissected into two sets E_0 and $E_1 \in \mathcal{F}$ such that $E_0 \cap E_1 = \phi$, $E_0 \cup E_1 = \Omega$, $\mu(E_0) = 0$ and $v(E_1) = 0$. The μ and v are said to be *mutually singular measures* and is denoted by $v \perp \mu$.

Theorem 12.1 (*Lebesgue decomposition and Radon-Nikodym (R.N.)*).   Let $(\Omega, \mathcal{F}, \mu)$ be a σ-finite measure space and v be a σ-additive σ-finite set function. Then there exists a unique decomposition $v = v_1 + v_2$ into σ-additive set functions v_1 and v_2 which are σ-finite and such that v_1 is singular with respoect to μ and $v_2 \ll \mu$. Further there is a a.e. $[\mu]$ finite real valued measurable function f on Ω such that

$$v_2(E) = \int_E f \, d\mu \text{ for all } E \in \mathcal{F}.$$ The function f is unique is the sense that if we also

have $v_2(E) = \int_E g \, d\mu$ for all $E \in \mathcal{F}$, then $f = g$ a.e. $[\mu]$.

Corollary (Radon-Nikodym, 1930).   Under the condition of the last theorem if $v \ll \mu$, then there exists finite real-valued function f on Ω such that $v(E) = \int_E f \, d\mu$

for all $E \in \mathcal{F}$.

Proof of Theorem 12.1   Since we can decompose Ω as a countable union of disjoint sets on each of which both μ and v are finite, without loss of generality we assume that μ and v are finite on Ω. We first prove that the decomposition is unique. Let $v = v_1 + v_2 = v_3 + v_4$ where v_1 and v_3 are singular and v_2 and v_4 are absolutely continuous (μ). Then $v_1 - v_3 = v_4 - v_2$. By definition there exist E_0 and $E_1 \in \mathcal{F}$ such that

$$\mu(E_0) = 0 \text{ and } (v_1 - v_3)(E) = (v_1 - v_3)(E \cap E_0) \tag{1}$$

for all $E \in \mathcal{F}$. But $v_4 - v_2$ is absolutely continuous and hence

$$(v_4 - v_2)(E \cap E_0) = 0 \tag{2}$$

Therefore $(v_4 - v_2)(E) = (v_1 - v_3)(E \cap E_0) = (v_4 - v_2)(E \cap E_0) = 0$ (by (2)) for $E \in \mathcal{F}$. Hence

$$v_1(E) = v_3(E) = v_2(E) = v_4(E) \text{ for all } E \in \mathcal{F}$$

To prove uniqueness of f, suppose that if f is not unique then there exists a finite

measurable g such that $v_2(E) = \int_E g \, d\mu = \int_E f \, d\mu \Rightarrow \int_E (f - g) \, d\mu = 0$ for all E

$\in \mathcal{F}$. Taking $E = \{\omega : f(\omega) > g(\omega)\}$ in the last equation we get $f = g$ a.e. $[\mu]$.

Now we state Hahn-Jordan decomposition (whose proof is given later on in Appendix II) without proof.

Hahn-Jordan decomposition. Let v be a σ-additive set function defined on a measurable space $(\Omega, \mathcal{F})$. Then there exist two non-negative σ-additive set functions v_+ and v_- such that

$$v(E) = v_+(E) - v_-(E) \text{ for all } E \in \mathcal{F}$$

where $v_+(E) = v(E \cap A) \geq 0$ and $v_-(E) = -v(E \cap N) \geq 0$ for $A \cup N = \Omega$, $A \cap N = \phi$, $A, N \in \mathcal{F}$.

Going back to the proof of the main theorem the Hahn-Jordon decomposition gives $v = v_+ - v_-$. Since the set function v is finite or σ-finite respectively iff both v_+ and v_- are and $v \ll \mu$ iff $v_+ \ll \mu$ and $v_- \ll \mu$ (see Problems 12.4 and 12.6). So se can assume v is a measure on $\mathcal{F}$. Let $\mathcal{X} = \{f : \Omega \to R^+, f$ is $1 - 1$ measurable function such that $v(E) \geq \int_E f \, d\mu$ for all $E \in \mathcal{F}\}$. Then $\mathcal{X}$ is non empty as $0 \in \mathcal{X}$.

Let $\alpha = \sup\limits_{f \in \mathcal{X}} [\int f \, d\mu]$ and let $\{f_n\}$ be a sequence in $\mathcal{X}$ such that $\lim \int f_n \, d\mu = \alpha$. If E is any measurable set, n is a fixed positive integer and $g_n = \max (f_1, \ldots, f_n)$, then we can prove by induction that E is the union of disjoint measurable sets E_i, $i = 1, \ldots, n$, such that $g_n = f_i$ on E_i, $i = 1, \ldots, n$. For, let $n = 2$ and let $E_1 = \{\omega : \omega \in E, f_1(\omega) \geq f_2(\omega)\}$, $E_2 = E \cap E_1^C$, then $E = E_1 \cup E_2$ has the desired property. Supposing the decomposition possible for n, let $g_{n+1} = \max (f_1, \ldots, f_{n+1}) = \max (g_n, f_{n+1})$. So $E = E_n \cup E_{n+1}$ where $g_{n+1} = f_{n+1}$ on E_{n+1}, $g_{n+1} = g_n$ on E_n and $E_n \cap E_{n+1} = \phi$. But then by the inductive hypothesis we have $E_n = \bigcup\limits_{i=1}^{n+1} E_i$ and $g_{n+1}(\omega) = f_i(\omega)$ for $\omega \in E_i$, $i = 1, \ldots, n + 1$. Now, since each $f_i \in \mathcal{X}$.

$$\int_E g_n \, d\mu = \sum_{i=1}^{n} \int_{E_i} f_i \, d\mu \leq \sum_{i=1}^{n} v(E_i) = v(E) \tag{1}$$

Hence $0 \leq g_n \in \mathcal{X}$ and $g_n \uparrow$ in n, so we write $\lim g_n = f_0$ a.e. Then (1) and monotone convergence theorem imply that $\int_E f_0 \, d\mu = \lim \int_E g_n \, d\mu \leq v(E)$, so $f_0 \in \mathcal{X}$. Hence $\alpha \geq \int f_0 \, d\mu \geq \int g_n \, d\mu \geq \int f_n \, d\mu$ and we must have

$$\alpha = \int f_0 \, d\mu \tag{2}$$

Let
$$v_2(E) = \int_E f_0 \, d\mu \tag{3}$$

Then $v_2 \ll \mu$. To prove Lebesgue decomposition define $v_1(E) = v(E) - v_2(E)$ and by definition of $\mathcal{X}$, $v_1(E) \geq 0$ for all $E \in \mathcal{F}$. Consider the σ-additive set function

$$\lambda_n = v_1 - 1/n\mu \tag{4}$$

Again by Hahn-Jordon decomposition for each fixed n there exist A_n and N_n such that $A_n \cup N_n = \Omega$, $N_n \cap A_n = \phi$ and $E \subset A_n \Rightarrow \lambda_n(E) \geq 0$ and $E \subset N_n \Rightarrow \lambda_n(E) \leq 0$. Then from (3) for $E \subset A_n$,

$$\nu(E) = \nu_1(E) + \nu_2(E) \geq \nu_2(E) + \frac{1}{n}\mu(E) \text{ (by (4))} = \int_E (f_0 + 1/n)\, d\mu \qquad (5)$$

Define $\quad g = f_0\, I_{N_n}(\cdot) + (f_0 + 1/n)\, I_{A_n}(\cdot) = f_0 + \frac{1}{n} I_{A_n}(\cdot) \leq f_0 + 1/n \qquad (6)$

By (5), and since $g \in \mathcal{X}$, integrating both sides of (6), $\int g\, d\mu \geq \alpha$ unless $\mu(A_n) = 0$ for all $n \geq 1$ (otherwise contradicting the maximality of α). Hence $\mu(A) = 0$ where $A = \bigcup_1^\infty A_n$. Further $\Omega - A \subset N_n$ for all $n \geq 1$, by definition of λ_n,

$$\lambda_n(\Omega - A) = \nu_1(\Omega - A) - \frac{1}{n}\mu(\Omega - A) \leq 0.$$

Due to σ-finiteness of μ, without loss of generality taking μ finite,

$$0 \leq \nu_1(\Omega - A) \leq \frac{1}{n}\mu(\Omega - A) \to 0 \text{ as } n \to \infty.$$

Hence $\nu_1(\Omega - A) = 0$ and $\nu_1(E) = \nu_1(E - A)$ for all $E \in \mathcal{F}$. So ν_1 singular.

Proof of corollary (Radon-Nikodym theorem). First assume ν is finite. Since $\nu \ll \mu$, the singular part ν in Lebesgue decomposition is identically zero and $\nu(E)$ $= \nu_2(E) = \int_E f_0\, d\mu$ for all $E \in \mathcal{F}$. Since $\int f_0\, d\mu \leq \nu(\Omega) < \infty$, there exists a finite valued measurable function f, also non-negative, such that $f = f_0$ a.e. $[\mu]$. The Hahn-Jordon decomposition gives $\nu = \nu_+ - \nu_-$, $\nu_+(E) \int_E f_1\, d\mu$, $\nu_-(E) = \int_E f_2\, d\mu$ where f_1 and f_2 are finite-valued non-negative measurable functions of which at least one is integrable. So, for $E \in \mathcal{F}$, $\nu(E) = \nu_+(E) - \nu_-(E) = \int_E f\, d\mu$ where the integral of $f = f_1 - f_2$ is well-defined.

Remarks.

   (i)   If ν is only σ-additive set function then *Radon-Nikodym* theorem still holds but f may not be finite.

   (ii) If moreover ν is finite, then f is integrable.

   (iii) If ν is a measure, then $f \geq 0$ a.e. $[\mu]$.

   (iv) Let μ and ν are σ-finite σ-additive set functions on $(\Omega, \mathcal{F})$ and suppose the $\nu \ll \mu$. Then the Radon-Nikodym derivative $d\nu/d\mu$, of ν with respect to μ, is any measurable function f such that $\nu(E) = \int_E f\, d\mu$ for each $E \in \mathcal{F}$, where $\int f\, d\mu = \int f\, d\mu_+ - \int f\, d\mu_-$. If μ is Lebesgue measure m then f is simply called the density of ν. If moreover ν is a probability measure, then f is called the probability density function (p.d.f.) of ν with respect to m.

12.2   CONDITIONAL EXPECTATION AND CONDITIONAL PROBABILITY

From a theoretical point of view, conditioning is a useful means of exploiting auxiliary information. From a practical vintage point, conditional probabilities reflect the change in unconditional probabilities due to additional knowledge from the data.

The latter is represented by a sub σ-field $\mathcal{A}$ of the basic σ-field $\mathcal{F}$ of events occurring in the underlying probability space $(\Omega, \mathcal{F}, P)$. Let X be a $\mathcal{F}$-measurable function of Ω whose integral exists, i.e. $|EX| \leq \infty$. Let $\mathcal{A} \subset \mathcal{F}$ be a sub σ-field, then,

$$\phi(B) = \int_B X \, dP, B \in \mathcal{A}$$

is the restriction of ϕ to sets of $\mathcal{A}$, denoted by $\phi \mid \mathcal{A}$. It is $\mathcal{A}$-measurable, $\phi \ll P$ (henceforth P-continuous) and σ-additive set function. Hence by Radon-Nikodym theorem, there exists a $\mathcal{A}$-measurable function $Y = d\phi/dP$ on Ω, called the conditional expectation of X given $\mathcal{A}$, denoted by $E(X/\mathcal{A})$. Y is defined by two conditions:

(i) Y is $\mathcal{A}$-measurable, and (ii) $\displaystyle \int_B Y \, dP = \int_B X \, dP$

$$= \int_B \left(\frac{d\phi \mid \mathcal{A}}{dP} \right) dP \tag{1}$$

for all $B \in \mathcal{A}$. In view of (i) and (ii), any $\mathcal{A}$-measurable function Z differs from $Y = E(X \mid \mathcal{A})$ on a set of probability zero also qualifies as $E(X \mid \mathcal{A})$. In other words, the conditional expectation $E(X/\mathcal{A})$ is only defined to within an equivalence, i.e. in any representative of a class of functions whose elements differ from one another only on sets of probability measure zero is an undesirable affair. We know that Radon-Nikodym derivative $E(X/\mathcal{A})$ is unique a.s. if $d\phi \mid \mathcal{A}/dP$ is σ-finite and we have also seen even without σ-finiteness condition there exists a a.s. finite $\mathcal{A}$-measurable function Z which can be served as a version of $E(X/\mathcal{A})$ but $E(Z)$ may not exists.

Let X be an $\mathcal{F}$-measurable function with $|EX| \leq \infty$ and $\{Y_t, t \in T\}$, $\{\mathcal{A}_t, t \in T\}$, non-empty families of r.vs. and σ-fields of sets, respectively. It is customary to define

$$E\{X \mid Y_t, t \in T\} = E\{X \mid \sigma(Y_t, t \in T)\}, E\{X \mid \mathcal{A}_t, t \in T\}$$

$$= E\{X \mid \sigma(\mathcal{A}_t, t \in T)\},$$

and in particular,

$$E\{X \mid Y_1, ..., Y_n\} = E\{X \mid \sigma(Y_1, ..., Y_n)\}, E(X \mid Y) = E(X \mid \sigma(Y)).$$

Since, by definition $E\{X \mid Y, ..., Y_n\}$ is $\sigma\{Y, ..., Y_n\}$-measurable, there exists a Borel measurable function g on R^n such that $E\{X \mid Y, ..., Y_n\} = g\{Y_1, ..., Y_n\}$. Conversely, if g is a Borel function of R^n such that $|Eg\{Y_1, ..., Y_n\}| \leq \infty$ and for every $B \in \sigma\{Y_1, ..., Y_n\}$

$$\int_B g\{Y_1, \ldots, Y_n\}\, dP = \int_B X\, dP,$$

then

$$g(Y_1, \ldots, Y_n) = E(X \mid Y_1, \ldots, Y_n) \text{ a.s.}$$

In particular, if $Y = I_B$ for some $B \in \mathcal{F}$, then $\sigma(Y) = \{\phi, B, B^C, \Omega\}$ and every version of $E(X \mid Y)$ must be constant on each of the events B, B^C, necessitating

$$E(X|Y)(\omega) = \begin{cases} \dfrac{1}{P(B)} \displaystyle\int_A x\, dP & \text{if } \omega \in B \\[3mm] \dfrac{1}{P(B^C)} \displaystyle\int_{B^C} x\, dP & \text{if } \omega \in (B)^C \end{cases} \tag{2}$$

where either of the constants on the right can be constructed as any number in $\overline{R}$ when the corresponding event B or B^C has probability zero.

12.2.1 Definition of Conditional Probability

For $A, B \in \mathcal{F}$ and $\mathcal{A} \subset \mathcal{F}$, define

$$P(A \mid \mathcal{A}) = E(I_A \mid \mathcal{A}) \text{ and } P(A \mid B) = E(I_A \mid I_B) \tag{3}$$

Then $P(A \mid \mathcal{A})$ is called the *conditional probability* of the event A given the sub σ-field $\mathcal{A}$ and according to previous section is $\mathcal{A}$-measurable function of Ω satisfying

$$\int_C P(A \mid \mathcal{A})\, dP = \int_C I_A\, dP, C \in \mathcal{A} \tag{4}$$

$P(A \mid B)$ is called the conditional probability of the event A given the event B and by (2), if $P(B) > 0$

$$P(A \mid B) = \frac{1}{P(B)} \int_B I_A\, dP = \frac{P(AB)}{P(B)}, \omega \in B \tag{5}$$

12.2.2 Definition of Conditional Probability Measure

If $\mu(\Omega)$ is non-zero and finite, considering $\mu(.)/\mu(\Omega)$ as a new set function, any finite measure may be viewed as a generalization of probability measure. Let B be an arbitrary but fixed subset of Ω and $B \in \mathcal{F}$. Consider the class of event

$$B \cap \mathcal{F} = \{B \cap A \mid A \in \mathcal{F}\} = \mathcal{F}_B, \text{ (say)} \tag{6}$$

This class is a σ-field of subsets of B. Hence $(B, \mathcal{F}_B)$ is a measurable space. Probability measure defined on this space is not normal, since $P(B)$ may not be unity, in general. However, as in previous discussion it can be normal, by defining the measure P_B as

$$P_B(B \cap A) = \frac{P(B \cap A)}{P(B)} \text{ on } \mathcal{F}_B, \text{ if } P(B) > 0.$$

Then $(B, \mathcal{F}_B, P_B)$ is a probability space. Domain of P_B is the class of events which are subsets of B. For a given B, we may view P_B to be a set function defined on members of $\mathcal{F}$. Then it is called *conditional probability measure, given B*, denoted by P_B or $P(\cdot \mid B)$. It is easy to check that P_B is non-negative, $P_B(\Omega) = 1$ and it is σ-additive on $\mathcal{F}$ and hence P_B is a probability measure on $(\Omega, \mathcal{F})$. Thus $(\Omega, \mathcal{F}, P_B)$ is a probability space. $P_B(A)$ is called conditional probability of A given B. $P_B(.)$ is called the *conditional probability* distribution on $(\Omega, \mathcal{F})$.

EXAMPLE 12.2   Suppose $\mathcal{A} = \{B, B^C, \phi, \Omega\} \subset \mathcal{F}$ and $X = I_A$, $A \in \mathcal{F}$, $(A \neq B)$. Then (1) becomes

$$\int_B X \, dP = \int_B I_A \, dP = P(AB) = \int_B E(X \mid \mathcal{A}) \, dP \tag{7}$$

$$\int_{B^C} X \, dP = P(AB^C) = \int_{B^C} E(X \mid \mathcal{A}) \, dP \tag{8}$$

Since $E(X \mid \mathcal{A})$ is $\mathcal{A}$-measurable, it is of the form $C_0 I_B + C_1 I_{B^C}$. Thus, $E(X \mid \mathcal{A}) = C_0$ for all $\omega \in \mathcal{A}$. Putting this in (8)

$$P(AB) = C_0 P(B) \Rightarrow C_0 = P(AB)/P(B) \text{ if } P(B) > 0$$

Similarly
$$C_1 = \frac{P(AB^C)}{P(B^C)} \text{ if } P(B^C) > 0. \text{ Thus,}$$

$$E(I_A \mid \mathcal{A})(\omega) = \begin{cases} P(AB)/P(B) & \text{if } P(B) > 0 \\ P(AB^C)/P(B^C) & \text{if } P(B^C) > 0. \end{cases}$$

So conditional expectation of I_A given $\mathcal{A}$ takes the value $P(A \mid B)$ if $\omega \in B$ and $P(AB^C)$ if $\omega \in B^C$. This function will be called the *conditional probability of A given $\mathcal{A}$* agreeing with our previous Definition (5).

12.2.3   Properties of Conditional Expectation

In our discussion to follow $X, X_1, X_2, \ldots$ are extended r.vs. on $(\Omega, \mathcal{F}, P)$, with all necessary expectations assumed to exist; $X : (\Omega, \mathcal{F}) \to (\Omega', \mathcal{F}')$ is a random object (measurable transformation) and $\mathcal{G}$ is a sub σ-field of $\mathcal{F}$ a.s. (or a.e.) if not stated will always mean with respect to $[P]$. $E(X \mid \mathcal{G}) : (\Omega, P) \to (\overline{R}, \mathcal{B})$.

The following are some simple consequences of the definition of conditional expectation:

$$E(X \mid \mathcal{G}) = X \text{ a.s. if } X \text{ is } \mathcal{G}\text{-measurable} \tag{1}$$

If Y is a constant k a.s., then

    (a) $E(Y \mid \mathcal{G}) = k$ a.s.
    (a′) $E(Y \mid X = x) = k$ a.s.

In particular, $E(1 \mid \mathcal{G}) = 1$ a.s. and $E(cX \mid \mathcal{G}) = cE(X \mid \mathcal{G})$ a.s. if $\mid EX \mid < \infty$ and c is a constant. If $X_1 \leq X_2$ a.s., then

    (b) $E(X_1 \mid \mathcal{G}) \leq E(X_2 \mid \mathcal{G})$
    (b′) $E(X_1 \mid X = x) \leq E(X_2 \mid X = x)$

In particular, $E(X \mid \mathcal{G}) \geq 0$ if $X \geq 0$

(c) $\mid E(X \mid \mathcal{G}) \mid \leq E(\mid X \mid \mid \mathcal{G})$

(c') $\mid E(X \mid Y = y) \leq E(\mid X \mid \mid Y = y)$

Additivity theorem for conditional expectation

Theorem 12.2 If $a, b, \in R$ and $aE(X_1) + bE(X_2)$ is well defined (not of the form $\infty - \infty$), then (a) $E(aX_1 + bX_2 \mid \mathcal{G}) = aE(X_1 \mid \mathcal{G}) + bE(X_2 \mid \mathcal{G})$ and (a') $E(aX_1 + bX_2 \mid X = x) = aE(X_1 \mid X = x) + bE(X_2 \mid X)$. In particular, $E(X_1 + X_2 \mid \mathcal{G}) = E(X_1 \mid \mathcal{G}) + E(X_2 \mid \mathcal{G})$ if $E(X_1^- + X_2^-) < \infty$ or $E(X_1^+ + X_2^+) < \infty$.

Proof. (a) If $A \in \mathcal{G}$

$$\int_A (aX_1 + bX_2)\, dP = \int_A aX_1\, dP + \int_A bX_2\, dP \text{ (by \quad additivity \quad of \quad integral)}$$

$$= \int_A aE(X_1 \mid \mathcal{G})\, dP + \int_A bE(X_2 \mid \mathcal{G})\, dP \text{ (by definition of conditional expectation).}$$

Thus $\int_A aE(X_1 \mid \mathcal{G})\, dP + \int_A bE(X_2 \mid \mathcal{G})\, dP$ is well defined, so again by additivity

theorem for integrals, $\int_A (aX_1 + bX_2)\, dP = \int_A [aE(X_1 \mid \mathcal{G}) + bE(X_2 \mid \mathcal{G})]\, dP$, as

desired.

(a') same as (a)

These properties assert roughly that if $TX = E(X \mid \mathcal{G})$, then T is linear, order preserving (monotone), and $T_1 = 1$ (constant preserving). That T is also idempotent, i.e. $T^2 = T$ follows from the theorem given below.

Theorem 12.3 If $\mathcal{F}_1, \mathcal{F}_2$ are σ-fields with $\mathcal{F}_1 \subset \mathcal{F}_2 \subset \mathcal{F}$ and $\mid (E(X) \mid < \infty$, then

$$E[E(X \mid \mathcal{F}_2) \mid \mathcal{F}_1] = E(X \mid \mathcal{F}_1) = E[E(X \mid \mathcal{F}_1) \mid \mathcal{F}_2] \tag{2}$$

Proof. Since $E[E(X \mid \mathcal{F}_2) \mid \mathcal{F}_1]$ is $\mathcal{F}_1$-measurable and for $B \in \mathcal{F}_1$

$$\int_B E[E(X \mid \mathcal{F}_2) \mid \mathcal{F}_1]\, dP = \int_B E(X \mid \mathcal{F}_2)\, dP = \int_B X\, dP,$$

the first equality of the theorem follows. Since $E(X \mid \mathcal{F}_1)$ is $\mathcal{F}_i$-measurable for $i = 1$ and hence $i = 2$, the second equality in (2) follows from (1).

Theorem 12.4 Let $\{X_n, n \geq 1\}$ be a sequence of r.vs. and $\mathcal{G}$ a σ-field of events and assume $X_n \geq Z, n \geq 1$ and $E(Z) > -\infty$.

(a) *(Monotone convergence theorem for conditional expectations)*

If $X_n \uparrow X$ a.s., then (1) $E(X_n \mid \mathcal{G}) \uparrow E(X \mid \mathcal{G})$ a.s. and
(1') $E(X_n \mid Y = y) \uparrow E(X \mid Y = y)$ a.s.

(b) *(Fatou's lemma for conditional expectations)*

(2) $\lim_{n \to \infty} \inf E(X_n \mid \mathcal{G}) \geq E(\lim_{n \to \infty} \inf X_n \mid \mathcal{G})$ a.s. and

(2') $\lim_{n \to \infty} \inf E(X_n \mid Y = y) \geq E(\lim_{n \to \infty} \inf X_n \mid Y = \dot{y})$ a.s.

Now assume $X_n \leq Z$, $n \geq 1$ and $E(Z) < \infty$.

   (c) If $X_n \downarrow X$ a.s.,

   (3) then $E(X_n \mid \mathscr{G}) \downarrow E(X \mid \mathscr{G})$ a.s.
   and (3′) $E(X_n \mid Y = y) \downarrow E(X \mid Y = y)$ a.s.

   (d) (4) $\lim\limits_{n \to \infty} \sup E(X_n \mid \mathscr{G}) \leq E(\lim\limits_{n \to \infty} \sup X_n \mid \mathscr{G})$ a.s.

   (4′) $\lim\limits_{n \to \infty} \sup E(X_n \mid Y = y) \leq E(\lim\limits_{n \to \infty} \sup X_n \mid Y = y)$ a.s.

   (e) (*Dominated convergence theorem for conditional expectations*)
   Let $\{X_n, \ n \geq 1\}$ be a sequence of r.vs. and $\mathscr{G}$ a σ-field of events. If
   $X_n \to X$ a.s. and $\mid X_n \mid \leq \mid Z \mid$ a.s., $n \geq 1$, then $E(X_n \mid \mathscr{G}) \to E(X \mid \mathscr{G})$ a.s.
   and $E(X_n \mid Y = y) = E(X \mid Y = y)$ a.s.

Proof. (a) (1) We have $0 \leq X_n^+ \uparrow X^+$.

$\int_A X_n^+ \, dP = \int_A E(X_n^+ \mid \mathscr{G}) \, dP$, $A \in \mathscr{G}$. By monotone property of conditional

expectations, $E(X_n^+ \mid \mathscr{G})$ increases to h, a $\mathscr{G}$-measurable function. By ordinary

monotone convergence theorem

$$\int_A X^+ \, dP = \int_A h \, dP \text{ and, since } h \text{ is } \mathscr{G}\text{-measurable, } h = E(X^+ \mid \mathscr{G}) \text{ a.s.}$$

Also $0 \leq X_n^- \leq Z^-$, which is integrable; hence similarly $E(X_n^- \mid \mathscr{G}) \to E(X^- \mid \mathscr{G})$.
Now $E(X_n \mid \mathscr{G}) = E(X_n^+ \mid \mathscr{G}) - E(X_n^- \mid \mathscr{G})$ (note that X_n^- is integrable) $\to E(X^+ \mid \mathscr{G})$
$- E(X^- \mid \mathscr{G})$ (by Theorem 12.3) (note that X^- is integrable).
   Proof of (1′) is same as (1).

(b) (2) Let $Y_n = \inf\limits_{k \geq n} X_k$. Then $Z \leq Y_n \uparrow Y = \lim\limits_{n \to \infty} \inf X_n$. By (a) $E(Y_n \mid \mathscr{G}) \uparrow E(Y$

$\mid \mathscr{G})$ a.s. But $Y_n \leq X_n$, so $E(Y \mid \mathscr{G}) = \lim\limits_{n \to \infty} E(Y_n \mid \mathscr{G}) = \lim\limits_{n \to \infty} \inf E(Y_n \mid \mathscr{G}) \leq \lim\limits_{n \to \infty}$
$\inf E(X_n \mid \mathscr{G})$ (by simple consequence (b))

   (2′) is same as (2).
   (c) (3) This follows from (a) upon replacing all extended r.vs. by their
negatives.
   (3′) is same as (3).

   (d) (4) $E(\lim \sup X_n \mid \mathscr{G}) = -E(\lim\limits_{n \to \infty} \inf (-X_n) \mid \mathscr{G})$

$$\geq - \lim\limits_{n \to \infty} \inf E(- X_n \mid \mathscr{G}) \text{ by (b)}$$

$$= \lim\limits_{n \to \infty} \sup E(X_n \mid \mathscr{G})$$

   (4′) is same as (4).
   By (b) and (c) $\lim\limits_{n \to \infty} \inf E(X_n \mid \mathscr{G}) \geq E(\lim\limits_{n \to \infty} \inf X_n \mid \mathscr{G}) = E(\lim\limits_{n \to \infty} X_n \mid \mathscr{G})$ (∵
$\lim X_n = X$ a.s. exists)

$$= E(\lim\limits_{n \to \infty} \sup X_n \mid \mathscr{G}) \geq \lim\limits_{n \to \infty} \sup E(X_n \mid \mathscr{G}) \geq \lim\limits_{n \to \infty} \inf E(X_n \mid \mathscr{G}) \qquad (1)$$

Hence $\lim_{n \to \infty} E(X_n \mid \mathcal{G})$ exists and all the quantities in (1) are equal to $E(X \mid \mathcal{G})$.

Therefore $\lim_{n \to \infty} E(X_n \mid \mathcal{G}) = E(X \mid \mathcal{G})$.

Theorem 12.5   Let X be a r.v. with $| EX | \leq \infty$ and $\mathcal{G}$ be a σ-field of events. If Y is a finite valued $\mathcal{G}$-measurable r.v. such that $| EXY | \leq \infty$, then $E(XY \mid \mathcal{G}) = YE(X \mid \mathcal{G})$ a.s.

Proof.   If $Y = I_B$, $B \in \mathcal{G}$ and $C \in \mathcal{G}$, we have $\displaystyle\int_C XY \, dP = \int_{C \cap B} E(X \mid \mathcal{G}) \, dP$

$= \displaystyle\int_C I_B E(X \mid \mathcal{G}) \, dP = \int_C Y E(X \mid \mathcal{G}) \, dP$ and $YE(X \mid \mathcal{G})$ is $\mathcal{G}$-measurable. Thus, the result holds for indicators.

Now let $Y = \displaystyle\sum_{j=1}^{n} y_j I_{B_j}$ be a simple function, with $B_j \in \mathcal{G}$. By Theorem 12.3,

$$E(XY \mid \mathcal{G}) = \sum_{j=1}^{n} y_j E(XI_{B_j} \mid \mathcal{G}) = \sum_{j=1}^{n} y_j I_{B_j} E(X \mid \mathcal{G}) = YE(X \mid \mathcal{G}).$$

Let $Y \geq 0$ a.s. be a $\mathcal{G}$-measurable function. Then there exists $\mathcal{G}$-measurable sequence of simple functions $0 \leq Y_n \uparrow Y$. By separate consideration of $X^{\pm}$ and $Y^{\pm}$ it may be supposed that $X \geq 0$ and $Y \geq 0$. Moreover, by the monotone convergence theorem for conditional expectations it may even by assumed that X and Y are bounded r.vs. Since $Y_n X^+ \uparrow YX^+$, where $0 \leq Y_n$ is a $\mathcal{G}$-measurable simple function $Y_n \uparrow Y$, by monotone convergence theorem for conditional expectations $E(YX^+ \mid \mathcal{G})$ $= \lim E(Y_n X^+ \mid \mathcal{G}) = \lim Y_n E(X^+ \mid \mathcal{G})$ a.s. $= YE(X^+ \mid \mathcal{G})$. Similarly $E(YX^- \mid \mathcal{G}) = YE(X^- \mid \mathcal{G})$. Therefore, $E(YX \mid \mathcal{G}) = YE(X \mid \mathcal{G})$. In general,

$$\begin{aligned} E(YX \mid \mathcal{G}) &= E(Y^+ X \mid \mathcal{G}) - E(Y^- X \mid \mathcal{G}) \\ &= Y^+ E(X \mid \mathcal{G}) - Y^- E(X \mid \mathcal{G}) \\ &= YE(X \mid \mathcal{G}) \, \text{a.s.} \end{aligned}$$

Conditional expectation and conditional probability

Theorem 12.6 (*Jensen's inequality for conditional expectations*).   Let $E \mid X \mid < \infty$, $E \mid Y \mid < \infty$ and g be a real valued continuous convex function on the real line such that $E (g^+ (Y)) < \infty$. If (i) $X = E(Y \mid \mathcal{G})$ a.s., or (ii) g is also non-decreasing and $X \leq E(Y \mid \mathcal{G})$ a.s. then $g(X) \leq E(g(Y) \mid \mathcal{G})$ a.s. In particular $E(g(X)) \geq g(E(X))$.

Proof.   Define

$$h(t) = \lim_{s \to t-} \frac{g(s) - g(t)}{s - t}$$

and hence h is finite non-decreasing on $(-\infty, \infty)$. Since the secant line of a convex function is always above the one-sided tangent line, $g(t) \geq g(s) + (t - s) \, h(s)$, $-\infty < s, t < \infty$, and hence

$$g(Y) \geq g[E(Y \mid \mathcal{G})] + [Y - E(Y \mid \mathcal{G})]h(E(Y \mid \mathcal{G})), \text{ a.s.}$$

Thus, if $B = [|E(Y \mid \mathcal{G})| \leq N]$, $N > 0$, then $h(E(Y \mid \mathcal{G}))$ is bounded on $B \in \mathcal{G}$ and $I_B g(Y) \geq I_B g(E(Y \mid \mathcal{G})) + I_B[Y - E(Y \mid \mathcal{G})] \, h(E(Y \mid \mathcal{G}))$, a.s. Hence by Theorem 12.5.

$I_B E(g(Y) \mid \mathcal{G}) = E[I_B g(Y) \mid \mathcal{G})] \geq I_B g(E(Y \mid \mathcal{G}))$, a.s. and since $I_B \to 1$ a.s. as $N \to \infty$,

$$E(g(Y) \mid \mathcal{G}) \geq g(E(Y \mid \mathcal{G})) \text{ a.s.} \tag{2}$$

$$X = E(Y \mid \mathcal{G}) \text{ a.s.} \Rightarrow E(g(Y) \mid \mathcal{G}) \geq g(X) \text{ a.s.} \tag{3}$$

Now g being non-decreasing (2) implies $E(g(Y) \mid \mathcal{G}) \geq g(X)$ a.s.

PROBLEMS AND COMPLEMENTS

12.1   Show that if μ and ν are measures such that $\nu \ll \mu$ and $\nu \perp \mu$, then ν is identically zero.

12.2   If μ and ν are two finite measures on a σ-field $\mathcal{F}$, then $\nu \ll \mu + \nu$, where $\mu + \nu$ is defined as $(\mu + \nu)(E) = \mu(E) + \nu(E)$ for all $E \in \mathcal{F}$.

12.3   If ν is only a σ-finite signed measure the Radon-Nikodym theorem still holds.

12.4   Show that the condition ν is σ-finite is necessary in the Radon-Nikodym theorem.

12.5   Show that if $\nu(E) = \displaystyle\int_E f \, d\mu$ for each $E \in \mathcal{F}$, where $f \geq 0$, f is measurable and $\mu \, [\omega : f(\omega) = \infty] > 0$, then ν is not σ-finite.

NOTE. This shows that $\nu \ll \mu$ and μ σ-finite does not imply that ν is σ-finite.

12.6   Show that the condition: μ σ-finite, is necessary in the Radon-Nikodym theorem.

12.7   Let μ be a signed measure on $(\Omega, \mathcal{F})$ and let ν be a finite-valued signed measure on $(\Omega, \mathcal{F})$ such that $\nu \ll \mu$; then given $\varepsilon > 0$ there exists $\delta > 0$ such that $|\nu|(E) < \varepsilon$ whenever $|\mu|(E) < \delta$.

NOTE.   The converse is true without finiteness conditon.

12.8   If μ and ν are signed measures on $(\Omega, \mathcal{F})$ and for all $\varepsilon > 0$, there exists $\delta > 0$ such that whenever $|\mu|(E) < \delta, |\nu|(E) < \varepsilon$, then show that $\nu \ll \mu$.

12.9   If ν_1 and ν_2 are σ-finite measures on $(\Omega, \mathcal{F})$ and $\nu_1 \ll \mu$, $\nu_2 \ll \mu$, then show that

$$\frac{d(\nu_1 + \nu_2)}{d\mu} = \frac{d\nu_1}{d\mu} + \frac{d\nu_2}{d\mu} [\mu].$$

12.10 If λ, μ and ν are σ-finite measures on $\mathcal{F}$ and $\nu \ll \mu$ and $\mu \ll \lambda$, show that

$$\nu \ll \lambda \text{ and } \frac{d\nu}{d\lambda} = \frac{d\nu}{d\mu} \cdot \frac{d\mu}{d\lambda} [\lambda].$$

NOTE.   The result is *the Chain rule* for *Radon-Nikodym derivaties*.

12.11 Two σ-finite measures μ and v are called *equivalent,* denoted $\mu \equiv v$, if each is absolutely continuous with respect to the other. Show that this is indeed an equivalence relation. If $(\Omega, \mathcal{F}, P)$ is a probability space, $X_i \geq 0$ is a $\mathcal{S}_1$ r.v. and $P_i(A) \int_A x_i \, dp$, $i = 1, 2$, and if $P\{X_1 = 0) \Delta(X_2 = 2)\} = 0$, then show that P_1 and P_2 are equivalent measures.

12.12 If v, μ and λ are σ-finite measures on a measurable space $(\Omega, \mathcal{F})$ with $v \ll \mu$ and $\mu \ll \lambda$, then $v \ll \lambda$ and

$$\frac{dv}{d\lambda} = \frac{dv}{d\mu} \frac{d\mu}{d\lambda} \text{ a.e.} [\lambda] \text{ (Chain rule)}.$$

[*Hint: Lemma.*   If μ be a σ-finite measure and $v \ll \mu$, then v is a σ-finite measure on $(\Omega, \mathcal{F})$. If X is a $\mathcal{F}$-measurable function whose integral $\int X \, dv_1$ exists, then for every $A \in \mathcal{F}$, $\displaystyle\int_A X \, dv_1 = \int_A X \frac{dv_1}{d\mu} d\mu$].

12.13 If Ω is the real line, $\mathcal{B}$ is the Borel field with respect to on the real line, μ is the Lebesgue measure on $\mathcal{B}$ and v is a measure on $\mathcal{B}$ such that

$$n((-\infty, a]) = \begin{cases} 0 & \text{if } a < 0 \\ 1 - e^{-a} & \text{if } a \geq 0 \end{cases}$$

then give the Radon-Nikodym derivative of v with respect to μ.

12.14 Let F_θ denote the d.f. of a normal r.v. with mean θ and variance 1 and μ_θ denote the associated Lebesgue-Stieltjes measure. Examine whether μ_θ, is absolutely continuous with respect to μ_0. If so, find the Radon-Nikodym derivative of μ_θ with respect to μ_0. Is μ_0 absolutely continuous with respect to μ_θ, $\theta \neq 0$?

12.15 If $E|Y| < \infty$ and $X : (\Omega, \mathcal{F}) \rightarrow (\Omega', \mathcal{F}')$ and $Z : (\Omega, \mathcal{F}) \rightarrow (\Omega'', \mathcal{F}'')$, show that if (X, Y) and Z are independent, then $E(Y \mid X, Z) = E(Y \mid X)$.

12.16 (Conditional Cauchy-Schwartz inequality). Prove that $E^2(|XY| \,|\mathcal{G}) \leq E(X^2 \mid \mathcal{G}) E(Y^2 \mid \mathcal{G})$.

13

Martingale

13.1 INTRODUCTION

The name Martingale was introduced into modern probability by Ville (1939), though Bernstein (1927, 1939, 1940, 1941) and Lévy (1937) used it without giving a name. Like probability theory it has its origin in gambling theory and it is often profitable to interpret results in terms of a gambling situation. For example if $X_1, X_2, \ldots$ is a sequence of r.vs., we may think of X_n as the fortune of a gambler after n trials in a succession of games and σ-field $\mathcal{F}_n$ as the information contained in the first n games. Having survived the first n games the gambler's expected fortune after game $n + 1$ is $E(X_{n+1} \mid \mathcal{F}_n)$. If this equals X_n, the game is "fair" since the expected gain on trial $n + 1$ is $E(X_{n+1} - X_n \mid \mathcal{F}_n) = X_n - X_n = 0$. If $E(X_{n+1} \mid \mathcal{F}_n) \geq X_n$, the game is "favourable" and if $E(X_{n+1} \mid \mathcal{F}_n) \leq X_n$, the game is "unfavourable". We are going to study sequences of this type of r.vs.; the results to be obtained will have significance outside the casino as well as inside.

Definition 13.1 Let $(\Omega, \mathcal{F}, P)$ be a probability space, $\{X_n\}$ a sequence of r.vs. on $(\Omega, \mathcal{F}, P)$ and $\mathcal{F}_1 \subset \mathcal{F}_2 \subset \ldots$ an increasing sequence of sub-σ-fields of $\mathcal{F}$; If (i) X_n is $\mathcal{F}_n$-measurable, (ii) $E \mid X_n \mid < \infty$, and (iii) $E(X_n \mid \mathcal{F}_{n-1}) = X_{n-1}$ a.s. then $\{X_n, \mathcal{F}_n\}$ is called a *martingale sequence*.

Note. Condition (iii) can be replaced by $E(X_n \mid \mathcal{F}_m) = X_m$ a.s. if $n > m$ (prove it!).

If on the other hand, $\mathcal{F}_n$ is a decreasing sequence of sub-σ-fields of $\mathcal{F}$ with $\{X_n\}$ satisfying (i), (ii) and (iv) $E(X_n \mid \mathcal{F}_{n+1}) = X_{n+1}$ a.s. then $\{X_n, \mathcal{F}_n\}$ is called a *reverse martingale sequence*. Clearly $\{(X_i, \mathcal{F}_i); 1 \leq i \leq n\}$ is a reverse martingale iff $\{(X_{n-i+1}, \mathcal{F}_{n-i+1}); 1 \leq i \leq n\}$ is a martingale. $\{X_n, \mathcal{F}_n\}$ is called a *submartingale* if equality in condition (iii) is changed to inequality $E(X_{n+1} \mid \mathcal{F}_n) \geq X_n$ a.s.; similarly a *supermartingale* if $E(X_{n+1} \mid \mathcal{F}_n) \leq X_n$ a.s. Again going back to the origin of martingale, we may quote from as early as Oxford Dictionary (1815) and Bachelier (1900): "It means a system for recouping losses by doubling the stake after each loss". To explain more clearly let us consider the following example:

A gambler has x dollars on trial n, he wins x dollars if the coin shows in head and $-x$ dollars if the coin comes up tails. Suppose that the gambler has infinite

178

capital. Strategy is to bet one dollar on trial one and for each $n \geq 1$ to double his bet on trial $n + 1$ if he losses on trial n and quit at trial n if he wins on trial n. Let Y_n, $n \geq 1$ be the sequence of his winnings. Then $\{Y_n, n \geq 1\}$ is a Markov process, i.e. $P[Y_n \in A | Y_{n-1}, \ldots, Y_1] = P[Y_n \in A | Y_{n-1}]$ a.s. for every measurable event A. Also $P[Y_n = 2^{n-1} | Y_{n-1} = -2^{n-2}] = 1/2 = P[Y_n = -2^{n-1} | Y_{n-1} = 2^{n-2}]$, $n \geq 2$ and $P[Y_n = 0 | Y_{n-1} = 0] = P[Y_n = 0 | Y_{n-1} = 2^{n-2}] = 1$.

Let $\mathcal{F}_n = \mathcal{B}(Y_1, Y_2, \ldots, Y_n)$. Then $\{Y_n, \mathcal{F}_n\}$ is a martingale difference sequence, i.e. $E(Y_n | \mathcal{F}_{n-1}) = 0$ a.s.

Hence the gambler is sure to win this game and go home one dollar richer. We see that the gambler wins one dollar by playing "fair". This is known as "bold strategy". Later on J.L. Doob in 1940's and early fifties made it well known (culminating in his masterpiece *Stochastic Process* (1953)—still today the best source for students to learn martingales) by discovering *Martingale convergence theorem*.

Remarks.   (1) In the definition of submartingale (supermartingale) condition (ii) $E | X_n | < \infty$, $n \geq 1$ can be replaced by the weaker condition (ii') $EX_n^+ < \infty$

($EX_n^- < \infty$), $n \geq 1$.

EXAMPLE 13.1   Prove the last result.

(2) $\{X_n, \mathcal{F}_n\}$ is a martingale iff $\int_A X_n \, dP = \int_A X_{n+1} \, dP$ for all $A \in \mathcal{F}_n$, $n \geq 1$.

This follows from the condition $E(X_{n+1} | \mathcal{F}_n) = X_n$ a.s. $[P] \Leftrightarrow \int_A E(X_{n+1} | \mathcal{F}_n) \, dP$

$= \int_A X_n \, dP$ for all $A \in \mathcal{F}_n$. Also $\int_A E(X_{n+1} | \mathcal{F}_n) \, dP = \int_A X_{n+1} \, dP$ by definition of conditional expectation (see Section 12.2). Similar results are true for submartingale and supermartingale. In particular a martingale is a constant expectation (i.e. $EX_n = c$, $n \geq 1$) process. Similarly, a submartingale is an increasing expectation process and a supermartingale is a decreasing expectation process.

(3) If $\{X_n, \mathcal{F}_n\}$ and $\{Y_n, \mathcal{F}_n\}$ are submartingales, so is $\{\max(X_n, Y_n), \mathcal{F}_n\}$. For $E(\max(X_{n+1}, Y_{n+1}) | \mathcal{F}_n) \geq E(X_{n+1} | \mathcal{F}_n) \geq X_n$ a.s. and similarly $E(\max X_{n+1}, Y_{n+1}) | \mathcal{F}_n) \geq Y_n$ a.s. Similarly, if $\{X_n, \mathcal{F}_n\}$ and $\{Y_n, \mathcal{F}_n\}$ are supermartingales, so is $\{\min(X_n, Y_n), \mathcal{F}_n\}$.

(4) More generally we have the following theorem regarding construction of submartingales from martingales or from other submartingales.

Theorem 13.1

(a) Let $\{X_n, \mathcal{F}_n\}$ be a submartingale, g a convex increasing function. If $g(X_n)$ is integrable for all $n \geq 1$, then $\{g(X_n), \mathcal{F}_n\}$ is a submartingale. In particular if $\{X_n\}$ is a submartingale, so is $\{X_n^+\}$.

(b) Let $\{X_n, \mathcal{F}_n\}$ be a martingale, g a convex function. If $g(X_n)$ is integrable for all $n \geq 1$, then $\{g(X_n), \mathcal{F}_n\}$ is a submartingale. In particular if $r \geq 1$, $\{X_n\}$ is a martingale and $| X_n |^r$ is integrable, $n \geq 1$, then $\{|X_n|^r\}$ is a submartingale.

Proof. By Jensen's inequality for conditional expectation $E(g(X_{n+1} \mid \mathcal{F}_n)) \geq g(E(X_{n+1} \mid \mathcal{F}_n))$. In (a) by the submartingale property $E(X_{n+1} \mid \mathcal{F}_n) \geq X_n$; hence $g(E(X_{n+1} \mid \mathcal{F}_n)) \geq g(X_n)$ since g is increasing. In (b) by martingale property $E(X_{n+1} \mid \mathcal{F}_n) = X_n$, therefore $g(E(X_{n+1} \mid \mathcal{F}_n)) = g(X_n)$. The result is proved.

Examples of martingales

(a) $X_n = X$, then $\{X_n\}$ is a martingale; If $X_1 \leq X_2 \leq \ldots$, then $\{X_n\}$ is a submartingale; If $X_1 \geq X_2 \geq \ldots$, then $\{X_n\}$ is a supermartingale.

(b) Let $\{Y_n\}$ be a sequence of independent r.vs. with $EY_j = \mu_j \neq 0$ and set

$$X_n = \prod_{j=1}^{n} (Y_j/\mu_j), \quad \mathcal{F}_n = \mathcal{B}(Y_1, \ldots, Y_n). \text{ Then } \{X_n, \mathcal{F}_n\} \text{ is a martingale.}$$

Proof. $E(X_{n+1} \mid \mathcal{F}_n) = E\left(\dfrac{X_n Y_{n+1}}{\mu_{n+1}} \mid Y_1, \ldots, Y_n \right) = X_n \, E(Y_{n+1}/\mu_{n+1}) = X_n$ a.s.

(c) Let Y be an integrable r.v. on $(\Omega, \mathcal{F}, P)$ such that $\mathcal{F}_1 \subset \ldots \subset \mathcal{F}_{n-1} \subset \mathcal{F}_n \subset \mathcal{F}$. Then $X_n = E(Y \mid \mathcal{F}_n)$ is a martingale with respect to $\mathcal{F}_n$ since $E(X_{n+1} \mid \mathcal{F}_n) = E(E(Y \mid \mathcal{F}_{n+1}) \mid \mathcal{F}_n) = E(Y \mid \mathcal{F}_n) = X_n, \; (\mathcal{F}_n \subset \mathcal{F}_{n+1})$.

(d) (*Polya urn scheme*) Let an urn contain b black and r red balls. Set $T_0 = b/(b + r)$. At each drawing a ball is drawn at random, its colour is noted and a ball of that colour are added to the urn. Let b_n be the number of black balls and r_n be the number of red balls after the nth drawing. Let T_n be the proportion of black balls after the nth drawing. Show that $\{T_n\}$ is a martingale.

Proof. By Markov property of T_n

$$\mathrm{E}[T_n \mid T_{n-1}, \ldots, T_1] = E[T_n \mid T_{n-1}] \text{ a.s.}$$

Therefore

$$E\left[T_n \mid T_{n-1} = \frac{b_{n-1}}{b_{n-1} + r_{n-1}} \right] = \frac{b_{n-1} + a}{b_{n-1} + a + r_{n-1}} \cdot \frac{b_{n-1}}{b_{n-1} + r_{n-1}}$$

$$+ \frac{b_{n-1}}{b_{n-1} + r_{n-1} + a} \cdot \frac{r_{n-1}}{b_{n-1} + r_{n-1}}$$

$$= \frac{b_{n-1}}{b_{n-1} + r_{n-1}} = T_{n-1}.$$

So $\{T_n, n \geq 1\}$ is a martingale.

(e) (*Branching process*) (i) Let X_n be a Markov chain with state space $S = \{0, 1, 2, \ldots\}$. The state at time n, denoted by X_n, is to represent the number of "offsprings" after n "generations". Take $X_0 = 1$, and define $X_{n+1} = \sum_{k=1}^{X_n} Y_k$, $n = 0, 1, 2, \ldots$, where Y_i are i.i.d. r.vs. with $P(Y_i = j) = p_j, j = 0, 1, 2, \ldots$ which is the probability that a given individual will produce exactly j "offsprings". Denote $p_{ij} = P(Y_1 + \ldots + Y_i = j), \; i = 1, 2, \ldots$ and $j = 0, 1, 2, \ldots$ and $p_{00} = 1$. Let $W_n = X_n/m^n$

and $m = E(Y_i) = \sum_{j=0}^{\infty} j p_j$, $i \geq 1$. If $0 < m < \infty$, then show that $\{W_n, \mathscr{F}_n\}$ is a martingale,

where $\mathscr{F}_n = \mathscr{B}(X_0, \ldots, X_n) = \mathscr{B}\left(\dfrac{X_1}{m}, \dfrac{X_2}{m^2}, \ldots, \dfrac{X_n}{m^n}\right)$ and $EW_n = 1$ for all $n \geq 1$.

Proof. $E(X_{n+1}/m^{n+1} \mid X_0 = i_0, \ldots, X_n = i_n) = \sum_{j=0}^{\infty} \dfrac{j}{m^{n+1}} P(Y_1 + \ldots + Y_{i_n} = j)$

$$= \sum_{j=0}^{\infty} p_{i_n j} / m^{n+1} = \frac{1}{m^{n+1}} E(Y_1 + \ldots + Y_{i_n}) = \frac{i_n m}{m^{n+1}} = i_n / m^n = \frac{X_n}{m^n}.$$

Hence $\{W_n, \mathscr{F}_n\}$ is a martingale. Since martingale is an expectation constant process, $E(W_n) = \ldots = E(W_1) = E(Y_1/m) = m/m = 1$.

(e) (ii) Let $G(s) = \sum_j p_j s^j$, $s \geq 0$ be the generating function of the offspring distribution, i.e. probability distribution of Y_i as in e(i). If $G(r) = r$, i.e. $p_1 = 1$ and $p_j = 0$ if $j \neq 1$, then show that $\{r^{X_n}, \mathscr{F}_n\}$ is a martingale.

Proof. $E(r^{X_{n+1}} \mid X_0 = i_0, \ldots, X_n = i_n) = \sum_{j=0}^{\infty} p_{i_n j} r^j$

$$= \sum_{j=0}^{\infty} r^j P(Y_1 + \ldots + Y_{i_n} = j) = E(r^{Y_1 + \ldots + Y_{i_n}}) = [E(r^{Y_1})]^{i_n} = [G(r)]^{i_n} = r^{i_n}$$

Therefore, $\qquad\qquad\qquad E(r^{Y_{n+1}} \mid \mathscr{F}_n) = r^{X_n}$ a.s.

(f) (*U-statistics*) $U_n = \dfrac{1}{\dbinom{n}{m}} \sum_{1 \geq i_1 < \ldots < i_n \leq n} h(X_{i_1}, \ldots, X_{i_m})$ is called a U-statistic

with a kernel $h(x_1, \ldots, x_m)$. Given a random sample $X_1, \ldots, X_n$, $n \geq m$ the corresponding U-statistic may be expresed as $U_n = E\{h(X_1, \ldots, X_m) \mid X(n)\}$, where $X_{(n)} = (X_{n1}, \ldots, X_{nn})$ is the order statistic and $\min\limits_{1 \leq j \leq n} X_j = X_{n1} \leq X_{n2} \leq \ldots \leq X_{nm} = \max\limits_{1 \leq j \leq n} X_j$.

Berk (1966) showed that if h is a symmetric kernel with $E_F(h) = \theta = \theta(F)$ and $E_F |h| < \infty$ then $\{U_n, \mathscr{F}_n\}$ is a reverse martingale with respect to $\mathscr{F}_n = \sigma\{X_{(n)}, X_{n+1}, X_{n+2}, \ldots\}$.

(g) Let $\{X_n, n \geq 1\}$ be independent integrable r.vs. with $E(X_n) = 0$, then for $k \geq 1$, $\{U_{k,n}, n \geq k\}$ is an integrable martingale where

$$U_{k,n} = \sum_{1 \leq i_1 < \ldots < i_k \leq n} X_{i_1} \ldots X_{i_k}, \quad n \geq k.$$

More generally, if $\{X_n, \mathscr{F}_n, n \geq 1\}$ are L_k r.vs. with $E(X_{n+1} \mid \mathscr{F}_n) = 0$, $n \geq 1$, then $\{U_{k,n}, \mathscr{F}_n, n \geq k\}$ is an integrable martingale with $E(U_{k,k}) = 0$. The results easily follow by taking conditional expectations.

(h)   Recall the definition: Random variables $\{X_i\}_{i \geq 1}$ are called interchangeable (or exchangeable) if the joint d.f. of every finite subset of k of these r.vs. is a symmetric function. Let $\{X_n, n \geq 1\}$ be interchangeable r.vs. and $h(\cdot)$, a symmetric Borel measurable function on R^m with $E \mid h(X_1, \ldots, X_m) \mid < \infty$. A sequence of so-called U-statistics is defined for any integer $m \geq 1$ by

$$U_{m,n} = \binom{n}{m}^{-1} \sum_{1 \leq i_1 < \ldots < i_m \leq n} h(X_{i_1}, \ldots, X_{i_m}), n \geq m.$$

Then $\{U_{m,n}, \mathcal{F}_n, n \geq m\}$ is a reverse martingale.

Proof.   If $\mathcal{F}_n = \sigma(U_{m,k}, k \geq n)$, then for $1 \leq i_1 < \ldots < i_m \leq n + 1$ and $A \in \mathcal{F}_{n+1}$,

$$\int_A h(X_{i_1}, \ldots, X_{i_m})dP = \int_A h(X_1, \ldots, X_m)dP \text{ by symmetry and exchangeability. This}$$

implies $E(h(X_{i_1}, \ldots, X_{i_m}) \mid \mathcal{F}_{n+1}) = E(h(X_1, \ldots, X_m) \mid \mathcal{F}_{n+1})$ a.s. for $1 \leq i_1 < \ldots < i_m \leq n + 1$, and hence

$$E(U_{m,n} \mid \mathcal{F}_{n+1}) = E(U_{m,n+1} \mid \mathcal{F}_{n+1}) = U_{m,n+1} \text{ a.s.}$$

So $\{(U_{m,n}, \mathcal{F}_n) \, n \geq m\}$ is a reverse martingale and hence if $U_n^* = U_{m,-n}$ and $\mathcal{F}_n^* = \mathcal{F}_{-n}, n \leq -m$ then $\{(U_n^*, \mathcal{F}_n^*) \, n \leq -m\}$ is a martingale.

(i) (*Likelihood ratio*)   Let $\{X_n\}_1^\infty$ be a sequence of r.vs. defined on the measurable space $\{\Omega, \mathcal{F}\}$. Assume that under the probability measure P on $\mathcal{F}(X_1, \ldots, X_n)$ has density p_n, and under the probability measure Q on $\mathcal{F}(X_1, \ldots, X_n)$ has density q_n. Define

$$L_n = \begin{cases} \dfrac{q_n(X_1, \ldots, X_n)}{p_n(X_1, \ldots, X_n)} & \text{if } p_n > 0 \\ 0 & \text{otherwise} \end{cases}$$

Show that if $\mathcal{F}_n = \sigma(X_1, \ldots, X_n)$, $\{L_n, \mathcal{F}_n\}$ is a supermartingale on $(\Omega, \mathcal{F}, P)$ and $0 \leq EL_n \leq 1$ for all $n \geq 1$.

Solution. $E(L_n) = \displaystyle\int_{\{p_n(\mathbf{x})>0\}} q_n(\mathbf{x})/p_n(\mathbf{x}) \cdot p_n(\mathbf{x}) \, d\mathbf{x}$

$$\leq \int_{R^n} q_n(\mathbf{x}) \, d\mathbf{x} = 1$$

Let $C = \{(X_1, \ldots, X_n) \in \mathcal{B}\}$, $\mathcal{B} \in (R^n)$. Then $\displaystyle\int_C L_{n+1} \, dP = E(L_{n+1}I_C)$; hence

$$\int_C L_{n+1} \, dP = \int_{\{\mathbf{x} \in B, p_{n+1}(\mathbf{x}')>0\}} q_{n+1}(\mathbf{x}')/p_{n+1}(\mathbf{x}')p_{n+1}(\mathbf{x}') \, d\mathbf{x}' \tag{1}$$

where

$$\mathbf{x}' = (x_1, \ldots, x_{n+1}) \in R^{n+1}, \mathbf{x} = (x_1, \ldots, x_n).$$

Now $p_n(\mathbf{x}) \int_{-\infty}^{\infty} p_{n+1}(\mathbf{x}') \, dx_{n+1}$; hence if $\mathbf{x} \in B$ and $p_n(\mathbf{x}) = 0$, then $p_{n+1}(\mathbf{x}') = 0$ except on a set of Lebesgue measure 0, so the integration of $q_{n+1}(\mathbf{x}')$ with respect to X_{n+1} in (1) will be 0, i.e.

$$\int_{\{\mathbf{x}\in B, p_n(\mathbf{x})=0, p_{n+1}(\mathbf{x}')>0\}} q_{n+1}(\mathbf{x}') \, dx' = 0.$$

Therefore the right hand side of (1) is

$$\int_{\{\mathbf{x}\in B, p_n(\mathbf{x})>0, p_{n+1}(\mathbf{x}')>0\}} q_{n+1}(\mathbf{x}') \, dx' \le \int_{\{\mathbf{x}\in B, p_n(\mathbf{x})>0\}} q_{n+1}(\mathbf{x}') \, dx'$$

$$= \int_{\{\mathbf{x}\in B, p_n(\mathbf{x})>0\}} q_n(\mathbf{x})/p_n(\mathbf{x}) \cdot p_n(\mathbf{x}) \, d\mathbf{x}$$

$$= E[I_A \, L_n] = \int_A L_n \, dP$$

Hence L_n is a supermartingale

13.2 HAJEK-RENYI INEQUALITY

Let $\{X_n\}$ be a sequence of independent r.vs. with $EX_n = 0$, var $(X_n) = \sigma_n^2 > 0$. Let $\{C_i\}$ be a non-increasing sequence of positive constants. Let $S_k = \sum_1^k X_i$, $1 \le k \le n$. Then for $\varepsilon > 0$,

$$P[\max_{j\le k\le n} C_k |S_k| > \varepsilon] \le 1/\varepsilon^2 \, \{C_j^2 \sum_{i=1}^j \sigma_1^2 + \sum_{i+1}^n C_i^2 \sigma_i^2\] \text{ holds.}$$

The proof of Hajek-Renyi inequality depends on the following lemma.

Lemma 13.1 Let $\{S_n, \mathscr{F}_n, n \ge 1\}$ be a submartingale, $\varepsilon > 0$ and $0 < a_n$ increasing and $\mathscr{F}_{n-1}$-measurable for $n \ge 1$. Then

$$\varepsilon P[\max_{1\le k\le n} S_k/a_k \ge \varepsilon] \le E S_1^+/a_1 + \sum_{k=2}^n E(S_k^+ - S_{k-1}^+)/a_k \tag{1}$$

for each $n \ge 1$.

Proof. $P[\max_{k\le n} S_k/a_k \ge \varepsilon] = P[\max_{k\le n} S_k^+/a_k \ge \varepsilon], \, \varepsilon > 0$ \qquad (2)

since $\varepsilon > 0$, $S_k^+ = \max(0, S_k) \ge 0$, the equality (2) holds in the case $\max_{k\le n} S_k \ge 0$. The case $\max_{k\le n} S_k < 0$ is trivial. Also note that $\{S_k^+, k \ge 1\}$ is again a submartingale. Thus, without loss of generality suppose $S_k \ge 0$ a.s. for $k \ge 1$. Let $Z_k + S_1/a_1 + \sum_2^k (S_i - S_{i-1})/a_i$. Then, $\{Z_k, k \ge 1\}$ is a non-negative submartingale $(Z_k \ge 0 + S_k/a_k \ge 0$ a.s.$)$. Let t be the first $n \ge k \ge 1$ such that $S_k/a_k \ge \varepsilon$ if such a k exists, otherwise let $t = n + 1$. Summation by parts shows that

$$Z_k = S_k/a_k + \sum_{i=1}^{k-1} (a_i^{-1} - a_{i+1}^{-1})S_i.$$

On $t = k$, $\varepsilon \leq S_k/a_k \leq Z_k$, thus on $t \leq n$, $Z_t \geq \varepsilon$.

Hence $\varepsilon P(\max_{1 \leq k \leq n} S_k/a_k \geq \varepsilon) \leq \varepsilon P(t \leq n)$

$$\leq E(Z_t I(t \leq n)) = \sum_{j=1}^{n} \int_{t=j} Z_j \, dP \leq \sum_{j=1}^{n} \int_{t=j} Z_n \, dP$$

(since $[t = j] \in \mathscr{F}_j$, $E(Z_n \mid \mathscr{F}_j) \geq Z_j$ if $j \leq n$)

$$= E(Z_n I(t \leq n)) \leq E(Z_n).$$

Since $Z_n \geq 0$ a.s., inequality (1) holds.

Proof of Hajek-Renyi inequality.   Since $EX_n = 0$ for all $n \geq 1$, $S_n = X_1 + \ldots + X_n$ is a martingale and $\{ C_n^2 S_n^2 \ n \geq 1\}$ is a non-negative submartingale.

$$\varepsilon^2 P[\max_{j \leq k \leq n} C_k |S_k| > \varepsilon] = \varepsilon^2 P[\max_{j \leq k \leq n} C_k^2 |S_k^2| > \varepsilon^2]$$

$$\leq C_j^2 E(S_j^2) + \sum_{i=j+1}^{n} E(C_i^2 S_i^2 - C_i^2 S_{i-1}^2)$$

$$= C_j^2 \sum_{i=1}^{j} \sigma_i^2 + \sum_{i=j+1}^{n} C_i^2 \sigma_i^2, \text{ since}$$

$$0 < a_i = 1/C_i^2 \text{ and } E(C_i^2 S_i^2 - C_i^2 S_{i-1}^2) = C_i^2 (\sum_{k=1}^{i} \sigma_k^2 - \sum_{k=1}^{i-1} \sigma_k^2) = C_i^2 \sigma_i^2.$$

Corollary 13.1   Let $\{T_n, \mathscr{F}_n, n \geq 1\}$ be a non-negative submartingale with $0 < a_n$ non-decreasing and $\mathscr{F}_{n-1}$-measurable for $n \geq 1$. Suppose for some $p \geq 1$, $ET_1^p/a_1^p < \infty$ and

$$\sum_{k=0}^{\infty} E(T_k^p - T_{k-1}^p)/a_k^p < \infty \tag{3}$$

Then SLLN holds, i.e. $\lim T_n/a_n = 0$ a.s.

Proof. $\{T_n^p, \mathscr{F}_n, n \geq 1\}$ is a non-negative submartingale. Using the lemma

$$\varepsilon^p P[\max_{j \geq n} T_k/a_k \geq \varepsilon] = \varepsilon^p P[\sup_{k \geq n} T_k^p/a_k^p \geq \varepsilon^p]$$

$$< ET_n^p/a_n^p + \sum_{k=n+1}^{\infty} E((T_k^p - T_{k-1}^p)/a_k^p).$$

By (3) and Kronecker's lemma

$$\sum_{k=2}^{n} E[(T_k^p - T_{k-1}^p)/a_n^p] = E(T_n^p/a_n^p) - E(T_1^p/a_n^p) \to 0$$

as $n \to \infty$. Thus $E(T_n^p/a_n^p) \to 0$. Hence $P[\sup_{k \geq n} T_k/a_k \geq \varepsilon] \to 0$ as $n \to \infty$. This implies $T_n/a_n \to 0$ a.s. as $n \to \infty$, since by (3) $\sum_{k=n+1}^{\infty} E(T_k^p - T_{k-1}^p)/a_k^p \to 0$ a.s. as $n \to \infty$.

Corollary 13.2 (*Chow 1960 and 1967*). Let $\{X_k\}$ be a martingale difference sequence. Then for $p \geq 2$, $S_n/n \to 0$ a.s. as $n \to \infty$ if $\sum_{k=1}^{\infty} E|X_k|^p/k^{1+p/2} < \infty$.

PROBLEMS AND COMPLEMENTS

13.1 (Halmos's optional skipping theorem). Let Y_n denote the fortune, after the nth game, of a gambler when he uses no skipping strategy and $\{X_n\}$ the fortune when he uses a skipping strategy. The r.v. $Z_n = 1$ if he bets in game n and $Z_n = 0$ if he passes game n. If the game is initially favourable (submartingale) or fair (martingale) show that it remains favourable or fair and no skipping strategy can increase the expected winning.

13.2 Let X_1 be uniformly distributed over [0, 1]. We successively define a sequence as follows. If it is given that $X_1 = x_1, \ldots. x_{n-1} = x_{n-1}$, then define X_n as a r.v. uniformly distributed on $[0, X_{n-1}]$. Show that the sequence $\{X_n\}$ is a supermartingale. Calculate $E(X_n)$.

***13.3** Let $\{Y_n\}$ be a Markov chain (MC) with transition matrix $p_{x,y}$, $x, y \in S$, the state space. A function f on S is called a right eigen function $\exists \lambda$ s.t. if $f(y) = \sum_{x \in s} p_{x,s} f(x)$. Show that $X_n = \lambda^{-n} f(Y_n)$ is a martingale if f is a right eigenfunction w.r.t. the eigenvalue and $E(f(Y_n)) < \infty$ for all n.

13.4 Let $\{Y_n)$ be a sequence of i.i.d. $N(0, 1)$ r.vs. Define $S_n = \sum_{k=1}^{n} Y_k$ and $X_n^\alpha = \exp[\alpha S_n - n\alpha^2/2]$, $\alpha \in \mathbb{R}$. Then show that (i) $\{X_n^\alpha\}$ is a martingale (ii) $\xi_n = \int X_n^\alpha \, dF(\alpha)$ is also a martingale where F is a d.f.

---

*Indicates difficult problem.

14

Measure Extension and Lebesgue-Stieltjes Measure

14.1 MEASURE EXTENSION AND CARATHÉODORY EXTENSION THEOREM

14.1.1 Introduction

It is extremely useful to extend the domain of definition of a probability measure or more generally a measure from a class of sets (e.g. semi-algebra or field) to a larger class of sets (normally σ-field). Actually from a non-negative set function defined on a semi-algebra (or a field) we construct a countably additive set function on a field and then extend this to a σ-field. For example, Lebesgue measure on the real line is defined on open intervals (or semi-closed intervals) as length and we use this to define outer measure and the Lebesgue measurable sets.

14.1.2 Outer Measure

Definition 14.1 By an *outer measure* μ^* we mean an extended real-valued set function defined on all subsets of Ω with the following properties:

1. $\mu^*(\phi) = 0$

2. μ^* is monotone: $A \subset B$ implies $\mu^*(A) \leq \mu^*(B)$

3. μ^* is countable σ-subadditive: $\mu^*(\overset{\infty}{\underset{n=1}{\cup}} A_n) \leq \overset{\infty}{\underset{1}{\Sigma}} \mu^*(A_n)$

Definition 14.2 A set E is said to be μ^*-measurable if *for every*

$$A \subset \Omega, \mu^*(A) = \mu^*(A \cap E) + \mu^*(A \cap E^C) \tag{1}$$

It is by subadditivity equivalent to $\mu^*(A \cap E) + \mu^*(A \cap E^C) \leq \mu^*(A)$. We shall prove later that the class of all μ^*-measurable sets is a σ-field.

Definition 14.3   A semi-algebra $\mathcal{L}$ is a non-empty class of subsets of Ω containing Ω such that for each $A \in \mathcal{L}$ there is a *finite partition* of A^C in $\mathcal{L}$ (i.e. $A^C = A_1 \cup A_2 \cup \ldots \cup A_k$, $A_i \cap A_j = \phi$ for $i \neq j$ and $A_i \in \mathcal{L}$) is closed under intersection and $\phi \in \mathcal{L}$. The class of all semi-closed intervals $(a, b]$, $-\infty \leq a \leq b \leq \infty$ and the class of all rectangles $[(a, b] \times (c, d]]$, $\infty \leq a \leq b \leq \infty$, $-\infty \leq c \leq d \leq \infty$ are simple examples of semi-algebra. (Note that the class α of all finite union of disjoint sets in $\mathcal{L}$ is the field generated by $\mathcal{L}$ (obviously $\alpha \subset a \; (\mathcal{L})$ and α is the field containing $\mathcal{L}$ and hence $\alpha \supset a \; (\mathcal{L})$.)

Definition 14.4   Let $\mathcal{G}_1$ and $\mathcal{G}_2$ be two classes of subsets of Ω with $\mathcal{G}_1 \subset \mathcal{G}_2$. If μ and ν are set functions defined on $\mathcal{G}_1$ and $\mathcal{G}_2$, respectively such that $\mu(A) = \nu(A)$ for all $A \in \mathcal{G}_1$, then ν is said to be an *extension* of μ to $\mathcal{G}_2$, and μ is the *restriction* of ν to $\mathcal{G}_1$, denoted by $\mu = \nu \mid \mathcal{G}_1$.

Theorem 14.1   Let μ be a non-negative additive set function on a field $\mathcal{A}$ and $\{A_n, n \geq 1\} \subset \mathcal{A}$.

    (a) If $\{A_n \; n \geq 1\}$ is a disjoint class of subsets and $\bigcup\limits_{n=1}^{\infty} A_n \subset A_0$, then

$$\mu(A_0) \geq \sum_{n=1}^{\infty} \mu(A_n),$$

    (b) If $A_0 \subset \bigcup\limits_{n=1}^{m} A_n$ for some $m = 1, 2, \ldots$, then $\mu(A_0) \leq \sum\limits_{n=1}^{m} \mu(A_n)$.

    Proof.   (a) is obvious if $\mu(A_0) = \infty$.

Let $\mu(A_0) < \infty$. Now $\left(A_0 - \bigcup\limits_{n=1}^{m} A_n\right) \in \mathcal{A}$, $m - 1, 2, \ldots$

Then

$$0 \leq \mu\left(A_0 - \bigcup_{n=1}^{m} A_n\right) = \mu(A_0) - \mu\left(\bigcup_{n=1}^{m} A_n\right) = \mu(A_0) - \sum_{1}^{m} \mu(A_n)$$

Therefore,

$$\mu(A_0) \geq \sum_{n=1}^{m} \mu(A_n) \to \sum_{1}^{\infty} \mu(A_n) \text{ as } m \to \infty.$$

    (b) $A_0 = \bigcup\limits_{n=1}^{m} (A_0 A_n) = A_0 A_1 \cup (A_0 A_2 \, A_1^C) \cup \ldots \cup (A_0 A_m \, A_{m-1}^C \ldots A_1^C)$

Therefore,

$$\mu(A_0) = \mu(A_0 A_1) + \mu(A_0 A_2 \, A_1^C) + \ldots + \mu(A_0 A_m \, A_{m-1}^C \ldots A_1^C)$$

$$\leq \mu(A_1) + \mu(A_2) + \ldots + \mu(A_m)$$

(since non-negative set function is monotone)

$$= \sum_{n=1}^{m} \mu(A_n).$$

Theorem 14.2   If μ is non-negative additive set function on a semi-algebra $\mathcal{L}$, then there exists a unique extension ν of μ to the field a generated by $\mathcal{L}$ such that ν is additive. Moreover, if μ is σ-additive on $\mathcal{L}$, so is ν on α.

Proof.   Since Q is the class of all finite union of disjoint sets in $\mathcal{L}$, every $A \in \mathcal{A}$ has a finite partition $\{A_n, 1 \leq n \leq m\}$ in $\mathcal{L}$. For such an $\mathcal{A}$, defined $v(A)$

$= \sum_{n=1}^{m} \mu(A_n)$. Then v is consistently defined on $\mathcal{A}$, since if $A \in \mathcal{A}$ has two distinct finite partitions

$$\{A_n, 1 \leq n \leq m\} \text{ and } \{B_j, 1 \leq j \leq k\} \text{ in } \mathcal{L}, \ v(A) = \sum_{n=1}^{m} \mu(A_n) = \sum_{n=1}^{m} \mu(AA_n)$$

$$= \sum_{n=1}^{m} \sum_{j=1}^{k} \mu(A_n \, B_j) = \sum_{j=1}^{k} \sum_{n=1}^{m} \mu(A_n \, B_j) = \sum_{j=1}^{k} \mu(AB_j) = \sum_{j=1}^{k} \mu(B_j).$$

It is easy to verify that v is additive on $\mathcal{A}$. The uniqueness of v follows from the fact that if v_1 is additive on $\mathcal{A}$ and $v_1 \mid \mathcal{L} = \mu$, then for any finite partition $\{A_n, 1 \leq n \leq m\}$ in $\mathcal{L}$ of $A \in \mathcal{A}$,

$$v_1(A) = \sum_{n=1}^{m} v_1(A_n) = \sum_{n=1}^{m} \mu(A_n) = v(A).$$

Suppose next that μ is σ-additive on $\mathcal{L}$, $\{A_n, n \geq 1\}$ is a σ-partition in $\mathcal{A}$ of $A \in \mathcal{A}$ and $\{C_n, 1 \leq n \leq m\}$ is a partition of A in $\mathcal{L}$. For each n, let $\{B_j, j_{n-1} < j \leq j_n\}$ be a finite partition of A_n in $\mathcal{L}$, where $j_0 = 0$. Then $(B_j, j \geq 1\}$ is a σ-partition of A in $\mathcal{L}$ and

$$\sum_1^{\infty} v(A_n) = \sum_{n=1}^{\infty} \sum_{j=j_{n-1}+1}^{j_n} \mu(B_j) = \sum_{j=1}^{\infty} \mu(B_j) = \sum_{j=1}^{\infty} \mu(AB_j)$$

$$= \sum_{j=1}^{\infty} \sum_{n=1}^{m} \mu(C_n B_j) = \sum_{j=1}^{\infty} \sum_{n=1}^{m} \mu(C_n B_j) = \sum_{n=1}^{m} \mu(C_n) = v(A).$$

Remarks.   (i) Recall that a measure μ on a class $\mathcal{A}$ of subsets of a space Ω is σ-finite if $\Omega = \bigcup_1^{\infty} A_n$ with $A_n \in \mathcal{A}$ and $\mu(A_n) < \infty$ for all $n \geq 1$.

(ii) For each subset $A \subset \Omega$ define a set function

$$v(A) = \inf_{\{S_n\}} \{\sum_1^{\infty} \mu(S_n) : \bigcup_1^{\infty} S_n \supset A, S_n \in \mathcal{L} \text{ for } n \geq 1\}$$

where μ is a measure defined on a class of sets. $\mathcal{L}$. If no such covering $\{S_n\}$ of $\mathcal{A}$-sets exists, take $v(A) = \infty$ in accordance with convention that the infimum over an empty set is ∞. This set function v defined above on the σ-algebra of all subsets of Ω is called the *"outer-measure induced by μ"*. Obviously v satisfies first two properties of an outer measure. If $v(A_n) = \infty$ for some n, then obviously $v(\bigcup_n A_n)$ $\leq \sum_n v(A_n)$. Otherwise cover each A_n by $\mathcal{L}$-sets $S_{n,k}$ satisfying $\sum_k \mu(S_{n,k}) < v(A_n)$ $+ \varepsilon/2^n$; then $v(\bigcup_n A_n) \leq \sum_{n,k} \mu(S_{n,k}) < \sum_n v(A_n) + \varepsilon$. Thus v is indeed an outer measure (since $\varepsilon > 0$ is arbitrary).

(iii) Any measure μ on a σ-field $\mathcal{F}$ is called *complete* if $B \subset A \in \mathcal{F}$ and $\mu(A) = 0$ imply $B \in \mathcal{F}$. A *complete measure space* is a measure space whose measure is complete. It follows from Prob. 10.5 that if $\mathcal{A}$ is a semi-algebra then a measure on $\sigma(\mathcal{A})$ is determined by its values on $\mathcal{A}$ if it is σ-finite there. Now we are in a position to state and prove the following must important theorem (the so-called, Carathéodory extension theorem) which shows how the measure of $\mathcal{A}$-set can be extended uniquely to a measure on $\sigma(\mathcal{A})$.

Theorem 14.3 (*Carathéodory extension*).   If μ is a measure on a semi-algebra $\mathcal{L}$ of subsets of Ω, then there exists a measure extension v of μ to $\sigma(\mathcal{L})$. Moreover, if μ is a probability measure or a σ-finite measure, then v is likewise and the extension is unique.

Proof.   For each subset $A \subset \Omega$, define

$$v(A) = \inf \left\{ \sum_1^\infty \mu(S_n) : \bigcup_1^\infty S_n \supset A, S_n \in \mathcal{L}, n \geq 1 \right\} \tag{1}$$

(note that v is infact an outer-measure) and

$$\mathcal{M} = \{ F \subset \Omega : v(FA) + v(F^C A) = v(A) \text{ for all } A \subset \Omega \}. \tag{2}$$

It follows from the definition that F is a $v(\mu^*)$-measurable set and $\mathcal{M}$ is the class of all μ^*-measurable sets. Obviously, for $A \subset B \subset \Omega$, $0 = v(\phi) \leq v(A) \leq v(B)$; moreover, $\Omega \in \mathcal{M}$ and $F^C \in \mathcal{M}$ whenever $F \in \mathcal{M}$. Hence v is indeed an outer-measure if we can show that v is countably subadditive. The proof is divided into seven steps.

Step (1): $\mu = v \mid \mathcal{L}$ *i.e.* μ *is the restriction of v to $\mathcal{L}$.* If $S \in \mathcal{L}$, then by (1) $v(S) \leq \mu(S) + \mu(\phi) + \mu(\phi) \ldots = \mu(S) \ldots$ (a), while if $\bigcup_{n=1}^{\infty} S_n \supset S$, where $S_n \in \mathcal{L}$, $n \geq 1$, then, since μ is a measure on $\mathcal{L}$, $\mu(S) \leq \sum^{\infty} \mu(S_n)$ and also by (1) $\mu(S) \leq v(S) \ldots$ (b). So by (a) and (b), $\mu(S) = v(S)$ if $S \in \overset{1}{\mathcal{L}}$.

Step 2: v is countably-subadditive, i.e. $A_0 \subset \bigcup_{n=1}^{\infty} A_n \subset \Omega$

$$v(A_0) \leq \sum_{n=1}^{\infty} v(A_n) \tag{3}$$

Without loss of generality we assume that $v(A_n) < \infty$, $n \geq 1$. For any $\varepsilon > 0$ and $n \geq 1$, choose $S_{n,m} \in \mathcal{L}$, $m \geq 1$, with

$$A_n \subset \bigcup_{m=1}^{\infty} S_{n,m}, \ \sum_{m=1}^{\infty} \mu(S_{n,m}) \leq v(A_n) + \varepsilon/2^n, \varepsilon > 0$$

Since

$$\bigcup_{n,m} S_{n,m} \supset \bigcup_n A_n \supset A_0, \text{ by (1)}$$

$$v(A_0) \leq \sum_{n,m} \mu(S_{n,m}) \leq \sum_{n=1}^{\infty} v(A_n) + \varepsilon$$

and letting $\varepsilon \to 0$, we obtain (3).

Step (3): $\mathcal{L} \subset \mathcal{M}$. Let $S \in \mathcal{L}$ and $A \subset \Omega$. By Step (2) and the definition of $\mathcal{M}$ it suffices to verify that

$$v(A) \geq v(SA) + v(S^C A) \tag{4}$$

Since if $v(A) = \infty$, the proof is trivial, we suppose $v(A) < \infty$. For any $\varepsilon > 0$, choose $S_n \in \mathcal{L}$, $n \geq 1$, such that $\bigcup\limits_{n=1}^{\infty} S_n \supset A$ and

$$v(A) + \varepsilon \geq \sum\limits_{n=1}^{\infty} \mu(S_n) \tag{5}$$

Since $\mathcal{L}$ is a semi-algebra, for each $n \geq 1$, there exists a finite partition of $S^C S_n$ in $\mathcal{L}$, say $\{S_{n,m}, m = 1, \ldots, m_n\}$. Then $\{SS_n, S_{n,m}, m = 1, 2, \ldots, m_n\}$ is a partition of S_n in $\mathcal{L}$, and by finite additivity of μ on $\mathcal{L}$

$$\mu(S_n) = \mu(SS_n) + \sum\limits_{m=1}^{m_n} \mu(S_{n,m}), \, n \geq 1 \tag{6}$$

But $\bigcup\limits_{n=1}^{\infty} \bigcup\limits_{m=1}^{m,n} S_{n,m} = \bigcup\limits_{n=1}^{\infty} S^C S_n \supset AS^C$ and

$$\bigcup\limits_{n=1}^{\infty} SS_n \supset SA, \text{ hence by (1)}$$

$$\sum\limits_{n=1}^{\infty} \sum\limits_{m=1}^{m_n} \mu(S_{n,m}) \geq v(AS^C), \, \sum\limits_{n=1}^{\infty} \mu(SS_n) \geq v(SA) \tag{7}$$

From (5), (6) and (7)

$$v(A) + \varepsilon \geq \sum\limits_{n=1}^{\infty} \mu(SS_n) + \sum\limits_{n=1}^{\infty} \sum\limits_{m=1}^{m_n} \mu(S_{n,m}) \geq v(SA) + v(S^C A) \tag{8}$$

Letting $\varepsilon \to 0$, we get (4).

Step (4): $\mathcal{M}$ is *a field* and for every $F \in \mathcal{M}$ and $A \subset \Omega$ and finite partition

$$\{F_n, n = 1, 2, \ldots, m\} \text{ of } F \text{ in } \mathcal{M}, \, v(AF) = \sum\limits_{n=1}^{\infty} v(AF_n) \tag{9}$$

Now for $A \subset \Omega$ and $F_1 \in \mathcal{M}$, $i = 1, 2, \ldots$

$$v(A) = v(AF_1) + v(A F_1^C) = v(AF_1 F_2) + v(AF_1 F_2^C)$$
$$+ v(A F_1^C F_2) + v(A F_1^C F_2^C) \tag{10}$$

Replacing A by $A(F_1 \cup F_2)$ in (10)

$$v(A(F_1 \cup F_2)) = v(A(F_1 F_2)) + v(AF_1 F_2^C) + v(A F_1^C F_2) \tag{11}$$

From (10) and (11), $v(A) = v(A(F_1 \cup F_2)) + v(A \, F_1^C F_2^C)$

$$= v(A(F_1 \cup F_2)) + v(A(F_1 \cup F_2)^C).$$

So by definition of $\mathcal{M}$, $F_1 \cup F_2 \in \mathcal{M}$. Obviously $\mathcal{M}$ is closed under complements and hence $\mathcal{M}$ is a field and moreover, if $F_1 F_2 = \phi$, (11) yields $v(A(F_1 \cup F_2))$

$= v(AF_1) + v(AF_2)$ which is (9) when $m = 2$ for general m, (9) follows by induction on μ.

Step (5): $\mathcal{M}$ *is a* σ*-field* and for every $F \in \mathcal{M}$ and $A \subset \Omega$ and countable partition $\{F_n, n \geq 1\}$ of F in $\mathcal{M}$,

$$v(AF) = \sum_{n=1}^{\infty} v(AF_n) \tag{12}$$

Let $\{F_n, n \geq 1\}$ be the countable partition of F and set $E_n = \bigcup_{1}^{n} F_i$. By Step (4), $F_n \in \mathcal{M}$ for every positive integer n and hence by definition of $\mathcal{M}$ for any $A \subset \Omega$,

$$v(A) = v(AE_n) + v(A E_n^C) \geq v(AE_n) + v(AF^C)$$

$$= \sum_{i=1}^{m} v(AF_i) + v(AF^C) \text{ (by (9)) for every } n \geq 1 \tag{13}$$

Hence by countable-subadditivity of v (see (3)) and since (13) is true for every $n \geq 1$.

$$v(A) \geq \sum_{n=1}^{\infty} v(AF_i) + v(AF^C) \geq v(AF) + v(AF^C) \tag{14}$$

and so equality holds throughout (14), yielding (12) upon replacement of A by AF.

Moreover, if $\{F_n, n \geq 1\}$ is any sequence of disjoint sets of $\mathcal{M}$ and $F = \bigcup_{n=1}^{\infty} F_i$, (14) is still valied and hence it is clear that $\mathcal{M}$ is a σ-field. It is obvious that if μ is finite or σ-finite, so is v. (In fact, for $A \subset \Omega$, $v(A) \leq v(\Omega) = \mu(\Omega) < \infty$ if μ is finite.)

If μ is σ-finite on $\mathcal{L}$, then there exist disjoint $A_i \in \mathcal{L}$ with $\bigcup_{1}^{\infty} A_i = \Omega$, $\mu(A_i) < \infty$ for all $i \geq 1$. If $A \in \mathcal{M}$, then $A = \bigcup_{i=1}^{\infty} (A_i \cap A)$ (a countable partition in $\mathcal{M}$, being closed under intersections) and hence $v(A) \sum_{i=1}^{\infty} v(A_i \cap A) \leq \sum_{i=1}^{\infty} v(A_i) = \sum_{i=1}^{\infty} \mu(A_i)$ with $\mu(A_i) < \infty$. Hence $v(A_i \cap A) < \infty$ for all $i \geq 1$.

Step (6): $\mathcal{M} \supset \sigma(\mathcal{L})$ and $\mu = v \mid \mathcal{L}$. First result follows from (3) and (5) and v is a measure on μ follows from (12) by taking $A = \Omega$ and $\mu = v \mid \mathcal{L}$ is just Step (1).

Step (7):   Uniqueness of v. Let v_1 be any measure extension of μ to $\sigma(\mathcal{L})$ and define $\mathcal{G} = \{A \mid A \in \sigma(\mathcal{L}) \text{ and } v(A) = v_1(A)\}$. If $\mu(\Omega) < \infty$, then $v(\Omega) = v_1(\Omega) < \infty$ and $\mathcal{G} \supset \mathcal{L}$. Now we shall see that $\mathcal{G}$ is a λ-system. Now $\Omega \in \mathcal{G}$. If $A_n \in \mathcal{G} \subseteq \sigma(\mathcal{L})$ and $A_n \uparrow A$, then $v(A_n) = v_1(A_n)$. Hence by continuity theorem of measure $v(A) = v_1(A)$. Therefore $A \in \mathcal{G}$. Also $\mathcal{G}$ is closed under proper difference. Obviously $\mathcal{L}$ is a π-system. Hence $\mathcal{G} \supset \sigma(\mathcal{L})$ and hence $v = v_1$ on $\sigma(\mathcal{L})$. If μ is only σ-finite on $\mathcal{L}$, then there exist sets $E_n \in \mathcal{L}$ with $\bigcup_{1}^{\infty} E_n = \Omega$ and $\mu(E_n) < \infty$, $n \geq 1$. Then as in the finite case $v = v_1$ on $E_n \cap \sigma(\mathcal{L})$ for all $n \geq 1$. Therefore $v = v_1$ on $\sigma(\mathcal{L})$.

Theorem 14.4   The outer measure v stipulated in (1) defines a complete measure extension of μ to $\mathcal{M}$, the σ-field of v-measurable sets.

Proof.   Let $B \subset D \in \mathcal{M}$ and $v(D) = 0$. Then

$$v(BA) + v(B^C A) \leq v(DA) + v(B^C A) \leq v(D) + v(B^C A) = v(B^C A) \leq v(A).$$

Since v is subadditive $v(BA) + v(B^C A) \geq v(A)$. Therefore $v(BA) + v(B^C A) = v(A)$ for all $A \subset \Omega$. Hence $B \in \mathcal{M}$. So v is a complete measure on $\mathcal{M}$.

14.2   CONSTRUCTION OF LEBESGUE-STIELTJES MEASURE

Let R be the real line and $\mathcal{B}$ be the Borel sets of R.

Definition 14.5   The measure μ-defined on $\mathcal{B}$ and finite for bounded sets is called a Borel measure on R. To each finite Borel measure we associate a function F by setting $F(x) = \mu(-\infty, x]$. The function F is called the cumulative distribution function (c.d.f.; in short distribution function—d.f.).

Theorem 14.5   If μ is a finite Borel measure on the real line its c.d.f. F is a monotone non-decreasing function which is continuous on the right. Moreover

$$\lim_{x \to -\infty} F(x) = 0.$$

The *proof* is given in Section 3.1.

Lemma 14.1   Let F be a non-decreasing, right continuous function on $\bar{R} = [-\infty, \infty]$ with $F(\infty) = F(\infty -)$ and $|F(t)| < \infty$, $|t| < \infty$.

If
$$\mu_F((a, b]) = F(b) - F(a), \quad -\infty \leq a \leq b \leq \infty,$$

$$\mu_F(\{-\infty\}) = 0 = \mu(\phi), \quad \mu(\bar{R}) = F(\infty) - F(-\infty),$$

then μ_F is countably additive on the semi-algebra $\mathcal{L}$ of right semi-closed intervals.

Proof.   To apply Theorem 14.1, consider $S \in \mathcal{L}$ with $S \neq \{-\infty\}$ or ϕ and let

$\{S_n\}_1^\infty$ be a countable partition of S in $\mathcal{L}$. Then $\mu_F(S) \geq \sum_1^\infty \mu_F(S_n)$.

To prove the reverse inequality, since $\mu_F([-\infty, b]) = \mu_F((-\infty, b])$ we suppose that $S = (a, b]$, $-\infty \leq a < b \leq \infty$. Moreover, by right continuity, it is enough to establish that

$$\mu_F((c, d]) \leq \sum_1^\infty \mu_F(S_n) \tag{1}$$

if $-\infty \leq a < c < d \leq b < d \neq \infty$.

Since $-\infty \notin (a, b)$, $S_n \neq \{-\infty\}$, $n \geq 1$ hence $S_n = (a_n, b_n]$, $-\infty \leq a_n \leq b_n \leq \infty$, $n \geq 1$.

For any $\varepsilon > 0$, set $I_n = (a_n, c_n + \delta_n)$, where $c_n = \min(b_n, d)$, $n \geq 1$, and choose $\delta_n > 0$ such that $F(c_n + \delta_n) < F(c_n) + \varepsilon/2^n$, $n \geq 1$ by the right continuity of F. Then $(c, d] \subset \bigcup_1^\infty I_n$ and by Heine-Borel theorem $(c, d] \subset [c, d] \subset \bigcup_1^\infty I_{nj}$ for some subsequence $\{n_j\}$ of $\{n\}$ for which $I_{nj} \neq \phi$, and hence by Theorem 14.1

$$\mu_F((c, d]) \le \sum_1^k (F(c_{n_j} + \delta_{n_j}) - F(a_{n_j}))$$

$$\le \sum_1^k (F(c_{n_j}) - F(a_{n_j})) + \sum_1^k \varepsilon 2^{-n_j}$$

$$\le \sum_1^k (F(c_{n_j}) - F(a_{n_j})) + \varepsilon$$

$$\le \sum_1^k \mu_F(S_j) + \varepsilon$$

Hence (1) is proved by letting $\varepsilon \to 0$.

Theorem 14.6 (*Converse of Theorem 14.5*).   Let F be a monotone non-decreasing point function which is continuous on the right. The there is a unique Borel measure μ_F such that for all a, b real numbers we have $\mu_F(a, b) = F(b) - F(a)$, $a < b$.

Proof.   Since it is convenient to work in the compact space $\overline{R}$, so we extend F to a map of $\overline{R}$ into $\overline{R}$ by defining $F(\infty) = \lim_{x \to \infty} F(x)$, $F(-\infty) = \lim_{x \to -\infty} F(x)$; the limit exist by monotonicity. Define

$$\mu_F(a, b] = F(b) - F(a) \quad a, b \in \overline{R}, \ a < b \tag{1}$$

and let

$$\mu_F([-\infty, b]) = F(b) - F(-\infty) = \mu_F(-\infty, b) \tag{2}$$

then μ is defined on all right semi-closed intervals of $\overline{R}$ (counting $[-\infty, b]$ as semi-closed). If $I_1, \ldots, I_k$ are disjoint right semi-closed intervals of $\overline{R}$, we define

$$\mu_F(\cup_1^k I_j) = \sum_{j=1}^k \mu_F(I_j) \tag{3}$$

Now take $\mathcal{L}(\overline{R})$ as the semi-algebra consists of right semi-closed intervals of $\overline{R}$, then by Theorem 14.2 μ_F admits a unique extension as a finitely additive set function to the field generated by $\mathcal{L}$. Then by Lemma 14.1 μ_F is countably additive on $\mathcal{L}(\overline{R})$. Let $\alpha(R)$ be the field of all finite disjoint union of right semi-closed intervals of R and by (3) extend μ_F to $\alpha(R)$. It follows from Theorem 14.2, μ_F is countably additive on $\alpha(R)$. By Theorem 14.3 μ_F can be extended to $\sigma(\mathcal{L})$. Since the class of Borel sets is the smallest σ-field containing $\mathcal{L}(R)$, we have an extension of μ_F to a Borel measure. Since $R = \bigcup_{n=-\infty}^{\infty} (n, n + 1]$ and $\mu_F(n, n + 1] = F(n + 1) - F(n) < \infty$, μ_F is indeed σ-finite. Thus, the extension is unique on $\mathcal{B}(R)$.

Remarks.   (i) μ_F need not be σ-finite on $\alpha(\overline{R})$, since the sets $(-n, n]$ do not cover $\overline{R}$.

(ii) μ_F (in the proof of Theorem 14.6) is called Lebesgue-Stieltjes measure associated with the d.f. F.

(iii) If
$$F(x) = \begin{cases} x & \text{if } 0 \le x \le 1 \\ 0 & \text{if } x \le 0 \\ 1 & \text{if } x \ge 1 \end{cases}$$

Then μ_F is the renowned Lebesgue measure on $[0, 1]$ generalizing the notion of length.

The Carathéodory extension of μ to the sets $\mathcal{M}$ if $\mathcal{M} = \mathcal{M}_F$, the μ^*-measurable sets which form a σ-field covering $\mathcal{B}(R)$, the Borel sets of R and this μ^*-measurable sets are called Lebesgue measurable sets which is complete by Theorem 14.4.

Exercise 14.1   Let $\Omega = R$

α = algebra generated by right semi-closed intervals $(a, b]$ or the algebra generated by open intervals of R.

$\mathcal{B}$ = σ-field generated by the semi-closed intervals of R.

Define $m^*(A) = \inf\limits_{A \subset \cup I_n \atop n} \sum\limits_n l(I_n)$, here I_n's are open intervals and $l(I_n)$ is the length of (I_n). In this case m^* is an outer measure and we get Lebesgue measure on $\mathcal{B}(R)$.

Exercise 14.2   If $m^*(E) = 0$, show that E is a Lebesgue measurable set.

Proof. $m^*(A \cap E) + m^*(A \cap E^C) \leq m^*(E) + m^*(A) = m^*(A)$

So $\qquad\qquad m^*(A \cap E) + m^*(A \cap E^C) = m^*(A)$ for $A \subset R$.

Hence E is m^*-measurable, i.e. E is a Lebesgue measurable set.

Exercise 14.3   Show that $m^*\{q\} = 0$ and hence Lebesgue measure of a countable set is zero, where $\{q\}$ is a singleton.

Proof. $l[a, b] = m^*(a, b] = m^*[a, b] = l(a, b]$. Therefore $m^*\{a\} = 0$. Now countable additivity of Lebesgue measure proves the result.

14.3   COMPARISON OF RIEMANN AND LEBESGUE INTEGRALS

In Section 3.3 we have seen by means of a counter example that the scope of Lebesgue integral is much wider than Riemann integral. In this section we prove that Lebesgue integration is in fact more general than Riemann integration and we obtain a precise criterion for Riemann integrability.

Theorem 14.7   Let f be a bounded real valued function on $[a, b]$. (a) The function f is Riemann integrable on $[a, b]$ iff f is continuous a.e. on $[a, b]$ (with respect to Lebesgue measure).

(b) If f is Riemann integrable on $[a, b]$, then f is Lebesgue integrable on $[a, b]$ and the two integrals are equal.

Proof.   The following facts about Riemann and Lebesgue integration are necessary before proving the theorem. Here $\Omega = [a, b]$, $\mathcal{F} = \mathcal{B}[a, b]$, the Lebesgue measurable subsets of $[a, b]$ and μ = the Lebesgue measure on $\mathcal{F}$. Let us construct upper and lower Darboux sums of f corresponding to the partition $P = \{a = x_0 < x_1 < \ldots < x_n = b\}$ of $[a, b]$ as follows:

Let $U(P) = \sum_{i=1}^{n} M_i(x_i - x_{i-1})$ and $L(P) = \sum_{1}^{n} m_i(x_i - x_{i-1})$ be the upper and lower sums respectively, where

$$M_i = \sup_{x \in (x_{i-1}, x_i]} f(x) \text{ and } m_i = \inf_{x \in (x_{i-1}, x_i]} f(x), \ i = 1, 2, \ldots, n.$$

Then

$$U(P) = \int_a^b \alpha(x) \, d\mu \text{ and } L(P) = \int_a^b \beta(x) \, d\mu \tag{1}$$

where $\alpha(x) = M_i$ if $x \in (x_{i-1}, x_i]$ and $\beta(x) = m_i$ if $x \in (x_{i-1}, x_i]$. The step functions $\alpha(x)$ and $\beta(x)$ are called upper and lower functions of f corresponding to the

partition P. Now consider $U(P_k) = \int_a^b \alpha_k \, d\mu, L(P_k) = \int_a^b \beta_k \, d\mu,$ then

$$\alpha_1 \geq \alpha_2 \geq \ldots \geq f \geq \ldots \geq \beta_2 \geq \beta_1 \tag{2}$$

where $P_1, P_2, \ldots, P_k$, be a sequence of partitions of $[a, b]$ such that P_{k+1} is a refinement of P_k for all $k \geq 1$ and such that $| P_k |$-max $(x_1 - x_{i-1}) \to 0$ as $k \to \infty$. Since α_k and β_k's are monotone functions and bounded below (above) by f, have limits $\phi(\Psi)$.

If $| f | < M$, then $| \alpha_k | < M$ and $| \beta_k | < M$ for all $k \geq 1$. Also

$$\int_a^b M \, d\mu = M\mu[a, b] = M(b - a) < \infty \tag{3}$$

Also note that f is continuous iff $\phi(x) = f(x) = \psi(x)$ \tag{4}

provided x is not an end point of any of the subintervals of P_ks.

Proof of the main results. (a) We know that a function is Riemann integrable on $[a, b]$ if $\lim_{k \to \infty} U(P_k) = \lim_{k \to \infty} L(P_k) = $ a finite number r, independent of the particular sequence of partitions $\{P_k\}$. By dominated convergence theorem

$$\lim_{k \to \infty} U(P_k) = \lim_{k \to \infty} \int_a^b \alpha_k \, d\mu = \int_a^b \phi d\mu \tag{5}$$

$$\lim_{k \to \infty} L(P_k) = \lim_{k \to \infty} \int_a^b \beta_k \, d\mu = \int_a^b \psi \, d\mu \tag{6}$$

Hence f is Riemann integrable iff

$$\int_a^b \phi \, d\mu = \int_a^b \psi \, d\mu = r \tag{7}$$

and r is the value of the Riemann integral $r_{a,b}(f)$.

Since $\psi \leq f \leq \phi$, $\phi - \psi \geq 0$ a.e. From (7) $\phi = \psi = \phi$ a.e.

So f is Riemann integrable on $[a, b] \Rightarrow f$ is continuous a.e. $[\mu]$. Conversely, assume f is continuous a.e., then $\phi = \psi = f$ a.e. Now ϕ and ψ are limits of simple functions (in particular measurable functions) and hence are measurable. Thus f

differs from a measurable function (namely ϕ or ψ) on a subset of set of measure zero, and therefore f is measurable in Lebesgue sense, since Lebesgue measure is complete. Since f is bounded, by (3) f is Lebesgue integrable and since $\phi = \psi = f$ a.e., we have

$$\int_a^b \phi \, d\mu = \int_a^b \psi \, d\mu \left(= \int_a^b f \, d\mu \right) \tag{8}$$

independent of the particular sequence of partitions. Therefore by the definition of Riemann integral f is Riemann integrable.

(b) If f is Riemann integrable then by part (a) f is continuous a.e. on $[a, b]$ and f is bounded which implies f is Lebesgue integrable on $[a, b]$. Using similar arguments equation (8) yields $r_{a,b}(f) = \int_a^b f \, d\mu$, as desired.

PROBLEMS AND COMPLEMENTS

14.1 Show that the Lebesgue measure m defined on $\mathcal{M}$, the class of Lebesgue measurable sets of $\mathcal{B}$, is σ-finite and complete.

14.2 Let $(\Omega, \mathcal{F}, \mu)$ be a measure space and $\bar{\mu}$ on $\bar{\mathcal{F}}$ be the completion of μ on $\mathcal{F}$. Show that if $D \in \bar{\mathcal{F}}$ there exists $B \in \mathcal{F}$ such that $\bar{\mu}(D \Delta B) = 0$.

14.3 If F is a continuous d.f. show that $\bar{\mu}_F([x]) = 0$ for each x.

14.4 Let $\bar{\mu}_F$ be the completion of Lebesgue-Stieltjes measure induced by a d.f. F constant on $R = (-\infty, \infty)$ except at a finite number of points. Then show that every subset of $\mathbb{R}$ is $\bar{\mu}_F$ measurable (i.e. outer measure, $\mu_F^* = \bar{\mu}_F$).

14.5 Show that if μ is a σ-finite measure on a field α, then the extension v of μ to α^*, the class of μ^*-measurable sets, is also σ-finite.

14.6 Let f be a bounded real-valued function on $[a, b]$ and F be a non-decreasing right continuous function on $[a, b]$ with corresponding Lebesgue-Stieltjes measure μ_F (defined on Borel subsets of $[a, b]$).
(a) Show that f is Riemann-Stieltjes integrable iff f is continuous a.e. $[\mu]$ on $[a, b]$.
(b) Show that if f is Riemann-Stieltjes integrable, then f is integrable w.r.t. the completion of the measure μ_F and the two integrals are equal.

14.7 Show that if $A \in \mathcal{B}(\mathbb{R}^k)$ and $a \in \mathbb{R}^k$, then $a + A \in \mathcal{B}(\mathbb{R}^k)$ and $m(a + A) = m(A)$, where m is Lebesgue measure, i.e. Lebesgue measure is translation invariant.

14.8 Let μ be a Lebesgue-Stieltjes measure on $\mathcal{B}(\mathbb{R}^k)$ such that $\mu(a + I) = \mu(I)$ for all $a \in \mathbb{R}^k$ and all (right semi-closed) intervals I in $\mathbb{R}^k$. Show that $\mu = cm$, where m is the Lebesgue measure and c a positive constant.

14.9 Let F be a function on $\mathbb{R}$ given by

$$F(x) = \begin{cases} 0 & \text{if } x < -1 \\ 1+x & \text{if } -1 \le x < 0 \\ 2+x^2 & \text{if } 0 \le x < 2 \\ 9 & \text{if } x \ge 2 \end{cases}$$

Find the Lebesgue-Stieltjes measure of the following sets:

(a) $\{2\}$, (b) $[-0.5,\ 3]$, (c) $(-1,\ 0] \cup (1,\ 2)$, (d) $[0,\ 0.5) \cup (1,\ 2]$, (e) $\{x : |x| + 2x^2 > 1\}$.

Measure on Infinite Product Space and Kolmogorov's Consistency

Theorem AI.1 An in Section 11.1 product measure and Fubini theorems can easily be generalized to n-dimensional product measure theorem formalizing the notion of an n-stage random experiment where the probability of an event associated with the nth stage depends on the result of the first $n - 1$ trials. It is convenient to have a single probability space which is adequate to handle n-stage experiments for n arbitrarily large (not fixed in advance). Such a space can be constructed if the product measure theorem can be extended to infinitely many dimensions. Our first task is to construct the product of infinitely many σ-fields. The infinite dimensional version of the product measure theorem is generally used for probability measures and therefore will be stated in that context. In fact the construction to be used to prove this theorem runs into trouble for non-probability measure due to the distinguished role played by the number 1 in infinite (and even finite) products.

Definition AI.1 Let $(\Omega_i, \mathscr{F}_i)$ be a sequence of measurable space and $\Omega = \prod_{i=1}^{\infty} \Omega_i$, the set of all sequences $\{\omega_1, \omega_2, \ldots\}$ such that $\omega_i \in \Omega_i$, $i \geq 1$. If $B^n \subset \prod_{i=1}^{n} \Omega_i$, define $B_n = \{\omega \in \Omega : (\omega_1, \omega_2, \ldots, \omega_n) \in B^n\}$. The set B_n is called the *cylinder with base B^n*; the cylinder is said to be measurable if $B^n \in \prod_{i=1}^{n} \mathscr{F}_i$. If $B^n = A_1 \times \ldots, \times A_n$, where $A_i \subset \Omega_i$ for each i, B_n is called a *rectangle, a measurable rectangle* if $A_i \in \mathscr{F}_i$ for each i.

A cylinder with an n-dimensional base may always be regarded as having a higher dimensional base. We have seen that the measurable cylinders form a field and finite disjoint unions of rectangles is a field (see Section 14.1). The σ-field generated by the measurable cylinders is called the *product of the σ-fields $\mathscr{F}_i$*, written $\mathscr{F} = \prod_{i=1}^{\infty} \mathscr{F}_i$; $\prod_{i=1}^{\infty} \mathscr{F}_i$ is also the σ-field generated by the measurable rectangles. If all $\mathscr{F}_i$ coincide with a fixed σ-field $\mathscr{F}$, then $\prod_{i=1}^{\infty} \mathscr{F}_i$ is denoted by $\mathscr{F}^{\infty}$, and if all Ω_j coincide with a fixed set B, $\prod_{i=1}^{\infty} \Omega_j$ is denoted by B^{∞}.

Theorem AI.2 Let $(\Omega_i, \mathscr{F}_i)$, $i \geq 1$ be a sequence of measurable spaces. Suppose that we are given an arbitrary probability measure P_1 on $\mathscr{F}_1$ and for each $i \geq 1$ and each $(\omega_1, \ldots, \omega_i) \in \Omega_1 \times \ldots \times \Omega_i$ we are given a probability measure $P(\omega_1, \ldots, \omega_i; .)$ on $\mathscr{F}_{i+1}$. Assume that $P(\omega_1, \ldots, \omega_i; A)$ is measurable: $\left(\prod_{j=1}^{i} \Omega_j, \prod_{j=1}^{i} \mathscr{F}_j \right) \to (R, \mathscr{B}(R))$ for each fixed $A \in \mathscr{F}_{i+1}$. If $B^n \in \prod_{j=1}^{n} \mathscr{F}_i$, define

$$P_n(B^n) = \int_{\Omega_1} dP(\omega_1) \int_{\Omega_2} P(\omega_1, d\omega_2) \ldots \int_{\Omega_{n-1}} P(\omega_1, \ldots, \omega_{n-2}, d\omega_{n-1})$$

$$\int_{\Omega_n} I_{B^n}(\omega_1, \ldots, \omega_n) P(\omega_1, \ldots, \omega_{n-1}, d\omega_n).$$

Then there is a unique probability measure P on $\mathscr{F}$ such that for all n, P agrees with P_n on n-dimensional cylinders, i.e. $P[\omega \in \Omega : (\omega_1, \ldots, \omega_n) \in B^n] = P_n(B^n)$ for all $n \geq 1$ and all $B^n \in \prod_{i=1}^{n} \mathscr{F}_i$.

(NOTE. P_n is a probability measure on $\prod_{i=1}^{n} \mathscr{F}_i$.)

Proof. Any measurable cylinder can be presented in the form $B_n = [\omega \in \Omega : (\omega_1, \ldots, \omega_n) \in B^n]$ for some n and some $B^n \in \prod_{i=1}^{n} \mathscr{F}_i$. Define $P(B_n) = P_n(B^n)$. We must show that P is well-defined on measurable cylinders. Suppose that B_n can also be expressed as $[\omega \in \Omega : (\omega_1, \ldots, \omega_n) \in A^m]$ where $A^m \in \prod_{i=1}^{n} \mathscr{F}_i$. Then we have to show that $P_n(B^n) = P_m(A^m)$ for some $m < n$. Then $(\omega_1, \ldots, \omega_n) \in A^m$ iff $(\omega_1, \ldots, \omega_n) \in B^n = A^m \times \Omega_{m+1} \times \ldots \Omega_n$ and the fact that $P = P(\omega_1, \ldots, \omega_j \ldots)$ is a probability measure. From the definition of P_n, $P_n(B^n) = P_m(A^m)$. Since P_n is a measure on $\prod_{i=1}^{n} \mathscr{F}_i$, P is finitely additive on the field $\mathscr{F}_0$ of measurable cylinders. Let $\{B_n\}$ be a sequence of measurable cylinders $\downarrow \phi$. Assume if possible $\lim_{n \to \infty} P(B_n) > 0$. Then for each $n > 1$, $P(B_n) = \int_{\Omega_n} f_{n_1}^{(1)}(\omega_1) \, dP(\omega_1)$, where $f_n^{(1)}(\omega_1)$

$$= \int_{\Omega_2} P(\omega_1, d\omega_2) \ldots \int_{\Omega_n} I_{B^n} P(\omega_1, \ldots, \omega_n) P(\omega_1, \omega_2, \ldots, \omega_{n-1}, d\omega_n).$$

Since $B_{n+1} \subset B_n$, $B_n^{n+1} \subset B^n \times \Omega_{n+1}$ and hence

$$I_{B^{n+1}}(\omega_1, \ldots, \omega_{n+1}) \leq I_{B^n}(\omega_1, \ldots, \omega_n).$$

Therefore, for fixed ω_1, $f_n^{(1)}(\omega_1) \downarrow g_1(\omega_1)$ as $n \to \infty$. By monotone or dominated convergence theorem for conditional probability (expectation) $P(B_n) \to \int_{\Omega_1} g_1(\omega_1) dP_1(\omega_1)$. If $\lim_{n \to \infty} P(B_n) > 0$, then $g_1(\omega_1^*) > 0$ for some $\omega_1^* \in B^1 \subset \Omega_1$.

This follows from the fact that if $\omega_1^* \notin B^1$, then $I_{B^n}(\omega_1^*, \omega_2, \ldots, \omega_n) = 0$ for all $n \geq 1$ and hence $f_n^{(1)}(\omega_1^*) = 0$ for all n and that implies $g_1(\omega_1^*) = 0$, a contradiction. Now for each $n > 2$

$$f_n^{(1)}(\omega_1^*) = \int_{\Omega_2} f_n^{(2)}(\omega_2) P(\omega_1^*, d\omega_2).$$

As before, $f_n^{(2)}(\omega_2) \downarrow g_2(\omega_2)$, say and hence

$$f_n^{(1)}(\omega_n^*) \rightarrow \int_{\Omega_2} f_n^{(2)}(\omega_2) P(\omega_1^*, d\omega_2).$$

Now $f_n^{(1)}(\omega_n^*) \rightarrow g_1(\omega_n^*) > 0$ implies $g_2(\omega_n^*) > 0$ for some $\omega_2^* \in B \subset \Omega_2$ and as before $(\omega_1^*, \omega_2^*) \in B^2$. Proceeding similarly we obtain $\omega_1^*, \omega_2^*, \ldots$ such that for each $n \geq 1$, $(\omega_1^*, \omega_2^* \ldots \omega_n^*) \in B^n$. That implies $(\omega_1^*, \omega_2^*, \ldots) \in \bigcap_{n=1}^{\infty} B_n = \lim B_n = \phi$, a contradiction. Hence P is continuous from above at ϕ. Continuity theorem for set function (see Theorem 2.7) implies that P is σ-additive on $\mathcal{F}_0$, and the Caratheodory extension (Theorem 14.3) extends P to a probability measure on $\prod_{i=1}^{\infty} \mathcal{F}_1$. Also by construction, P agrees with P_n on n-dimensional cylinders. This proves the existence of the desired probability measure P. If Q is another such probability measure, then $P = Q$ on measurable cylinders, hence $P = Q$ on $\mathcal{F}$ by the uniqueness of Caratheodory extension.

The classical product measure theorem can be extended as follows:

Corollary AI.1   For each $i \geq 1$, let $(\Omega_i, \mathcal{F}_i, P_i)$ be an arbitrary probability space. Then there is a unique probability measure P on $\mathcal{F} = \prod_{i=1}^{\infty} \mathcal{F}_i$ such that $P(\omega \in \Omega : \omega_1 \in A_1, \ldots, \omega_n \in A_n) = \prod_{i=1}^{n} P(A_i)$ for all $n \geq 1$ and all $A_i \in \mathcal{F}_i$, $i \geq 1$.

NOTE. P is called the *product* of the P_i and is written as $P = \prod_{i=1}^{\infty} P_i$.

Proof.   In previous theorem, take $P(\omega_1, \ldots, \omega_i, B) = P_{i+1}(B)$ $B \in \mathcal{F}_{i+1}$. Then $P_n(A_1 \times \ldots \times A_n) = \prod_{i=1}^{n} P(A_i)$ and hence the probability measure P has the desired properties. If Q is another such probability measure, the $P = Q$ on the field of finite disjoint unions of measurable rectangles; hence $P = Q$ on $\mathcal{F}$ by the Caratheodory extension.

Is it always possible to define a sequence of r.vs. $\{X_n\}_1^{\infty}$ on some probability space $(\Omega, \mathcal{F}, P)$ such that the joint d.fs. of all finite subsets of r.vs. $X_1, \ldots, X_n$ coincide with d.fs. $F_{1,\ldots,n}$ assigned a priori? The answer is yes provided the assignment is not internally contradictory.

A family $\{F_{1,\ldots,n}(x_1, \ldots, x_n)\}$ of n-dimensional d.fs. defined for all $n \geq 1$ will be called *consistent* if for all $n \geq 1$,

$$F_{1,\ldots,n}(x_1, \ldots, x_n) = \lim_{x_{n+1} \to \infty} F_{1,\ldots,n+1}(x_1, \ldots, x_{n+1}) \tag{1}$$

Theorem AI.3 (*Kolmogorov consistency theorem*).   Let $\{F_{1,...,n}, n \geq 1\}$ be a consistent family of d.fs. Then there exists a probability measure P on $(R^\infty, \mathcal{F}^\infty)$ such that the d.fs. of the coordinate r.vs. $X_1, ..., X_n$ on $(R^\infty, \mathcal{F}^\infty, P)$ coincide with the preassigned d.fs. $F_{1,..., n}$, such that if

$$X_k(\omega) = \omega_k, \ k = 1, 2, ... \text{ for } \omega = (\omega_1, \omega_2, ...) \in R^\infty \tag{2}$$

then for all $n \geq 1$,

$$P(X_1 \leq x_1, ..., X_n \leq x_n) = F_{1,...,n}(x_1, ..., x_n) \tag{3}$$

Proof.   If A_n is an n-dimensional set of $\mathcal{B}^n$, define

$$P_n(A_n) = \int_{A_n} \cdots \int dF_{1,...,n}(x_1, ..., x_n) \tag{4}$$

Therefore $\{(R^n, \mathcal{B}^n, P_n)\}$ is a sequence of probability spaces. Then

$$P_{n+1}[\prod_{i=1}^{n}(a_i, b_i] \times R] = \lim_{\substack{a_{n+1} \to -\infty \\ b_{n+1} \to \infty}} P_{n+1}[\prod_{i=1}^{n+1}(a_i, b_i]]$$

$$= \lim_{\substack{a_{n+1} \to -\infty \\ b_{n+1} \to \infty}} \Delta_{n+1}^{a,b} F_{1,...,n+1} = \Delta_n^{a,b} F_{1,...,n} = P_n[\prod_{i=1}^{n}(a_i, b_i]]$$

by condition 1. Hence, $P_{n+1}(A_n \times R) = P_n(A_n)$ for all $A_n \in \mathcal{B}^n$, $n \geq 1$ and so there is a probability measure P on $(R^\infty, \mathcal{B}^\infty)$ such that for all $A_n \in \mathcal{B}^n$ $P(A_n \times R \times R \times ...) = P_n(A_n)$, $n \geq 1$. The coordinate functions defined by (2) are r.vs. and by (2) and (4) $P(X_1 \leq x_1, ..., X_n \leq x_n) = P(\omega : \omega_1 \leq x_1, ..., \omega_n \leq x_n) = F_{1,...n}(x_1, ..., x_n)$.

Corollary AI.2.   If $\{X_n\}_1^\infty$ be a sequence of r.vs. on some probability space $(\Omega, \mathcal{F}, P)$, then there exists a sequence of coordinate r.vs. $\{Y_n\}_1^\infty$ on the same probability space $(R^\infty, \mathcal{F}^\infty, P^*)$ such that the joint d.fs. of $(X_1, ..., X_n)$ and $(Y_1, ..., Y_n)$ are identical for all $n \geq 1$.

Hahn-Jordan Decomposition

In this section we allow measures to take negative values and then in the Hahn-Jordan decompositions show how in the study of such measures we may keep to the non-negative measures so far we have mostly discussed.

Definition AII.1   *An extended real-valued* set function v defined on a measurable space $(\Omega, \mathscr{F})$ is said to be a *signed measure* if

    (i) v takes at most one of the values ∞, $-\infty$

    (ii) $v(\phi) = 0$

    (iii) v is countably additive on $\mathscr{F}$.

Definition AII.2   A set P is called *positive* (with respect to v) if, for every $A \in \mathscr{F}$, $P \cap A \in \mathscr{F}$ and $v(P \cap A) \geq 0$; similarly N is called *negative* if, for every $A \in \mathscr{F}$, $N \cap A \in \mathscr{F}$ and $v(N \cap A) \leq 0$.

NOTE.   The null set ϕ is both positive and negative.

Theorem AII.1 (*Hahn decomposition*).   If v is a signed measure, then there exists two disjoint sets P and N such that $P \cup N = \Omega$ and P is positive and N is negative with respect to v.

NOTE.   The sets P and N are said to form Hahn decomposition of Ω with respect to v.

Proof.   Since v assumes at most one of the values $+\infty$ and $-\infty$, we may assume that $-\infty < v(A) \leq \infty$ for all $A \in \mathscr{F}$.

Since the difference of two negative sets, and a disjoint, countable union of negative sets are obviously negative, it follows that every countable union of negative sets is negative. Let $\alpha = \inf \{v(N) : N \in \mathscr{F}$ and N is a negative set$\}$. Let $\{N_n\}$ be a sequence of negative sets such that each $N_n \in \mathscr{F}$ and $\lim_{n \to \infty} v(N_n) = \alpha$; if $N = \bigcup_{n=1}^{\infty} N_n$, then N is a negative set, $N \in \mathscr{F}$ and $v(N)$ is minimal. We shall prove that the set $P = N^C$ is a positive set. Suppose that, on the contrary, $P_0 \subset P$, $P_0 \in \mathscr{F}$ and $v(P_0) < 0$. The set P_0 cannot be a negative set with $v(N \cap P_0) < v(N)$, which is imposible. Let k_1 be the smallest positive integer such that $v(P_1) \geq 1/k_1$;

for a measurable set $P_1 \subset P_0$. Since $v(P_0) < 0$, $v(P_0) < \infty$ and $v(P_1) < \infty$, $v(P_0 - P_1)$ $= v(P_0) - v(P_1) \leq v(P_0) - 1/k_1 < 0$, and hence the same argument applied to P_0 is applicable to $P_0 - P_1$. Similarly, k_2 is the smallest positive integer such that $v(P_2)$ $\geq 1/k_2$, for a measurable set $P_2 \subset P_0 - P_1$. Repeat the same procedure infinite number of times. Since v is finite valued for measurable subsets of P_0, we must have

$$\lim_{n \to \infty} \frac{1}{k_n} = 0.$$

It follows that every measurable subset N^* of $N_0 = P_0 \cap (\overset{\infty}{\underset{j=1}{\cup}} P_j)^C$, we have $P(N^*)$ ≤ 0, i.e. N_0 is a negative set. Since $N_0 \cap N = \phi$, and since $v(N_0) = v(P_0) - \overset{\infty}{\underset{j=1}{\Sigma}} v(P_j)$ $\leq v(P_0) < 0$, this contradicts the minimality of N, and hence $v(P_0) \geq 0$, i.e. P is a positive set.

NOTE.   Hahn decomposition is as such not unique but it is unique to the extent if P_1, N_1 and P_2, N_2 are Hahn decompositions of Ω with respect to v, then $v((P_1 \Delta P_2)E) = 0$, for all $E \in \mathcal{F}$.

Proof. $P_1 - P_2 = P_1 \cap N_2 \Rightarrow$ it is both a positive and a negative set with respect to v and hence $v((P_1 - P_2)E) = 0$, $E \in \mathcal{F}$. Similarly $v((P_2 - P_1)E) = 0$, $E \in \mathcal{F}$, and so $v((P_1 \Delta P_2)E) = 0$, $E \in \mathcal{F}$.

Definition AII.3   It follows from Hahn decomposition theorem that the equations

$$v^+(A) = v(A \cap P) \text{ and } v^-(A) = -v(A \cap N)$$

unambiguously define two set functions v^+ and v^- on $\mathcal{F}$, called respectively, the *upper variation* and the *lower variation* of v. The set function $|v|$, defined for every $A \in \mathcal{F}$ by $|v|(A) = v^+(A) + v^-(A)$, is the *total variation* of v. It can be shown that $|v(A)| = \sup \overset{n}{\underset{i=1}{\Sigma}} |v(A_i)|$ where E_i's are disjoint and $A = \overset{\infty}{\underset{i=1}{\cup}} A_i$. This justifies the term "total variation".

Theorem AII.2 (*Jordan decomposition*).   The upper, lower, and total variations of a signed measure v are measures and $v(A) = v^+(A) - v^-(A)$, $A \in \mathcal{F}$. If v is finite or σ-finite, then so also are v^+ and v^-; at least one of the measures v^+ and v^- is always finite. Moreover v^+ and v^- are mutually singular (written as $v^+ \perp v^-$) and this decomposition is unique.

Proof.   The variation of v are clearly non-negative; if for any

$$A \in \mathcal{F}, A = \overset{\infty}{\underset{i=1}{\cup}} A_i, A_n \in \mathcal{F} \text{ and } \mu(A_n) < \infty, n \geq 1,$$

by Hahn decomposition theorem the same is true for v^+ and v^-. The equation $v = v^+ - v^-$ follows from definitions of v^+ and v^-. v take on at most one of the values $+ \infty$ and $- \infty \Rightarrow$ at least one of the set function v^+ and v^- is always finite. Countable additivity of v^+ and v^- and $|v|$ is clear. Now $P \cap N = \phi$ and $v(P \cap N) = 0$. Hence,

$$v^+(N) = v(P \cap N) = 0, v^-(N^C) = - v(N^C \cap N) = - v(\phi) = 0 \qquad (1)$$

Hence $v^+ \perp v^-$. Let $v = v_1 - v_2$ be any other decomposition of v into mutually singular measures. Then $\Omega = A \cup B$, where $B = A^C$ and $v_1(B) = v_2(A) = 0$. Let $D \subset A$, then $v(D) = v_1(D) - v_2(D) = v_1(D) \geq 0$, so A is a positive set with respect to v. Similarly B is a negative set. For each $E \in \mathcal{F}$ we have $v_1(E) = v_1(E \cap A)$ $= v(E \cup A)$ and $v_2(E) = -v(E \cup B)$, so every such decomposition of v is obtained from a Hahn decomposition of Ω, as in (1). So it is enough to show that if A, B and A', B' are two Hahn decompositions then the measures obtained as in (1) are the same. We have $v(A \cup A') = v(A \cap A') + v(A \triangle A') = v(A \cap A')$ by Hahn decomposition theorem. For each $E \in \mathcal{F}$, as $A \cup A'$ is a positive set we have

$$v(E \cap (A \cap A')) \leq v(E \cap A) \leq v(E \cap (A \cup A'))$$

and
$$v(E \cap (A \cap A')) \leq v(E \cap A') \leq v(E \cap (A \cup A')).$$

Therefore, $v(E \cap A) = v(E \cap A')$ and v^+ defined by (1) is unique. But then $v^- = v^+ - v$ is also unique.

NOTE. The Hahn decomposition is of the space and is not unique whereas the Jordan decomposition is of the signed measure and is unique. Upper and lower variations of a signed measure v can be defined in an alternative way as

$$v^+(A) = \sup \, [v(B) : B \in \mathcal{F}, B \subset A]$$

$$v^-(A) = -\inf \, [v(B) : B \in \mathcal{F}, B \subset A] \tag{2}$$

Exercise AII.1   Show that (2) $\Leftrightarrow$ (1).

Proof of (2) $\Rightarrow$ (1).

For $B \in \mathcal{F}$ and $B \subset A$,

$$v(B) = v(B \cap N) + v(B \cap N^C)$$

$$\leq v(B \cap N^C) \; (N \text{ being the negative set} \Rightarrow v(B \cap N) \leq 0)$$

$$\leq v(B \cap N^C) + v((A - B)N^C)$$

$$= v(A \cap P) \; (N^C = P)$$

Taking supremum over B on both sides $v^+(A) \leq v(A \cap P)$. Also by the definition of v^+, $v(A \cap P) \leq v^+(A)$. Hence $v^+(A) = v(A \cap P)$. Similarly $v^-(A) = -v(A \cap N)$. To prove (1) $\Rightarrow$ (2), $A \cap N \subset A$ and N being negative we have seen in the proof of Hahn decomposition theorem $v(N)$ is minimal, it follows that

$$v(A \cap N) = \inf \, \{v(B) : B \in \mathcal{F}, B \subset A\}$$

Similarly
$$v(A \cap P) = \sup \, \{v(B) : B \in \mathcal{F}, B \subset A\}.$$

Proof of lemma (Jordan decomposition) in Chapter 3.

Let
$$p_k = P(r_k) = F(r_k) - F(r_k^-)$$

Define
$$F_d(x) = \sum_{r_k \leq x} p(r_k), \, x \in R \tag{1}$$

and
$$F_c(x) = F(x) - F_d(x) \tag{2}$$

Now as in Theorem 3.1 F_d is non-decreasing, non-negative and right continuous. By (2), F_c is also non-negative and right continuous. It is also left continuous and non-decreasing, since for $x_n < x_1$,

$$F_c(x_1) - F_c(x_n) = F(x_1) - F(x_n) - \sum_{x_n < r_k \leq x_1} p(r_k)$$

$$= F_c(x_1^-) - F_c(x_n) - \sum_{x_n < r_k < x_1} p(r_k) \geq 0 \text{ and } \to 0 \text{ as } x_n \uparrow x_1.$$

If possible suppose F'_c and F'_d be the other such decompositions of F, then

$$F = F_c + F_d = F'_c + F'_d$$

and hence $F_c - F'_c = F_d - F'_d$ but $F_c - F'_c$ is continuous and $F_d - F'_d$ is a step function. This contradiction leads to $F_c - F'_c = F_d - F'_d = 0$. Hence $F_c = F'_c$ and $F_d = F'_d$ and the decomposition is unique.

The proof of the lemma is complete.

Proof of Theorem 3.3.   Now F_c can be further split into a differentiable (absolutely continuous) part and a singular part in a unique manner by Lebesgue decomposition theorem given in Appendix I. Norming appropriately we can view these three parts as d.fs. of r.vs. and F which has been split up as the convex combination (mixture) of these three types of d.fs.

Large Sample Theory

Large Deviations

Introduction.   The problem of finding the rate at which the probability that either the sample mean (the partial sum) exceeds some preassigned large number (some preassigned sequence of numbers) is known as large deviations. There have been phenomenal developments in the theory of large deviations over the last two decades. Most significant are the works of Chernoff, Cramer, Bahadur, Donsker and Vardhan who have developed a powerful machinery in a series of papers to deal with many old and new problems in probability where precise estimates of large deviation (away from the central part of the distribution) probabilities play an important role.

Let $X_1, X_2, \ldots, X_n$ be i.i.d. r.vs, with mean zero and variance 1 and $S_n = X_1 + \ldots + X_n$ be the nth partial sum. Weak law of large numbers then asserts that $S_n/n \xrightarrow{P} 0$ as $n \to \infty$, that is, given $a > 0$,

$$P(|S_n/n| > a) \to 0 \text{ as } n \to \infty$$

In particular $P_n(a) = P(S_n/n > a) \to 0$ as $n \to \infty$, that is,

$$P_n(a) = P(S_n/\sqrt{n} > \sqrt{n}\, a) \to 0 \text{ as } n \to \infty$$

or

$$P(S_n/\sqrt{n} > x_n) \to 0 \text{ as } n \to \infty \text{ where } x_n = O(\sqrt{n}).$$

The event $\{S_n/n > a\}$ is known as large deviation event and the problem of large deviations is to find the rate at which $P_n(a)$ goes to zero as $n \to \infty$. Applications of large deviation principle can be found in testing of hypothesis, information theory and statistical mechanics.

Note that $P_n(a) \sim [1 - \Phi(x_n)]$ by CLT if $x_n = \sqrt{n}a$

$$\sim 1/x_n\, \Phi(x_n)$$

$$= 1/x_n\, \frac{1}{\sqrt{2\pi}} e^{-x_n^2/2} \text{ for large } n$$

Then we sould expect

$$P_n(a) \sim \frac{1}{\sqrt{na}\sqrt{2\pi}} e^{-na^2/2}$$

or

$$1/n \log P_n(a) \sim - a^2/2 \text{ for large } n.$$

Unfortunately, this is not true. Indeed we shall show that there is a function $I(a)$ depending on the distribution of X, such that

$$1/n \log P_n(a) \to -I(a) \text{ as } n \to \infty \text{ which is known as Chernoff's theorem.}$$

Chernoff's theorem.   The function $I(a)$ is nown as a *large deviation rate* function or simply a rate function.

To prove Chernoff's theorem we shall first prove some auxilliary results.

Let $M(t)$ and $C(t) = \log M(t)$ be the m.g.f. and cumulant generating function (c.g.f.) of X respectively. Assume that $M(t) < \infty$ for $t \in T$. Define

$$I(y) = \sup_{s \in T} [ys - C(s)] \tag{1}$$

Let $a > 0$ be a number such that there is a $t \in T$ and $C'(t) = a$. Then by differentiation from (1) we get

$$I(a) = at \pm C(t).$$

Let $Y = X - a$ and F be the d.f. of Y. Define a new r.v. Z with d.f. G given by

$$dG(z) = e^{I(a)+tz} \, dF(z).$$

Note that G is a proper d.f. because

(a) $$\int_{-\infty}^{\infty} dG(z) = e^{I(a)} \int e^{tz} dF(z) = e^{I(a)} E(e^{tY})$$

$$= e^{I(a)} e^{-at} M(t) = 1$$

(b) The m.g.f. of Z is given by

$$E(e^{uZ}) = \int e^{uz} \, dG(z) = e^{I(a)} E(e^{(u+t)Z})$$

$$= e^{-au} \frac{M(t+u)}{M(t)}$$

(c) The c.g.f. of Z is given by

$$- au + C(u + t) - C(t)$$

Hence $E(Z) = - a + C'(t) = 0$.

Let σ^2 be the variance of Z and let H_n be the d.f. of $(Z_1 + Z_2 + \ldots + Z_n)/$ $(\sqrt{n} \, \sigma)$. By CLT $H_n(x) \to \Phi(x)$ as $n \to \infty$. Also note that $dF(z) = e^{-I(a)} \cdot e^{-tz} \cdot dG(z)$.

Proof of Chernoff's theorem.   (This proof is due to R.R. Bahadur)

$$P(S_n/n > a) = P(Y_1 + \dots + Y_n > 0)$$

$$= \int_{y_1 + \dots + } \int_{y_n > 0} dF(y_1) \dots dF(y_n)$$

$$= e^{-nI(a)} \int_{y_1 + \dots + } \int_{y_n > 0} e^{-t(y_1 + \dots + y_n)} dG(y_1) \dots dG(y_n)$$

$$= e^{-nI(a)} \int_0^\infty e^{-\sqrt{n}\sigma t x} dH_n(x)$$

$$= e^{-nI(a)} \cdot J_n$$

where

$$J_n = \int_0^\infty e^{-\sqrt{n}\sigma t x} dH_n(x)$$

$$= H_n(x) e^{-\sqrt{n}bx} \Big|_0^\infty - \int_0^\infty H_n(x)(-\sqrt{n}b) e^{-\sqrt{n}bx} dx \, (b = \sigma t)$$

$$= H_n(0) + \sqrt{n}b \int_0^\infty H_n(x) e^{-\sqrt{n}bx} dx$$

$$= \sqrt{n}b \int_0^\infty [H_n(x) - H_n(0)] e^{-\sqrt{n}bx} dx$$

Let $\varepsilon > 0$ be given. Note that

$$1 \ge J_n = \sqrt{n}b \int_0^\infty [H_n(x) - H_n(0)] e^{-\sqrt{n}bx} dx$$

$$\ge [H_n(\varepsilon) - H_n(0)] \int_\varepsilon^\infty \sqrt{n}b e^{-\sqrt{n}bx} dx$$

$$= [H_n(\varepsilon) - H_n(0)] e^{-\sqrt{n}b\varepsilon}$$

Therefore $0 \ge 1/n \log J_n \ge - \sqrt{n}\, b\varepsilon/n + 1/n \log [H_n(\varepsilon) - H_n(0)]$. Hence $\lim\limits_{n \to \infty} 1/n \log J_n = 0$.

Therefore $1/n \log P_n(a) \to - I(a)$ as $n \to \infty$.

Remark.   Bahadur and Ranga Rao (1960, AMS) actually showed that

$$P(S_n/n > a) \sim t^{-1}(2\pi \, nC''(t))^{-1/2} \cdot e^{-nI(a)}, \, a = C'(t).$$

The following large deviation formulation of the large deviation principle is due to Vardhan (1966).

(i)   The rate function I is a finite non-negative valued function from $X \to [0, \infty]$, $X \subseteq R$. $I(x) \ge 0$, follows from taking $t = 0$.

(ii) I is lower semi-continuous on X, that is

$$\underline{\lim} \, I(x_n) \ge I(x) \text{ if } x_n \to x$$

(being the supremum of continuous functions)

(iii) I has compact level sets, that is for any $b > 0$, $K = \{x : I(x) \le b\}$ is compact.

$K \subset [b - \log M(-1), b + \log M(1)]$. Hence K is bounded and therefore compact.

NOTE. $(3) \Rightarrow (2)$ in the above definition.

Large Deviation Principle (LDP)

A sequence $\{P_n, n \ge 1\}$ of probability measures of $(R, \mathscr{B})$ is said to satisfy the LDP with function I if

1. $\displaystyle\lim_{n\to\infty} \sup 1/n \, \log P_n(F) \le -\, I(F)$, for all closed sets F

2. $\displaystyle\lim_{n\to\infty} \inf \, 1/n \, \log P_n(G) \ge -\, I(G)$, for all open sets G.

Remark.   Denote by $I(A) = \displaystyle\inf_{x\in A} I(x)$, note that $I(X) = 0$.

Some more properties of Rate Function I

(i) I is a convex function

(ii) $I(a) = 0$ iff $a = E(X)$

(iii) $I(x) \downarrow$ in $(-\infty, a)$ and $I(x) \uparrow$ in (a, ∞).

Proof.   (i) $\lambda I(x_1) + (1 - \lambda) I(x_2) \ge \lambda(tx_1 - \log M(t)) + (1 - \lambda)(tx_2 - \log M(t))$
$= t(\lambda x_1 + (1 - \lambda) x_2) - \log M(t)$ for all t.
Hence $\lambda I(x_1) + (1 - \lambda) I(x_2) \ge I(\lambda x_1 + (1 - \lambda)x_2)$. Hence I is convex.

(ii) By Jensen's inequality
$$E(\log (e^{tX})) \le \log E(e^{tX})$$
$$ta \le \log M(t) \text{ for all } t$$
Therefore $ta - \log M(t) \le 0$ for all t.
Hence $I(a) = 0$.

(iii) Choose $1 > \lambda > 0$ such that for $a < x_1 < x_2$,

$$x_1 = \lambda a + (1 - \lambda)x_2 \quad \left(1 - \lambda = \frac{x_1 - a}{x_2 - a}\right).$$

Since I is convex, $I(x_1) \le \lambda I(a) + (1 - \lambda)I(x_2) \le I(x_2)$.
Similarly for $x_1 < x_2 < a$, choose $1 > \lambda > 0$ such that

$$x_2 = \lambda a + (1 - \lambda)x_1$$

Hence $\qquad\qquad I(x_2) \le \lambda I(a) + (1 - \lambda)I(x_1) \le I(x_1)$.

Kolmogorov's strong law for i.i.d. r.vs. $X_1, X_2, \ldots$ with mean 0 states that
$\displaystyle\lim_{n\to\infty} S_n/n = 0$ a.s.

If $a < b$ and $0 \notin [a, b]$, then $\lim P[S_n/n \in (a, b)] = 0$. Cramer (1937) studied the problem of finding the rate at which the above probability tends to zero. Of course one cannot say very much if nothing more than the existence of the first moment is assumed. If one assumes the existence of the moment generating function, then one has the following theorem of Cramer generalizing Chernoff's theorem.

Cramer's theorem.   Let $X_1, \ldots, X_n, \ldots$ be i.i.d. r.vs. with $M(t) = E(e^{tX_1}) < \infty$ for all real t. Let P_n be the probability measure associated with S_n/n. Then P_n satisfies the LDP with rate function

$$I(x) = \sup_t \, [xt - \log M(t)].$$

Proof. We want to show

$$\overline{\lim} \ 1/n \ \log P_n(F) \le - I(F) \text{ for closed set } F. \text{ Let } a = EX.$$

If F is empty or $a \in F$ the result is trivial.

Case (1): $F \subset (a, \infty)$. Let $y_1 = \inf \{x : x \in F\}$. Then

$$I(F) = I(y_1) = \sup \, [ty_1 - C(t)]$$

Note that $ty_1 - C(t) \le ty_1 - ta = t(y_1 - a) \le 0$ if $t < 0$ $(y_1 - a \ge 0)$

Therefore, $$I(y_1) = \sup_{t \ge 0} \, [ty_1 - C(t)].$$

Now $$P_n(F) \le P(\overline{X}_n \ge y_1) \le e^{-ty_1} E(e^{t\overline{X}_n}), t \ge 0 = e^{-ty_1} [M(t/n)]^n.$$

Therefore, $$1/n \ \log P_n(F) \le - \, [t/n \ y_1 - \log M(t/n)], \ t \ge 0.$$

Hence $$1/n \ \log P_n(F) \le - \, [t/n \ y_1 - \log M(t/n)], \ t \ge 0$$

$$\le - \sup_{t \ge 0} \, [ty_1 - \log M(t/n)] = I(y_1).$$

Therefore $\overline{\lim} \ 1/n \ \log P_n(F) \le - I(F)$.

Case (2): $F \subset (-\infty, a)$

Similarly we can show that

$$\overline{\lim} \ 1/n \ \log P_n(F) \le - I(y_2) \text{ where } y_2 = \sup \{y : y \in F\}.$$

Case (3): Let F be a closed set not containing a such that $F_1 = (-\infty, a) \cap F \ne \phi$, $F_2 = F \cap (a, \infty) \ne \phi$.

Therefore $$P_n(F) \le P_n(F_1) + P_n(F_2) \le 2 \max (P_n(F_1), P_n(F_2)).$$

Hence $1/n \ \log P_n(F) \le 1/n \ \log 2 + 1/n \ \log (\max (P_n(F_1), P_n(F_2)).$

Therefore $$\overline{\lim} \ 1/n \ \log P_n(F) \le - \min (I(F_1), I(F_2)) = - I(F).$$

Let G be non-empty and open. We want to show that

$$\liminf_{n \to \infty} 1/n \ \log P_n(G) \ge \inf_{x \in G} I(x).$$

Let $x \in G$ such that $I(x) < \infty$.

Case 1: There is no t such that $I(x) = st - C(t)$. Assume that $x > a$.

Since $I(x) = \sup_{t \geq 0} [st - C(t)]$, there exists $t_n \to \infty$

such that $\qquad\qquad\qquad xt_n - C(t_n) \to I(x).$

Note that $\qquad\qquad \infty > e^{-I(x)} = \lim_{n \to \infty} e^{\log M(t_n) - t_n x}$

$$= \lim_{n \to \infty} \int e^{(y-x)t_n} dF(y)$$

$$= \lim_{n \to \infty} \int_x^\infty e^{(y-x)t_n} dF(y) \,(\text{since } P[Y < x] = 0).$$

Therefore $\qquad\qquad\qquad e^{-I(x)} = P(X = x).$

Hence $P_n(G) \geq P_n(\{x\}) \geq P[(X = x)]^n = e^{-nI(x)}$

Therefore $\liminf_{n \to \infty} 1/n \, \log P_n(G) \geq -I(x).$

A similar argument applies to $x < a$.

Case 2: $I(x) = xt_0 - C(t_0)$ for some t_0. Assume that $x > a$. Therefore $x = M'(t_0)/M(t_0) = C'(t_0)$.

Since G is open there exists a $\delta > 0$ such that $(x - \delta, x + \delta) \subset G$. Now

$$P_n(G) \geq P_n\,(x - \delta, x + \delta) = \int_{|S_n/n - x| < \delta} \prod_1^n dF(x_i)$$

$$= \int_{|S_n/n - x| < \delta} e^{-t_0 S_n}(M(t_0))^n \cdot e^{t_0 S_n}/[M(t_0)]^n \prod_1^n dF(x_i)$$

Hence $\qquad P_n(G) \geq e^{-nt_0(x+\delta)}(M(t_0))^n \int_{|S_n/n - x| < \delta} \prod_1^n D\tilde{F}(x_i)$

where $\tilde{F}(x) = e^{t_0 x}\, dF(x)/(M(t_0))$ is the d.f. of some r.v. X^*

Therefore $P_n(G) \geq \exp\,[-n(t_0(x + \delta) - C(t_0))]\, P\,(|\overline{X}_n^* - x| < \delta).$

Therefore $\underline{\lim}\, 1/n\, \log P_n(G) \geq - t_0(x + \delta) + C(t_0) + \underline{\lim}\, 1/n\, \log P(\overline{X}_n^* - x| < \delta)$

$\geq - I(x) - t_0\delta$ for small $\delta > 0$.

Therefore $\qquad\qquad \underline{\lim}\, 1/n\, \log P_n(G) \geq - I(x).$

A similar argument works for $x < a$.

Hence $\underline{\lim}\, 1/n\, \log P_n(G) \geq - I(G).$

Remarks.   (1) Chernoff's theorem is a special case of Cramer's theorem.

Proof. Let A^0, $\overline{A}$ denote the interior and closure of A. Then $-I(A_0) \leq \underline{\lim}$ $1/n \log P(S_n/n \in A^0) \leq \underline{\lim}$ $1/n \log P(S_n/n \in A) \leq \overline{\lim}$ $1/n \log P(S_n/n \in A)$ $\leq \lim$ $1/n \log P(S_n/n \in A \leq -I(\overline{A})$. Thus if $I(A^0) = I(\overline{A}) = I(A)$ we have

$$\lim_{n \to \infty} 1/n \log P(S_n/n \in A) = -I(A).$$

In particular take $A = [a, \infty]$. Assume $E(X) = 0$. If $a > 0$, $I(A) = I(a)$.

Therefore $\qquad \lim 1/n \log P(S_n/n \geq a) = -I(a)$.

[2] If $d_0 = \inf_{x \in (a,b)} I(x)$, $d_1 = \inf_{x \in [a,b]} I(x)$ then given $\varepsilon > 0$, there exists n_0 such that for $n \geq n_0$

$$e^{-n(d_0+\varepsilon)} \leq P[S_n/n\varepsilon(a,b)] \leq e^{-n(d_1+\varepsilon)}$$

i.e. if $d_0 = d_1$ and $EX = m \notin [a, b]$ then the probability decays exponentially fast and the function I gives the rate of decay in the exponent.

(3) Cramer's theorem deals with deviations of S_n of order n from the mean $nE(X)$.

References

Ash, R.B. (1972): *Real Analysis and Probability,* Academic Press, New York.

Bhatt, B.R. (1985): *Modern Probability Theory,* 2nd ed., Wiley Eastern, New Delhi.

Billingsley, P. (1968): *Cónvergence of Probability Measures,* Wiley, New York.

Billingsley, P. (1979): *Probability and Measure,* Wiley, New York.

Brieman, L. (1968): *Probability,* Addison-Wesley, Reading, Mass.

Cantelli, F.P. (1932): *Una teoria astratta del calcola delle Probabilita,* Ist Ital. Attuari 3.

Chandra, T.K. (1989), *Bull. Cal Math. Soc.* Vol. 81, No. 3, pp. 227–231.

Chow, Y.S. (1960): A Martingale Inequality and the Law of Large Numbers, *Proc. Amer. Math. Soc.* Vol. 11, pp. 107–111.

Chow, Y.S. and Tiecher, H. (1988): *Probability Theory,* 2nd ed., Springer-Verlag, New York.

Chung, K.L. (1951): *The strong law of large numbers, Proc. 2nd Berkeley Symp. Stat. and Prob.* pp. 341–352.

Chung, K.L. (1965): *A Course in Probability Theory,* Harcourt Brace, Jovanovich, New York.

Cramer, H. (1946): *Mathematical Methods of Statistics,* Princeton University Press, Princeton.

Cramer, H. (1970): *Random Variables and Probability Distributions,* Cambridge Univ. Press, London.

De Finetti, B. (1937): "La Prevision; ses lois logiques, ses sources subjectives", *Annales de l'Institut Heuri Poincare,* Vol. 7, pp. 1–68.

Doob, J.L. (1953): *Stochastic Processes,* Wiley, New York.

Dynkin, E.B. (1961): *Theory of Markov Processes,* Prentice-Hall, Englewood Cliffs, New Jersey.

Etemade, N. (1983): On the laws of large numbers for nonnegative random variables. *J. Mult. Anal.* 13, pp. 187–193.

Feller, W. (1950): *An Introduction to Probability Theory and its Applications,* Vol. I, 3rd ed., Wiley, New York.

Feller, W. (1966): *An Introduction to Probability Theory and its Applications,* Vol. II, Wiley, New York.

Glivenko, J. (1936): *Stieltjes Integral* (in Russian).

Gnedenko, B.V. and Kolmogorov (1954): *Limit Distribution for Sums of Independent Random Variables,* Addison-Wesley, Reading, Mass.

Hagg, J. (1924, 1928): *Sur un probleme general de probabilites et. ses diverses applications, Proc. Inst. Congr. Math,* Toronto, pp. 629–674.

Hajek, L. and Renyi, A. (1955): Generalization of an inequality of Kolmogorov, *Acta Math. Hung.* Vol. 6, pp. 281–283.

Halmos, P.R. (1958): *Measure Theory,* Van-Nostrand, Princeton.

Hsu, P.L. and Robbins, H. (1947): Complete convergence and the law of large numbers, *Proc. Nat. Acad. Sci.* (USA), 33, pp. 25–31.

Khintchine, A. and Kolmogorov, A.N. (1924): *Uber Konvergenz von Reichen, derem Glieder durch den zufall bestimmt Werden, Rec. Math. (Mat Sbornik)* Vol. 32, pp. 1173–1205.

Kolmogorov, A.N. (1950): *Foundations of Probability,* Chelsea, New York.

Lamperti, J. (1966): *Probability,* Benjamin, New York.

Lévy, P. (1937): *Theori de l'addition des variables aleatories,* 2nd ed., Gauthier-Villars, Paris, 1954.

Lindeberg, L. (1922): Eine neue Herleitung des Exponentialgesetzes in der wahrscheinlichkeitsrechnung, *Math, Zeit.* Vol. 15, pp. 211–225.

Loéve, M. (1963): *Probability Theory,* Van-Nostrand, Princeton.

Luckacs, E. and Laha, R.G. (1964): *Applications of Characteristic Functions,* Griffin, London.

Monroe, M.E. (1953): *Introduction to Measure and Integration,* Addison-Wesley, Reading, Mass.

Renyi, A. (1970): *Foundations of Probability,* Holden-Day, San Francisco.

Scheffe, H. (1947): "A useful convergence theorem for probability distribution" *Ann Math. Stat.* Vol. 18, pp. 434–438.

Serfling, R.J. (1980): *Approximation Theorems of Mathematical Statistics,* Wiley, New York.

Tucker, H.G. (1967): *A Graduate Course in Probability,* Academic Press, New York.

von Mises, R. (1939): *Uber aufteilungs and Besetzungs Wahrs cheinlichkeiten, Revue de la Faculte des Sciences de Universite d'Istambul,* N.S. Vol. 4, pp. 145–163.

Author Index

Subject Index